Bioactive Compounds - Biosynthesis, Characterization and Applications

Edited by Leila Queiroz Zepka, Tatiele Casagrande do Nascimento and Eduardo Jacob-Lopes

Published in London, United Kingdom

IntechOpen

Supporting open minds since 2005

Bioactive Compounds - Biosynthesis, Characterization and Applications
http://dx.doi.org/10.5772/intechopen.91131
Edited by Leila Queiroz Zepka, Tatiele Casagrande do Nascimento and Eduardo Jacob-Lopes

Contributors
Mohd Rehan, Achouak Madani, Nassima Mazouni, Mohammed Nedjhioui, Sunita Sharma, Mariane Bittencourt Fagundes, Roger Wagner, Kapil Singh Narayan, Reenu Kashyap, InnocentMary IfedibaluChukwu Ejiofor, Maria Goretti Igbokwe, Merugu Chandra Surya Rao, Dokka Venkata Swami, P. Ashok, Satya Prakash Nanda, Banavath Babu Rao, Mohd Abdul Gani, Shama M, Yakubu Jibira, Paul Atawuchugi, Iyakutye Jacob Nte, Holinshead Holly Gunn, Utkarsh Ghate, Hema Kulkarni, Stella Omokhefe Bruce, Felix A. Onyegbule, Zeinab Amanzadeh, Poournima Patil, Suresh Killedar, Mahmoud Ghorbani, Ricardo Santos Aleman, Leila Queiroz Zepka, Tatiele Casagrande do Nascimento, Eduardo Jacob-Lopes

First published in London, United Kingdom, 2021 by IntechOpen
IntechOpen is the global imprint of INTECHOPEN LIMITED, registered in England and Wales, registration number: 11086078, 5 Princes Gate Court, London, SW7 2QJ, United Kingdom
Printed in Croatia

British Library Cataloguing-in-Publication Data
A catalogue record for this book is available from the British Library

Additional hard and PDF copies can be obtained from orders@intechopen.com

Bioactive Compounds - Biosynthesis, Characterization and Applications
Edited by Leila Queiroz Zepka, Tatiele Casagrande do Nascimento and Eduardo Jacob-Lopes
p. cm.
Print ISBN 978-1-83969-269-7
Online ISBN 978-1-83969-270-3
eBook (PDF) ISBN 978-1-83969-271-0

For EU product safety concerns:
IN TECH d.o.o., Prolaz Marije Krucifikse Kozulić 3, 51000 Rijeka, Croatia, info@intechopen.com or visit our website at intechopen.com.

Meet the editors

Prof. Dr. Leila Queiroz Zepka is currently an associate professor in the Department of Food Technology and Science, Federal University of Santa Maria, Brazil. She has more than fifteen years of teaching and research experience. She has published more than 550 scientific publications/communications, including 15 books, 50 book chapters, 100 original research papers, 380 research communications in national and international conferences, and 12 patents. She is a member of the editorial board of five journals and acts as a reviewer for several national and international journals. Her research interests include microalgal biotechnology with an emphasis on microalgae-based products.

Dr. Tatiele Casagrande do Nascimento is currently a young researcher at the Department of Food Science and Technology, Federal University of Santa Maria, Brazil. She has a doctorate in Food Science and Technology. Her research interests include foods and health outcomes with an emphasis on natural pigments.

Prof. Dr. Eduardo Jacob-Lopes is currently an associate professor at the Department of Food Technology and Science, Federal University of Santa Maria, Brazil. He has more than fifteen years of teaching and research experience. He has coordinated and is coordinating more than fifty research projects and/or technological developments financed by public funding agencies and private initiatives. He has published more than 600 scientific publications/communications, including 15 books, 60 book chapters, 120 original research papers, 400 research communications in national and international conferences, and 13 patents. He is a member of the editorial board of ten journals and acts as a reviewer for several national and international journals. His research interests include bioprocess engineering and sustainable engineering with an emphasis on microalgal biotechnology.

Contents

Preface

Biosynthesis is a process of successive enzymatic reactions necessary to originate a specific biologically active natural metabolite. The knowledge of biosynthetic pathways and the factors that influence their regulation and accumulation is essential to monitor and optimize the mass production of these valuable metabolites, as they can occur through different pathways. Numerous natural sources synthesize a range of bioactive compounds with beneficial effects on human health, mainly antioxidant, anti-inflammatory, antimicrobial, hepatoprotective, immunomodulatory, anticancer, and dyslipidemia control effects. Among them are terpenic structures, phenolic compounds, and alkaloids, in addition to polysaccharides, peptides, and amino acids.

This book is a compilation of chapters on bioactive compounds with proven activities. It provides a valuable information source of bioactive compounds biosynthesized, which can be used for further development of bioproducts by industry. The volume is divided into the following sixteen chapters:

Chapter 1: "Introductory Chapter: An Overview on Bioactive Compounds with Focus in the Biosynthesis, Characterization and Applications"

Chapter 2: "Pharmacological Role of Biosynthetic Products"

Chapter 3: "Plants' Bioactive Metabolites and Extraction Methods"

Chapter 4: "Biosynthesis of Natural Products"

Chapter 5: "The Need to Use Microorganisms and Their Biosynthesized Bioactive Metabolites for Biological and Medical Activities"

Chapter 6: "Biosynthesis of Diverse Class Flavonoids *via* Shikimate and Phenylpropanoid Pathway"

Chapter 7: "Biosynthesis of the Immunomodulatory Molecule Capsular Polysaccharide A from *Bacteroides fragilis*"

Chapter 8: "Flavonoids: Understanding Their Biosynthetic Pathways in Plants and Health Benefits"

Chapter 9: "Sterols Biosynthesis in Algae"

Chapter 10: "Arginine Metabolism: An Enlightening Therapeutic Attribute for Cancer Treatment"

Chapter 11: "Electro-Spinning and Electro-Spraying as Innovative Approaches in Developing of a Suitable Food Vehicle for Polyphenols-Based Functional Ingredients"

Chapter 12: "Scope, Nutritional Importance and Value Addition in Palmyrah (*Borassus flabellifer L.*): An Under Exploited Crop"

Chapter 13: "Phenolic Compounds"

Chapter 14: " Polyphenols, Spices and Vegetarian Diet for Immunity and Anti-Inflammatory Drug Design"

Chapter 15: "Colon Available Bioactive Compounds Exhibits Anticancer Effect on *In-Vitro* Model of Colorectal Cancer"

Chapter 16: "Cysteine in Broiler Poultry Nutrition"

These chapters cover recent advances in knowledge about biosynthesized bioactive metabolites, including distribution, regulatory mechanisms, accumulation induction, environmental conditions, the impact of the use of advanced technologies, methods of obtaining, and bioactivity. This book is a valuable reference for researchers seeking to improve their knowledge of naturally produced bioactive metabolites.

Leila Queiroz Zepka, Tatiele Casagrande do Nascimento and Eduardo Jacob-Lopes
Food Science and Technology Department,
Federal University of Santa Maria,
Santa Maria, RS, Brazil

Chapter 1

Introductory Chapter: An Overview on Bioactive Compounds with Focus in the Biosynthesis, Characterization and Applications

Tatiele Casagrande do Nascimento, Eduardo Jacob-Lopes and Leila Queiroz Zepka

1. Overview

Natural compounds have been used globally for thousands of years [1]. However, currently, a world trend towards health has considerably boosted the search for natural alternatives for health promotion [2]. Several naturally synthesized compounds have bioactive functions and have been explored for different applications, especially in the food and pharmaceutical industry. These substances are chemical structures that perform specialized functions at the biological level [3].

According to a consensus established by [4], bioactive compounds are naturally occurring essential and non-essential compounds that can positively influence human health. Nutritionally, they have also been called nutraceuticals since 1979 because when ingested, they provide health benefits beyond basic nutrition [4].

Bioactive compounds make up a highly heterogeneous set of molecules with different chemical structures and distributions in nature [5]. Broadly, these metabolites are divided into three main groups: terpenes and terpenoids, phenolic compounds and alkaloids [3]. Among them, carotenoids, sterols, flavonoids are frequent examples.

Most of the bioactive terpenes investigated are tetraterpenes with C40 skeletons [6]. formed from eight isoprenoid units (C5) and characterized by a central sequence of conjugated double bonds [7]. Similarly, sterols also belong to the group of terpenes, and they are triterpenes (C30) with a basic structure consisting of a tetracyclic ring and a C17 side chain [8]. The phenolic compounds flavonoids have low molecular weight, are consisting of 15 carbon atoms, organized in the basic configuration C6-C3-C6 [9]. In contrast, alkaloids are usually heterocyclic organic compounds (basic pH) that contain nitrogen atoms [10].

In addition to the main groups, other molecules have been shown some bioactivity, such as polysaccharides, amino acids and peptides, indicating that the diversity of bioactive compounds is comprehensive and is in a growing process of exploration and investigation in various sources.

Microorganisms, plants and animals offer many bioactive products of great interest for application in the food and pharmaceutical industry [3, 11]. According to [12], more than 80% and 30% of the active compounds used in food and medicine, respectively, are obtained from natural sources.

 IntechOpen

Terpenoids generally constitute the largest and most diverse class of secondary metabolites in natural products. For example, it is estimated that more than 1200 natural carotenoids have been characterized from different sources, including plants, fruits, vegetables and microorganisms [6]. Sterols are present in most living organisms, including vertebrates, invertebrates, plants, fungi, and bacteria [13]. According to [14], sterols that occur plant, animal, and microbial are called phytosterols, zoosterols, and mycosterols.

On the other hand, phenolic compounds (including flavonoids) are widely distributed in the plant kingdom, are present in fruits, leaves, seeds and glycosylated in other parts of the plant [15]. Most known alkaloids are isolated from plants. However, they have also been reported in microorganisms, marine organisms, and terrestrial animals [10].

Regardless of the source, bioactive compounds must be obtained (isolated or extract) from some extraction technique. They are conventionally solvent extracted, considering important aspects such as solvent-compound affinity, extraction time and temperature. However, emerging technologies such as ultrasound, pulsed electric field, enzymatic digestion, extrusion, microwave, ohmic heating, supercritical fluids are increasingly used due to greater sustainability and efficiency [3, 12].

In terms of biosynthesis, bioactive compounds can be formed in different ways. Terpenes are biosynthesized via the cytosolic mevalonic acid (MVA) pathway and the methylerythritol phosphate (MEP) pathway [16]. The biosynthesis of phenolic compounds involves several pathways, the shikimic acid pathway, phenylpropanoid and flavonoid pathways [9]. In contrast, the shikimic acid pathway is the main route involved in alkaloid biosynthesis [17].

The benefits of these compounds are a consequence of several proven bioactive properties, mainly antioxidant, anti-inflammatory and antimicrobial effects [8].

In general, most bioactive compounds have a marked antioxidant capacity due to their ability to capture reactive species [8]. Furthermore, they improve endogenous antioxidant defenses in vivo, allowing an attractive therapeutic approach against oxidative stress and related diseases [18].

The role of bioactive in inflammatory processes is evidenced by reducing signalers such as pro-inflammatory cytokines, chemokines, interleukins, inducible enzymes (cyclooxygenase-2 and inducible nitric oxide synthase) and inflammatory mediators (prostaglandins, leukotrienes and thromboxane). These pathological events are associated with the development and progression of most chronic diseases, such as type II diabetes mellitus, obesity, neurodegenerative disorders, cardiovascular diseases and cancer [19].

The antimicrobial activity of bioactive compounds has been reported for different microorganisms [20]. This activity is often associated with phenolic compounds, and it is believed that they use active redox metals from the microbial cell, causing an imbalance in the redox state and consequently cell death [21].

In addition to these mentioned properties, several other effects are associated with bioactive compounds, including anticancer, neuroprotection, hepatoprotective, immunomodulatory activities and dyslipidemias control [18, 22–25].

The chapters presented in this book provide a reliable compilation of biosynthesized active compounds with proven activities that can contribute to the development of products by industry.

Author details

Tatiele Casagrande do Nascimento, Eduardo Jacob-Lopes and Leila Queiroz Zepka*
Department of Food Science and Technology, Federal University of Santa Maria, Santa Maria, RS, Brazil

*Address all correspondence to: zepkaleila@yahoo.com.br

References

[1] Y. Fu, J. Luo, J. Qin, and M. Yang, "Screening techniques for the identification of bioactive compounds in natural products," J. Pharm. Biomed. Anal., vol. 168, pp. 189-200, 2019, DOI:10.1016/j.jpba.2019.02.027.

[2] J. Fanzo et al., "A research vision for food systems in the 2020s: Defying the status quo," Glob. Food Sec., vol. 26, p. 100397, Sep. 2020, DOI:10.1016/j.gfs.2020.100397.

[3] J. Bonilla, F. C. Vargas, T. G. de Oliveira, G. L. da Aparecida Makishi, and P. J. do Amaral Sobral, "Recent patents on the application of bioactive compounds in food: a short review," Curr. Opin. Food Sci., vol. 5, pp. 1-7, Oct. 2015, DOI:10.1016/J.COFS.2015.05.012.

[4] B. HK et al., "Bioactive compounds: Definition and assessment of activity," Nutrition, vol. 25, no. 11-12, pp. 1202-1205, Nov. 2009, DOI:10.1016/J.NUT.2009.04.023.

[5] J. M. Carbonell-Capella, M. Buniowska, F. J. Barba, M. J. Esteve, and A. Frígola, "Analytical methods for determining bioavailability and bioaccessibility of bioactive compounds from fruits and vegetables: A review," Compr. Rev. Food Sci. Food Saf., vol. 13, no. 2, pp. 155-171, 2014, DOI:10.1111/1541-4337.12049.

[6] J. Yabuzaki, "Carotenoids database: Structures, chemical fingerprints and distribution among organisms," Database, vol. 2017, no. 1, pp. 1-11, 2017, DOI:10.1093/database/bax004.

[7] A. S. Fernandes, T. C. do Nascimento, E. Jacob-Lopes, V. V. De Rosso, and L. Q. Zepka, "Introductory Chapter: Carotenoids – A Brief Overview on Its Structure, Biosynthesis, Synthesis, and Applications," Prog. Carotenoid Res., pp. 1-16, 2018, DOI:10.5772/intechopen.79542.

[8] F. J. Barba, M. J. Esteve, and A. Frígola, "Bioactive components from leaf vegetable products," Stud. Nat. Prod. Chem., vol. 41, pp. 321-346, 2014, DOI:10.1016/B978-0-444-63294-4.00011-5.

[9] V. M. Patil and N. Masand, "Anticancer potential of flavonoids: Chemistry, biological activities, and future perspectives," Stud. Nat. Prod. Chem., vol. 59, pp. 401-430, Jan. 2018, DOI:10.1016/B978-0-444-64179-3.00012-8.

[10] R. Verpoorte, "ALKALOIDS," Encycl. Anal. Sci. Second Ed., pp. 56-61, Jan. 2005, DOI:10.1016/B0-12-369397-7/00010-8.

[11] R. P. Sinha and D.-P. Häder, "Introduction," Nat. Bioact. Compd., pp. 1-17, 2021, DOI:10.1016/b978-0-12-820655-3.00001-x.

[12] Q. Ren et al., "Recent advances in separation of bioactive natural products," Chinese J. Chem. Eng., vol. 21, no. 9, pp. 937-952, Sep. 2013, DOI:10.1016/S1004-9541(13)60560-1.

[13] N. B. Myant, "The distribution of sterols and related steroids in nature," Biol. Cholest. Relat. Steroids, pp. 123-159, Jan. 1981, DOI:10.1016/B978-0-433-22880-6.50010-2.

[14] M. H. Gordon, "FATS | classification," Encycl. Food Sci. Nutr., pp. 2287-2292, Jan. 2003, DOI:10.1016/B0-12-227055-X/00701-X.

[15] P. M. Angelo and N. Jorge, "Compostos fenólicos em alimentos – Uma breve revisão," Rev. Inst. Adolfo Lutz, vol. 66, no. 1, pp. 1-9, 2007.

[16] E. Oldfield and F.-Y. Lin, "Terpene biosynthesis: Modularity rules," Angew.

Chemie Int. Ed., vol. 51, no. 5, pp. 1124-1137, Jan. 2012, DOI:10.1002/ANIE.201103110.

[17] H. O. Gutzeit and J. Ludwig-Muller, Plant natural products: Synthesis, biological functions and practical applications, 1st ed. Wiley-Blackwell.

[18] R. R. Pujari, N. S. Vyawahare, and P. A. Thakurdesai, "Neuroprotective and antioxidant role of Phoenix dactylifera in permanent bilateral common carotid occlusion in rats," J. Acute Dis., vol. 3, no. 2, pp. 104-114, Jan. 2014, DOI:10.1016/S2221-6189(14)60026-3.

[19] A. F. Mendes, M. T. Cruz, and O. Gualillo, "Editorial: The physiology of inflammation—The final common pathway to disease," Front. Physiol., vol. 9, p. 1741, Dec. 2018, DOI:10.3389/FPHYS.2018.01741.

[20] S. MA, H. SH, S. K, and Y. JS, "Antibacterial Properties and Effects of Fruit Chilling and Extract Storage on Antioxidant Activity, Total Phenolic and Anthocyanin Content of Four Date Palm (*Phoenix dactylifera*) Cultivars," Molecules, vol. 21, no. 4, Apr. 2016, DOI:10.3390/MOLECULES21040419.

[21] H. Taleb, S. E. Maddocks, R. K. Morris, and A. D. Kanekanian, "Chemical characterisation and the anti-inflammatory, anti-angiogenic and antibacterial properties of date fruit (*Phoenix dactylifera* L.)," J. Ethnopharmacol., vol. 194, no. May, pp. 457-468, 2016, DOI:10.1016/j.jep.2016.10.032.

[22] C. Tenkerian, M. El-Sibai, C. F. Daher, and M. Mroueh, "Hepatoprotective, Antioxidant, and Anticancer Effects of the *Tragopogon porrifolius* Methanolic Extract," Evid. Based. Complement. Alternat. Med., vol. 2015, 2015, DOI:10.1155/2015/161720.

[23] X. Ji et al., "Bioactive compounds from herbal medicines to manage dyslipidemia," Biomed. Pharmacother., vol. 118, p. 109338, Oct. 2019, DOI:10.1016/J.BIOPHA.2019.109338.

[24] H. A. Adeola et al., "Bioactive compounds and their libraries: An insight into prospective phytotherapeutics approach for oral mucocutaneous cancers," Biomed. Pharmacother., vol. 141, p. 111809, Sep. 2021, DOI:10.1016/J.BIOPHA.2021.111809.

[25] A. Gupta, R. Kumar, R. Ganguly, A. K. Singh, H. K. Rana, and A. K. Pandey, "Antioxidant, anti-inflammatory and hepatoprotective activities of Terminalia bellirica and its bioactive component ellagic acid against diclofenac induced oxidative stress and hepatotoxicity," Toxicol. Reports, vol. 8, pp. 44-52, Jan. 2021, DOI:10.1016/J.TOXREP.2020.12.010.

Chapter 2

Pharmacological Role of Biosynthetic Products

Yakubu Jibira and Paul Atawuchugi

Abstract

A product such as hydrated carbons (carbohydrates), lipids, proteins, and nucleic acids play significant roles in plants metabolically. But there are other natural organic products manufactured by plants, some of the products are complex molecules, which are not primary metabolites. These biosynthetic products have possesses a variety of therapeutic merits in drug discovery. Some biosynthetic products show numerous appreciable therapeutic effects making them beneficial for trimming down polypharmacy and as viable candidates for the management of chronic diseases such as diabetes and hypertension in patients. The chapter discusses the pharmacological role of some biosynthetic products from plants and animals.

Keywords: alkaloids, phytosterols, biosynthesis, amino acids, pharmacological, terpenes

1. Introduction

Natural products are mostly produced by a living organism. A product such as carbohydrates, lipids, proteins, and nucleic acids play significantly impact on the primary role in in plants in terms of their metabolic reactions. Moreover, other natural organic compounds have also been known to be produced by plants, with which some of them are complex molecules, which are not primary metabolites. Different organisms may produce the same compounds through different pathways (e.g., convergent evolution), even if they are widely separated phylogenetically. The same organism may produce some compounds via over one biosynthetic path. There may be over one path available, such as in a changed linear process or metabolic grid. Even if the same compound is present in two different organisms, they may be formed via different pathways. This, however, is more likely for metabolites with simple structures. It derives the major precursors from the metabolism of carbohydrate (sugars), protein (amino acids), and lipid (fatty acid). The pathway derived biosynthetically for aromatic amino acids is an integral source of aromatic compounds such as flavonoids, phenols, and some alkaloids. Glycolysis yields an important metabolite such as acetyl-CoA and also via the beta-oxidation of fatty acids and also used in the tricarboxylic acid cycle (TCA) in the manufacture of organic acids, which are also starting materials for secondary metabolites. Also, acetyl-CoA plays important role in synthesizing terpenes, which forms a distinct class of metabolites.

2. Animal biosynthetic product

Animals contain many unique small molecules, including bioactive secondary metabolites. The compounds are protective, offensive or involved in communication [1]. Most of this product is also biologically effective, and they include the following.

2.1 Terpenes

Terpenes occur widely in nature. They are a large and varied class of hydrocarbons, which are produced by a wide variety of plants and by some animals. Terpenes are biosynthetically derived from isoprene units with the molecular formula C_5H_8. In bacteria and plants, isoprene precursor's dimethylallyl pyrophosphate and isopentenyl pyrophosphate can be made either via the mevalonate or deoxyxylulose phosphate pathways, but in animals' mevalonate is the source of these precursors [2]. The mevalonate pathway is the universal source of the terpenoid C5 precursor's isopentenyl pyrophosphate (IPP) and dimethylallyl pyrophosphate (DMAPP). In this pathway, three molecules of acetyl-CoA are condensed to 3-hydroxy-3-methylglutaryl-CoA with subsequent reduction to mevalonate, which is converted to IPP and DMAPP. Terpenes can exist as hydrocarbons or have oxygen-containing compounds such as hydroxyl, carboxyl, ketone, or aldehyde groups. After chemical modification of terpenes, it refers to the resulting compounds as terpenoids [3].

In the biochemical pathway of terpenoid synthesis, prenyltransferases take part in the condensation of activated forms of isoprene units. They link IPP with an isopentenyl diphosphate isomer of DMAPP in a "head-to-tail" manner. The linear chains of phenyl diphosphates that are formed in the reaction may also be changed through dimerization or cyclization by terpene synthase, forming terpenoids with new functions. They classify the resulting terpenes in order of size into hemiterpene, monoterpenes, sesquiterpenes, diterpene, triterpene, tetra terpenes, and polyterpenes. Hemiterpene comprises a single isoprene unit and changed into oxygen-containing derivatives called hemiterpenoids [3].

We have found terpenoids to be useful in the prevention and therapy of several diseases, including cancer, and also to have antimicrobial, antifungal, antiparasitic, antiviral, anti-allergenic, antispasmodic, antihyperglycemic, anti-inflammatory, and immunomodulatory properties [3].

2.2 Carnitine

Vaz and Wanders, [4] have described that carnitine plays a vital role in energy production in living organisms in terms of metabolism since it enables activated fatty acids to enter the mitochondria, where broken down via β-oxidation. Carnitine is present in all animal species, including other multiple micro-organisms and plants. They maintain its homoeostasis through endogenous synthesis, absorption from dietary sources and efficient tubular reabsorption by the kidney. Animal tissues contain relatively high amounts of carnitine, varying between 0.2 and 6 μmol/g, with the highest concentrations in heart and skeletal muscle. Apart from the diet being the primary source of carnitine, most mammals can synthesize carnitine internally [4].

Carnitine synthesis is from two amino acids; precisely lysine and methionine. Lysine serves as the carbon backbone of carnitine and the 4-N-methyl groups emanate from methionine [5]. Proteins in mammals contain N'trimethyl-lysine (TML) residues. N-methylation of these lysine residues occurs as a post-translational event in proteins such as calmodulin, myosin, actin, cytochrome c and histones. This reaction mostly catalysed by specific methyltransferase, which uses S-adenosylmethionine as a methyl donor. TML is the first metabolite of carnitine biosynthesis is released through the action of Lysosomal hydrolysis of these proteins. TML is first hydroxylated on the 3-position by TML to yield 3-hydroxy TML (HTML). Aldolytic cleavage of HTML yields 4-trimethylaminobutyraldehyde (TMABA) and glycine, a reaction catalysed by HTML aldolase (HTMLA; EC 4.1.2.'X'). Dehydrogenation of TMABA by TMABA dehydrogenase (TMABA-DH; EC 1.2.1.47) results in the formation of 4-Ntrimethylaminobutyrate (butyrobetaine). In the irrevocable step, butyrobetaine is hydroxylated on the 3-position by γ-butyrobetaine dioxygenase to yield carnitine [4]. Carnitine has an important role in the transport of activated long-chain fatty acids from the cytosol to the mitochondrial matrix where β-oxidation takes place. Second, they involve carnitine in transferring the products of peroxisomal β-oxidation, including acetyl-CoA, to the mitochondria for oxidation to CO2 and H2O in the Krebs cycle.

Pharmacologically, carnitine has possessed several effects on bone mass, male infertility, cognitive support, [6] metabolic function improvement, [7] neuroprotective effects in Alzheimer's dementia [8] and Parkinson's disease, [9] protection against oxidative stress damage, [10] congestive heart failure, hypertrophic heart disease, and peripheral arterial disease treatment [11]. They have also shown it to improve peripheral vasodilator activity [12].

From [13] the ability of an agent to control endothelial dysfunction is through upsetting the balance between the production of a vasodilator such as nitric oxide and vasoconstrictor, such as endothelin-1 substances in response to physical or chemical stimuli. This is because it is through that imbalance that leads to endothelial dysfunction, which is one of the initial steps in atherogenesis [14]. They relate endothelial dysfunction to cardiovascular risk factors such as arterial hypertension, dyslipidemia, diabetes, and obesity [15, 16]. They have therefore investigated carnitine gas in rats' models to cause an endothelium-dependent dilatation in arteries, since endothelial nitric oxide seemed to be the main mediator of vasodilatation [17]. For the cardioprotective ability of carnitine to serve as a cardioprotective agent, isolated rat working heart through different antioxidant mechanisms in most of the cases [18]. In terms of insulin resistance, this is where a 20 weeks treatment with carnitine in animal obese models resulted in a reduction in body weight, abdominal adiposity, plasmatic insulin, and liver triglyceride content [19]. Several, studies have been conducted on the antioxidant properties of carnitine and was due mainly to a reduction in both lipid peroxidation and free radical generation [20]. A decreased expression of inducible NO synthase and protein nitration, and inhibition of tubular necrosis and neutrophil infiltration in transplanted kidneys [21] . For antioxidant study [22] also revealed that carnitine and its derivatives fundamental mechanisms as participating in redox signalling that affects transcription factors (Nrf2, PPARα, NF-κB, etc.) and activating the vitagene network.

2.3 Pheromone

Pheromone plays a key role in sexual communication and reproduction in many insects such as the moth's species. In some insects, the pheromone is biosynthesized and released by specialized sex pheromone glands (PGs) that are along the intersegmental membrane between the 8th and 9th abdominal segments of females. Although the general pathway of sex pheromone synthesis in most species has not been established and the molecular mechanisms remain poorly understood [23]. Although de novo synthesis is more prevalent in the species studied to date, there are multiple examples of pheromone components derived from host precursors. Sometimes, such as leucine, used as starting material for fatty-acid derived sex pheromone biosynthesis by Holomelina spp. (Lepidoptera: Arctiidae), the putative plant-derived precursor is extensively elaborated by a typically de novo pathway. In other cases, it converts a highly elaborated host precursor to a pheromone component through a simple chemical transformation. While they originally reported the utilization of host precursors for pheromone biosynthesis in some insects, subsequent studies showed that pheromone biosynthesis was only or partially de novo [23].

The analysis of different pheromone glands in different species has revealed the occurrence of unusual fatty acids that have been proposed as precursors of pheromone components. Many lepidopteran sex pheromones are produced by P-oxidation steps with desaturase systems. However, only a few studies have a combination of different fatty acyl intermediates been used to show experimentally a pheromone biosynthetic pathway [23].

From [24] sex pheromone biosynthesis in moths begins with a palmitic or stearic acid moiety that is synthesized de novo in the PG through modifications of the fatty acid biosynthetic pathway. Through a series of enzymatic reactions such as desaturation, chain-shortening reaction, reduction, acetylation, and oxidation, it then converts the palmitic or stearic acids to the final pheromone components in a step-wise manner. Therefore, different enzymes are likely to be involved in the different reactions, and to date, the genes encoding 4 different classes of enzymes that are essential for this pathway have been functionally identified desaturases (Des), fatty acid reductases (FARs), fatty acid transport proteins (FATPs), and acyl-CoA-binding proteins (ACBPs) [24].

Pheromones, a chemical or blend of chemicals released by an organism that causes a specific behavioral or physiological reaction in one or more conspecific individuals are important mediators of communication for bacteria, plants, and animals in these environments. Pheromone systems of insects have proved to be some of the richest intellectual sources for the nascent science of chemical ecology the composite pheromones can be classed into six behaviorally functional groups: sex, aggregation, dispersal (spacing or epideictic), alarm, recruitment (trail), and maturation [24].

Pheromones are noted to function as opposite-sex attractants, same-sex repellents, and mother-infant bonding attractants and as menstrual cycle modulators [25]. A review article by [26] concluded that pheromones have aphrodisiac activity.

2.4 Melanin

Melanin is an abundant biological pigment that is present in mammalian which is located in areas such as skin, hair, eyes, ears and the nervous system. Birds' feathers, squid's ink, insects, plants and many other biological systems have been also known to contain melanin. It classifies melanin into three groups: eumelanin, pheomelanin and allomelanins. We term nervous system melanins as neuromelanin. Eumelanin colours commonly present as black or brown in animals [27].

Amino acid tyrosine is needed in the production of melanin, but its actions are catalyzed enzymatically by tyrosinase. In their primary biosynthetic pathway, tyrosine is hydroxylated to form the catecholamine 3,4-dihydroxyphenylalanine (DOPA), which is then oxidized to form 3,4-dioxyphenylalanine (dopa quinone) before cyclization to 5,6-indole quinones and their subsequent polymerization to form melanin [27].

Mason-Raper pathway can also produce melanin, and it started by the usage of tyrosine to form dihydroxyphenylalanine by tyrosinase through an oxidation reaction. The product (dihydroxyphenylalanine) formed is then oxidized by the same enzyme to dopa quinone, which rearranges spontaneously to leuco dopa chrome and then to dopa chrome. An unusual trait of dopa chrome that of decolorizing slowly if held under a vacuum allowed the identification of the subsequent intermediate 5,6-dihydroxy indole-2 carbolic acid, which loses its carboxyl group to become 5,6-dihydroxy dole. Upon exposure to air 5,6-dihydroxy indole is oxidized to indole 5,6-quinone and then to melano chrome, a purple compound that polymerizes to melanin [27].

Similar to the biosynthesis of eumelanin, melanin known as pheomelanin is biologically synthesized, except that it incorporates a precursor containing sulphur in the structure.

It has revealed melanin from natural sources to exhibit a broad spectrum of biological activities such as UV radiation protection, enzymatic lysis, and damage by oxidants, and resistance to drugs by pathogens, protection of insects against bacteria and antiviral protection. They have shown melanin to chelate metal ions and to act as a physiological redox buffer [27]. With this in mind, several studies have pharmacologically shown that melanin is indeed effective as an antioxidant, [28–30] Immunomodulatory and enhancement, [31–34] hepatoprotective, [35] anticarcinogenic effects, [36] and anti-inflammatory [37].

Studies conducted in *Nigella sativa* L. by [38, 39] which contains a lot of melanin in its seeds had the potential of treating imbalanced cytokine production and unconcern cancer and other immunotherapies. This was possible because melanin induced TNF-alpha, IL-6 and VEGF mRNA expression.

Melanins from various sources exhibit significant antioxidant activity, melanin protects pigmented cells and adjacent tissues by adsorbing potentially harmful substances, which are then slowly released in nontoxic concentrations, Besides, melanin could also interact with orally administered drugs and a vehicle for drug delivery, Melanin extracted from different species of tea displays protective effects against hydrazine-induced liver injury, a remarkable anti-inflammatory effect of melanin has been reported [34].

2.5 Cholesterol

The most ubiquitous sterol in the animal system is present as cholesterol. But plant lack cholesterol notwithstanding, they contain structurally similar other sterol and similar biosynthetic pathway exist both in plants and animals and some prokaryotes. In humans, Normal healthy adults synthesize cholesterol at a rate of approximately 1 g/day and consume approximately 0.3 g/day. A relatively constant level of cholesterol in the body (150–200 mg/dL) is maintained primarily by controlling the level of de novo synthesis, which is partly regulated in part by the dietary intake of cholesterol [40].

Cholesterol is biosynthesized from a 2-carbon metabolic intermediate, acetyl-CoA hooked end to end involving several enzymatic reactions and finally gets converted into the 27-carbon molecule of cholesterol. Metabolism (catabolism) of lipids, carbohydrates and proteins leads to the formation of AcetylCoA. The process of cholesterol synthesis has five major steps where the conversion of Acetyl-CoAs to 3-hydroxy-3-methylglutaryl-CoA (HMG-CoA) is the first. HMG-CoA is converted to mevalonate, followed by the formation of an isoprene-based molecule, isopentenyl pyrophosphate (IPP), with the concomitant loss of CO2 from mevalonate. It, therefore, converts the IPP to squalene, where cholesterol is formed from squalene as the last step. The reaction is repeated with the units of Acetyl-CoA. Two moles of acetyl-CoA are condensed in a reversal of the thiolase reaction, forming acetoacetyl-CoA. Acetoacetyl-CoA and the third mole of acetyl-CoA are converted to HMG-CoA by the action of HMG-CoA synthase. HMG-CoA is converted to mevalonate by HMG-CoA reductase, HMGR (it binds this enzyme in the endoplasmic reticulum, ER). HMGR requires NADPH as a cofactor, and two moles of NADPH are consumed during the conversion of HMG-CoA to mevalonate.

The reaction catalyzed by HMGR is the rate-limiting step of cholesterol biosynthesis. Mevalonate is then activated by three successive phosphorylations, yielding 5-pyrophosphomevalonate. Phosphorylation mevalonate and successive reactions maintain their solubility since otherwise, these are insoluble in water [40].

Cholesterol from both diet and synthesis is utilized in the formation of membranes and the synthesis of the steroid hormones and bile acids, regulating membrane fluidity and permeability as a cell membrane structural component, formatting lipid rafts with sphingolipid to mediate cell-to-cell recognition, adhesion, and communication [40].

3. Biosynthesis of plant origin

Different plants not only synthesize different aromatic secondary metabolites but also synthesize varying amounts of them at specific times and in specific subcellular compartments. One would expect that regulation of the differential biosynthesis of sometimes very complex molecular structures might involve regulation of the supply of the precursors influencing the rate-limiting step for carbon flow through the shikimate pathway. Recent data on transgenic potatoes give some sign that this is indeed the case [41].

3.1 Biosynthesis of terpenoid compounds

Some terpenoids play an important role in plant growth and development, such as gibberellin, as plant hormones regulate plant development. Other terpenoids play a role in the interaction between plants and the environment, such as participating in plant defence systems as phytoalexins and interspecies competition as interspecific sensing compounds. Terpenoids make up one of the largest and structurally diverse groups of naturally occurring compounds [42]. Mevalonic acid is mostly employed as a terpenoid synthetic racemate; however, mevalonic acid dimethoxy acetal may be resolved as its quinine salt. The acetates of the individual enantiomers of mevalonolactone are much less soluble than the racemate, and these may determine the purity and chirality of biosynthetic mevalonate. The steric course of many terpenoid biosynthetic processes has been followed using the stereospecifically deuterated and tritiated 2R,3R- [2-'H]-, 2S,3R- [2-3H]-, 3R,4R- [4–3 H] -, 3R,4S- [4- HI-, and 3R,5R- [S 3 H]-mevalonate. Several routes for preparing 5S-[5-3H] mevalonate have been described. [l-3H] Isopentenal is a substrate for liver alcohol dehydrogenase and this affords 1 s- [l-3H] isopentenyl, which may then be converted into mevalonic acid. Alternatively, [5-3H] mevalonic acid can be reduced enzymatically with mevalonate reductase to afford 5S-[5-3H] mevalonic acid. The hemithioacetal of mevalonate and coenzyme-A is reduced by 3-hydroxy-3-methylglutaryl COA reductase." The steric course of this reduction is now known and they can also adapt this to afford 3R,5S- [5-3H] mevalonic acid. They have reported a mevalonate kinase assay [13]. An enzyme system capable of forming the mono- and pyro-phosphates of mevalonic acid, isopentenyl pyrophosphate and dimethylallyl pyrophosphate has been isolated 'From orange-juice vesicles and shown to convert isopentenyl pyrophosphate and dimethylallyl pyrophosphate into linalool [42].

3.1.1 Pharmacology of terpenoids

Antitumour, [43] anti-inflammatory, antibacterial effects, cardiovascular effects, antimalarial, hypoglycemic effect, and transdermal absorption promotion have been investigated to give such pharmacological activities. Other activities apart from the mentioned above include insect resistance, immunoregulation, antioxidation, antiaging, and neuroprotection have also been known [3].

3.1.2 Works done on terpenoids

They have conducted several studies on terpenoids and their outcomes are promising. For antitumour activity, they have shown that perillyl alcohol which contains terpenoids plays a preventive and therapeutic role in cancer. The results showed that the administration of perillyl alcohol in rats can significantly reduce the incidence and multiplicity of colonic invasive adenocarcinoma caused by the injection of carcinogen azomethane [43]. Paeoniflorin which is a monoterpene glycoside compound isolated from the root of *Paeonia lactiflora* gave significant levels of anti-inflammatory activity. It could also dose-dependently inhibit the production of inflammatory factor nitric oxide (NO), interleukin-6 (IL-6) and tumour necrosis factor-alpha (TNF-α) induced by lipopolysaccharides (LPS) [44]. Besides, menthol, which is a cyclic monoterpene, has been shown to have antibacterial activity [45], they found that menthol showed significant inhibitory activity of biofilm when studying the effects of plant-derived terpenoids on *Candida albicans*. Several studies have also shown that terpenoids, particularly artemisinin which is a sesquiterpene lactone compound isolated from *Artemisia annua* Linn, has an effective antimalarial activity and currently in use [3]. The latest findings suggest that tanshinone IIA (TS) which is a terpenoid can prevent the formation of atherosclerosis and the damage and hypertrophy of the heart. This is because of its ability to inhibit the oxidation of low-density lipoprotein and the expression of proinflammatory factors, and TS also has certain activity and potential to stabilize atherosclerotic plaque [46].

3.2 Biosynthesis of pinocembrin

In plant propolis, pinocembrin is one of the most abundant flavonoid, and it could also be commonly found in a most plants. Biological synthesis plays a significant part in synthesizing pinocembrin owing to its increased yield and low cost in production. They can extract pinocembrin from product of nature but that methodology is of high production cost and reduced yield. Biological synthesis from microorganisms features the advantages of low cost and large product yield, which compensate for the lack pinocembrin natural sources [47]. *Escherichia coli* has been known to be used in the production of pinocembrin. Recently, efficient way of producing pinocembrin has been the main goal of most researchers. Biological production of pinocembrin mostly requires that one needs to supplement an expensive phenylpropanoid starting materials, resenting a key problem in previous studies. Genetic engineering is now the breakthrough in the synthesis of pinocembrin biosynthesis, where there is the usage of bacteria to construct the synthesis of pinocembrin from glucose. To manufacture the flavonoid precursor (2S)-pinocembrin directly from glucose, four-vectors have been assembled, 3-deoxy-D-arabinoheptulosonate-7-phosphate synthase, chorismite mutase/pre-phenate dehydratase, phenylalanine ammonia-lyase (PAL), 4-coumarate: CoA ligase (4CL), chalcone synthase (CHS), chalcone isomerase (CHI), malonate synthetase, malonate dehydratase, and malonate carrier protein. Pinocembrin synthesis from glucose can be achieved through adjustment of other corresponding parameters in the synthetic pathways [47].

3.2.1 Pharmacological activities

Pharmacologically, pinocembrin can exhibit anti-inflammatory, [48] antioxidant, [49] antibacterial [50] and neuroprotective activities [51]. They mainly use

Pinocembrin for the treatment of ischemic stroke. Moreover, current studies have shown that pinocembrin may have an effect of reversing Parkinson's disease (PD) and Alzheimer's disease (AD) in affected patients [52]. It has also been shown to exhibit anti-pulmonary fibrosis and vasodilating effects. Pinocembrin undergoes several pathways to perform its pharmacological effects. In addition, pinocembrin could also ease blood brain barrier (BBB) disruption and neurological injury by interfering and reducing the levels of inflammatory factors and reactive oxygen species (ROS). Also, pinocembrin is known to preserve mitochondrial integrity via the activation of the signal-regulated kinase/nuclear factor erythroid 2-related factor 2 (Erk1/2-Nrf2) pathway extracellularly [53]. Pinocembrin also attenuates apoptosis by affecting the p53 pathway, influencing the Bax-Bcl-2 ratio and cytochrome C release [54].

3.2.2 Works done on pinocembrin

The brain of rats has been demonstrated to be protected against apoptosis and oxidation induced by ischemia–reperfusion both in vivo and in vitro. Pinocembrin attenuates blood–brain barrier injury induced by global cerebral ischemia–reperfusion in rats [55]. Further research work is also showing that Pinocembrin has the potential of giving positive outcomes in the treatment of ischemic stroke. This is because it can significantly cause a reduction in the regions of cerebral infarction in rats and cerebral ischemia and reduce the level of cerebral oedema and apoptosis of the cells in the nerve [47]. Shen et al., [47] concluded that pinocembrin exhibits several effects on, Parkinson's disease, ischemic stroke, solid tumors, Alzheimer's disease, and some other diseases because of possibility of releasing inflammatory factors by halting several signaling pathways, such as PI3K/AKT and MAPK. Antioxidant role is also known in pinocembrin because of its ability to reduce the release of NO, nNOS, ROS and iNOS.

3.3 Biosynthesis of polyphenols

Polyphenols constitute an integral class of key secondary metabolites with multiple phenolic hydroxyl groups including flavonoids, stilbenes, phenolic acids, and tannins (hydrolysable and condensed) [56] synthesized mainly by a metabolic pathway termed phenylpropanoid [57].

The biological synthetic routes of polyphenols involve the phenylpropanoid and shikimic acid metabolism pathways [58]. In Salvia species, polyphenols found are mainly reduced by the phenylpropanoid metabolic pathway [57–60] and most derivatives have synonymous basic structures [61].

Tyrosine and Phenylalanine are precursor compounds of the phenylpropanoid metabolic pathway, and their biosynthetic pathways make up two (2) parallel branches of this pathway involving five rate-limiting enzymes [62, 63]. These enzymes consist of phenylalanine ammonia-lyase (PAL); which is a key regulatory enzyme in plant metabolism, cinnamic acid-4-hydroxylase (C4H) and 4-coumarate: coenzyme A (COA) ligase (a peculiar regulatory enzyme in the phenylalanine branch), rosmarinic acid synthase (a key enzyme in catalytic synthesis) and tyrosine aminotransferase (the initial key enzyme and rate-limiting enzyme in tyrosine metabolism pathway), [64].

Caffeic acid derivatives are phenolic acids derived from Salvia species which are mostly produced via esterification of caffeic acid with danshensu [65, 66]. Caffeic acid originates from the class of phenyl propionic acid, [67] and it is the fundamental structural unit of phenolic acids [68]. Phenylalanine is the precursor compound of caffeic acid, which helps in the production of caffeic acid through the action of C4H and PAL enzymes.

It plays a key role in the metabolic pathway of phenylpropanoid and is a precursor compound of rosmarinic acid [69, 70]. Studies have speculated that, catalysis of caffeoyl CoA in the main synthetic route of rosmarinic acid, caffeic acid. Subsequently, caffeoyl CoA and 4-hydroxyphenyl acetic acid is catalysed by hydroxycinnamoyl-CoA: hydroxyphenyllactate hydroxycinnamoyl transferase (rosmarinic acid synthase (RAS)) to earn caffeoyl-40-hydroxy phenylacetic acid (caffeoyl-40-HPLA).

The reaction is finally catalyzed by CYP98A14 to rosmarinic acid [71]. Rosmarinic acid is employed in the formation of Salvianolic acid E under the action of enzymes and other reactions which are then transformed into salvianolic acid B and other compounds. This observation seeks to infer that rosmarinic acid is the core constituent unit of a series of complex phenolic acids, such as salvianolic acids [72, 73]. Complex phenolic compounds formation is based mostly on rosmarinic acid synthesis [61].

The phenylpropanoid metabolic pathway is an important upstream pathway for producing flavonoids such as anthocyanins, isoflavonoids, and flavonoids [74].

3.3.1 *Pharmacology of polyphenols*

They exhibit numerous pharmacological activities, such as anti-cardiovascular, [75] anti-oxidation [76], anti-tumor [77] anti-inflammatory [78]. Other pharmacological activities exhibited by polyphenols include; anti-hypertensive (caffeic acid, chlorogenic acid and salvianolic acid A), memory and cognitive impairment improvement (rosmarinic acid, total salvianolic acid), hypoglycemic (Salvianolic acid B), antiviral (Protocatechuic aldehyde, Magnesium lithospermate B, rosmaric acid), prevents and treats cancer (Danshensu, Protocatechualdehyde) [79].

3.3.2 *Works done on polyphenols*

Chang et al., [80] have illustrated that phenolic acids exhibit better antioxidant properties because of their mechanism of action including, inhibition of free radical generation, free radical scavenging and lipid peroxidation. A study showed that rosmarinic acid, danshensu and caffeic acid were as effective as the positive control (quercetin), which scavenged 2,2-diphenyl-1-picrylhydrazyl (DPPH) radicals in a concentration-dependent way. On the other hand, ferulic acid was less effective [81]. Other studies have shown that danshensu and salvianolic acid B showed greater scavenging activities against HO·, –, DPPH, O2-, and 2,20-and-bis (3-ethylbenzothiazoline-6-sulphonic acid (ABTS) than the other constituents. Among them, the half-maximal inhibitory concentration (IC50) of salvianolic acid B and danshensu were not significantly different, and there was no obvious difference in the scavenging actions of hydroxyl radicals [82]. Thus, hydroxyl radical scavengers are phenolic acids are.

Studies on inhibition of spontaneous lipid peroxidation in liver tissue of polyphenols in mice lead to the formation to the following order after testing all the seven polyphenols. Rosmarinic emerged the most efficacious followed by caffeic acid, protocatechuic aldehyde, chlorogenic acid, ferulic acid, and danshensu. 3-hydroxycinnamic acid was studied in hydrogen peroxide -induced liver lipid peroxidation in rodents' model. Also, other in vitro studies confirmed the antioxidants effects of these phenolic acids [83].

Salvianolic acid B was found to be effective in the reduction of myocardial ischemia–reperfusion injury. Ischemia–reperfusion model was established by ligating the left circumflex artery in Sprague–Dawley (SD) rats against myocardial ischemia–reperfusion injury and the concentration and apoptotic index of the plasma level of myocardial enzymes (cardiac troponins (CTn) I and creatine kinase-MB (CKMB)), endothelin (ET), superoxide dismutase (SOD), nitric oxide (NO), malondialdehyde (MDA), and histological changes of the heart were determined. The outcome was observed that salvianolic acid B significantly increased the plasma levels of CTn I, CKMB, MDA, and ET contents; decrease in T-SOD and NO contents; reduction in infarct size; and improved myocardial ultrastructure. It was concluded that salvianolic acid B has an impact against conditions such as myocardial ischemia–reperfusion injury via the regulation of reducing oxidative stress, active oxygen metabolism, and myocardial apoptosis [84]. A study conducted by [85] also stated that salvianolic acid which is a phenolic could also be effective in alleviating ischemic-reperfusion injury. It could also prevent myocardial ischemia–reperfusion injury through an increased glucose condition by adjusting the NADPH oxidase 2 (Nox2)/reactive oxygen species (ROS)/ phosphorylated-c-Jun N-terminal kinase 2 (p-JNK2)/NF-κB pathway to reduce transient receptor potential cation channel, member 6 (TRPC6)/$Ca2+$ influx, subfamily C [86].

Anti-thrombotic effects which are through its actions on blood rheology has also been known to be a possible action of salvianolic acid in *S. miltiorrhiza*. It also acts as to prevent platelet aggregation by targeting P2Y1 or P2Y12 receptors, which are novel target receptors required for anti-platelet aggregation. Salvianolic acid B experimentally, only antagonizes the action of platelet P2Y12 receptors, while salvianolic acid A and C are P2Y12 and P2Y1 receptor inhibitors [87].

Other studies by [88] have also illustrated that caffeic acid can inhibit platelet-mediated thrombosis by P-selectin expression, repression of ADP-induced platelet aggregation, ATP release, and $Ca2+$ mobilization and attenuate the activation of ERK, p38, JNK, and integrin αIIbβ3. It could also increase cAMP expression levels.

Better antiplatelet and anti-thrombotic therapeutic efficacy have also been demonstrated in Danshensu when compared to other constituents. Several contributions have been made through its ability to selectively suppress balancing the ratio of thromboxane A2 (TXA2)/prostacyclin (PGI2) and the expression of cyclooxygenase (COX)-2 [89].

Several studies have been conducted on the anti-liver injury activity of polyphenols and from [90] salvianolic acid B protects liver cells by enhancing lysosome-associated membrane protein 1 (LAMP1) expression and preserving lysosomal membrane integrity through scavenging ROS. Salvianolic A has also been concluded by [91, 92] that it protects acute hepatic injury after inducing mice models with concanavalin A. The outcome of the liver function markers, alanine aminotransferase (ALT) and serum aspartate aminotransferase (AST) showed that salvianolic acid A significantly reduced ConA-induced AST and ALT activity. Also, there was a reduction in the hepatotoxic cytokine levels, such as interferon (IFN)-γ and tumor necrosis factor (TNF)-α; improvement in the increased NF-κB level and cleaved caspase-3; and reversal of B-cell lymphoma-extra-large (Bcl-xL) expression [79].

Wang et al. [90] concluded that salvianolic acid B seems to be effective and safe, and it could develop this natural component into a potential therapeutic agent for the management of glioma. This is because of its inhibitory actions on the human glioma U80 cells, which its initiation leads to p38-activation-mediated ROS production.

Reviews conducted by Rasouli et al. [93] throws more light on the importance of polyphenols from foods. This because they are beneficial in most health conditions including pernicious human diseases (HDs). Also, people who followed a specific diet particularly polyphenol-rich diets are of lower risk of several ranges of chronic diseases, such as diabetes, obesity, cancer, and heart disease.

3.4 Biosynthesis of berberine

Berberine is a quaternary ammonium salt and it's from a group of isoquinoline alkaloids termed 2,3–methylenedioxy-9,10-dimethoxyprotoberberine chloride; C20H18NO4+. It is highly concentrated in the roots, stem bark, rhizomes and of numerous plants including *Rhizoma coptidis, Tinospora cordifolia, Coptis chinensis, Hydrastis canadensis, Berberis vulgaris, Berberis aquifolium, Berberis aristata, Arcangelisia flava* and *Cortex rhellodendri* [94].

Berberine comprises of several derivatives such as berberrubine, jatrorrhizine thalifendine, and demethyleneberberine [94].

3.4.1 Pharmacological activity

Pharmacologically, almost all parts of the plant have been shown to have several properties. These pharmacological properties of berberine includes antiemetic, antipyretic, tonic, antimicrobial, antipruritic, antiarrhythmic, sedative, antioxidant, anti-inflammatory, hypotensive, antinociceptive, anticholinergic and cholagogue actions, and it has been used sometimes like dysentery cholecystitis, cholelithiasis, leishmaniasis, jaundice, malaria and gall stones [95]. Berberine has been used for treating diarrhoea and gastrointestinal disorders for a long time [96, 97]. It has multiple pharmacological effects including; antimicrobial activity against 54 microorganisms, [98] inhibition of intestinal ion secretion and smooth muscle contraction, inhibition of ventricular tachyarrhythmia, reduction of inflammation, stimulation of bile secretion and bilirubin discharge [99]. Moreover, [100] have reviewed that berberine to possess pharmacological activities such as insulin secretion promotion, insulin resistance reduction, increased insulin secretion, inhibiting gluconeogenesis in the liver, stimulating glycolysis in peripheral tissue cells, reducing intestinal absorption of glucose, and regulating lipid metabolism, and modulating gut microbiota. Also, it is significant in the treatment of diabetic nephropathy, diabetic neuropathy, and diabetic cardiomyopathy because of its anti-inflammatory and antioxidant activities, in inflammatory diseases.

3.4.2 Works done on berberine

Antidiabetic activity of berberine has been conducted by Pang et al., [101] His review article highlighted several mechanisms with which berberine act to improve glucose control. A bibliometric review conducted between 1985–2018 also outlines that berberine has significant antibacterial and antipyretic effects and is a commonly used drug for treating infectious diarrhoea and amoebiasis. Berberine has significant antipyretic and antibacterial effects and its commonly used for the treatment of diarrhea associated with infections.

Zhao et al., [102] also added., that berberine has an improved activity in improving nonalcoholic fatty liver disease by halting glucogenesis and comprehensively regulating lipid metabolism, and its effect on inhibiting lipogenesis in the liver was much stronger. He also suggested that weight loss may partly mediate the improvement and would be a drug of choice for NAFLD patients and glucose metabolic disorder. But they, therefore, require future clinical trials to confirm these effects.

The review article by Zhu et al., [103] threw more light on the mechanism of action of berberine. It was stated that berberine potentially works through an increase in insulin sensitivity, LDLR mRNA stabilization, improvement of mitochondrial function, regulation of adenosine monophosphate-activated protein kinase (AMPK) pathway, alleviation of oxidative stress, and regulation of gut microenvironment being the major targets of berberine in the treatment of NAFLD.

Also, the reduction of DNA methylation and that of proprotein convertase subtilisin/Kexin 9 (PCSK9) expression is also involved in the pharmacological mechanisms of berberine involved in the management of NAFLD. Several mechanisms such as the immunologic mechanism in relation to the treatment of NAFLD, drug combinations, development of berberine derivative, delivery routes, and drug dose can be considered for further research.

From Cicero and Baggioni, [104] a good deal of preclinical evidence supports the role of berberine in the management of cerebral ischemia, Alzheimer's disease, anxiety, mental depression, and schizophrenia. However, most of these data have been obtained purely through experimental models [105]. Of particular interest is the potential antidepressant activity of berberine, it was found to inhibit the immobility time in mice in both tail suspension test and forced swim test. These two antidepressant models all gave effects in a dose-independent manner [106]. Regarding the bioactivities of berberine reported, monoamine oxidase (MAO)-A activity is noted to be inhibited. From Kong et al., [107] (MAO)-A is an enzyme needed to catalyze the deamination of catecholamines oxidatively, and thus inhibiting degradation of these neurotransmitters. Levels of norepinephrine, serotonin and dopamine, neurotransmitters are increased due to induction by MAO-A enzyme after acute and chronic administration of berberine in mice [107]. Under Kulkarni and colleague's data, Arora and Chopra [106] revealed the protective antidepressant-like activity of berberine against the reserpine-induced biogenic amine depletion (a monoamine depletion) mostly employed in the induction of depression in animals.

However, to there is therefore no conclusive data on the evaluation of the potential antidepressant effects of berberine in higher mammals such as humans [108].

The review article by Td et al., [100] has shown that berberine has the potential of reversing neurodegenerative effects in diseases such as Alzheimer, Huntington's disease, Parkinson's and because of its antioxidant activity. They have also conducted antiviral activity on berberine by Warowicka et al., [94] and he observed berberine could regulate the MPK/mTOR, MEK–ERK, and NF-κB signaling routes, which are needed for viral replication. It is deduced that it provides adequate supports to the host immune response, thus leading to viral clearance. Berberine and its derivatives might promise agents to be considered in future in the fight against the recent pandemic SARS-CoV-2, which is the causative agent responsible for causing COVID-19.

3.5 Biosynthesis of aristolochic acid

Roots and rhizomes of most Aristolochia species contain mixtures of nitro phenanthrene-carboxylic acids. The key acid is the aristolochic acid I. The aristolochic acids; a group of substituted 10-nitro-1-phenanthrene acids have been known to occur in most species of the genus Aristolochia, and other members of the family Aristolochiaceae [109, 110]. Aristolochic acids emanate from aporphines through oxidative cleavage of the hetero ring. They stated that the results were consistent with the observation that the aristolochic acids rises from aporphines via oxidative cleavage of the hetero ring. Because the pathway from tyrosine to the product runs through dopa, which is an amino acid that undergoes reversible transamination in plants [111].

In all cases, aristolochic acid I and the structurally related alkaloid magnoflorine could be shown in the roots and rhizomes. The biogenetic relationship with the aporphine alkaloids was due to both the structure of aristolochic acid and its occurrence with magnoflorine. Aporphine skeleton could yield aristolochic acids through oxidative cleavage of the heterocyclic ring. Benzylisoquinoline norlaudanosoline is a key intermediate in the biosynthetic pathway, which can be formed from tyrosine or a biochemical equivalent. Aporphine skeleton is also produced from norlaudanosoline through the phenol oxidation and dienol-benzene rearrangement [110].

Aristolochiaceae family of plants produces aporphine alkaloid, 4,5-dioxoaporphine, which is referred to as possible intermediates of the precursors of aristolochic acids and aristo lactams [112].

3.5.1 Pharmacological activities of aristolochic acid

It has been reported by Okhale et al., [112] that Aristolochic acid antirheumatics, as diuretics, in the treatment of oedema, to facilitate childbirth, in wound healing and for less common conditions such as cough, hemorrhoids, and asthma. Aristolochic acids also possess antibacterial [113, 114] antifungal, antiviral, and antitumor effects and in more recently, have been used in the pharmaceutical industry as convention usage [115, 116]. Herbal preparations with active constituents being aristolochic acids have been used for different illnesses such as urinary tract infection, hepatitis, vaginitis, oedema, upper respiratory tract infection, eczema, bronchitis, headache, oral ulcer, neuralgia, dysmenorrhea, arthralgia, hypertension, cerebrovascular accident, heart failure and pneumonia [117].

3.5.2 Works done on aristolochic acid

Aristolochia species that contain aristolochic acid is Aristolochia triangularis. Oliveira et al., [118] had revealed several studies such as the antiproliferative effect, and his conclusions show its outcomes are very desirable.

Several studies have reported that aristolochic acid is the potential of causing carcinogenic effects in humans [119]. Nephrotoxic effects of the renal cortex and further damage to the liver and bladder when much of it is ingested; likely because of the formation of bulky chemical DNA adducts. AA is dA-AA formation is the most abundant and mutagenic form of DNA adduct associated with. In exons 2–11 of TP53, mutation results from bulky chemical DNA adducts, primarily of A: T base pairs [120, 121] have also conducted potential nephrotoxic effects of aristolochic acids and also proposed possible molecular mechanism of such effect. This is through induction of oxidative/nitros active stress and mitochondrial dysfunction, apoptosis induction, inflammatory responses induction, and fibrosis. A pharmacokinetic study conducted by [122] also added up that Aristolochic nephrotoxicity comprises dose-dependent and progressive tubular damage, even though significant changes in the morphology of glomeruli was not seen.

3.6 Biosynthesis of canthaxanthin

Canthaxanthin is a carotenoid, and it's one class identified to possess a lot of colouration. Canthaxanthin is biologically synthesized from the precursor, β-carotene, ketolase enzyme (BKT in algae and CRTW in bacteria) serves as the enzyme for the reaction. The allylic 4-position to a carbonyl group is oxidized in the β-ring, producing echinenone as intermediate product. The same enzyme sequentially transforms the 40-carbon atom in the second β-ring to a carbonyl [123, 124]. This is also a substrate for the synthesis of another keto carotenoid astaxanthin which is of commercial interest. The enzyme β-carotene hydroxylase introduces hydroxyl (OH) groups into the canthaxanthin rings at positions 3,30, resulting in astaxanthin formation [125].

3.6.1 Pharmacological activities of canthaxanthin

Antioxidant and free radical scavenging properties [125]. Canthaxanthin alters the onset of many diseases such as cataracts, atherosclerosis, multiple sclerosis, age-related macular degeneration, and cancer [126].

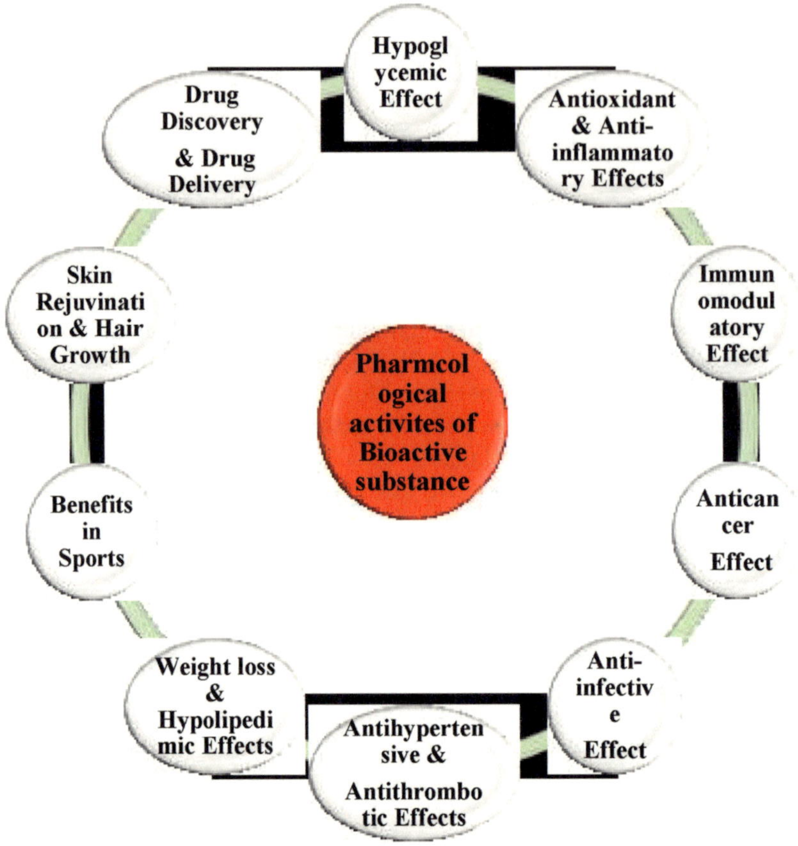

Figure 1.
A summary of biological activities and health-promoting properties of biological metabolite.

3.6.2 Works done on canthaxanthin

Studies have shown that canthaxanthin show significantly higher free radical scavenging and antioxidant properties than other xanthophylls and carotenes, and they can also scavenge reactive oxygen species and quench singlet oxygen. Also, it has been revealed in in vivo experiments that the supplementation of canthaxanthin led to a significant decrease in lipid peroxidation to preventing induced rats liver DNA damage [126]. They have also found it in cell detoxification of lipopolysaccharides. With respect to uses of canthaxanthin in humans, less comprehensive studies have been conducted to elucidate its effects and safety. It is popularly employed as natural skin-tanning agent, and cosmetics (**Figure 1**).

4. Conclusion

It employs most of these bioactive products in complementary medicine, and others are under clinical study. They can source bioactive products from plants, animal and some micro-organisms. They are produced mostly via a sequence of physiological processes or synthesized exogenously. Many biosynthetic products exhibit one therapeutic effect. These several biosynthetic metabolites have many pharmacological effects, thus making them useful candidates for drug discovery.

Author details

Yakubu Jibira* and Paul Atawuchugi
Kwame Nkrumah University of Science and Technology, Kumasi, Ghana

*Address all correspondence to: jyakubu2@st.knust.edu.gh; yakjab21@gmail.com

References

[1] Fewer DP and Metsa-Ketela (2020). A pharmaceutical model for the molecukar evolution of microbial natural products. The FEBS Journal. Blackwell Publishing Ltd 287 (7) : 1429-1449

[2] Walsh, C. T., and Tang, Y. (2017) *Natural Product Biosynthesis: Chemical Logic and Enzymatic Machinery*, Royal Society of Chemistry, London

[3] Yang, W., Chen, X., Li, Y., Guo, S., Wang, Z., Yu, X., 2020. Advances in Pharmacological Activities of Terpenoids.

[4] Vaz, F.D.M., Wanders, R.J.A., 2002. Carnitine biosynthesis in mammals 429, 417-429.

[5] Tanphaichitr, V., Horne, D. W. and Broquist, H. P. (1971) Lysine, a precursor of carnitine in the rat. J. Biol. Chem. *246*, 6364-6366

[6] Pekala, J., Bodkowski, R., Nowakowski, P., 2011. L-Carnitine - Metabolic Functions and Meaning in Humans Life.

[7] Ramsay RR, Zammit VA. Carnitine acyltransferases and their influence on CoA pools in health and disease. Mol Aspects Med. 2004;25: 475-493.

[8] Shenk JC, Liu J, Fischbach K, et al. The effect of acetyl-Lcarnitine and R-alpha-lipoic acid treatment in ApoE4 mouse as a model of human Alzheimer's disease. J Neurol Sci. 2009;283:199-206

[9] Zhang H, Jia H, Liu J, et al. Combined R-alpha-lipoic acid and acetyl-L-carnitine exerts efficient preventative effects in a cellular model of Parkinson's disease. J Cell Mol Med. 2010;14:215-225.

[10] Vanella A, Russo A, Acquaviva R, et al. L -propionyl-carnitine as superoxide scavenger, antioxidant, and DNA cleavage protector. Cell Biol Toxicol. 2000; 16:99-104.

[11] Arsenian MA. Carnitine and its derivatives in cardiovascular disease. Prog Cardiovasc Dis. 1997; 40:265-286.

[12] Paulson DJ, Traxler J, Schmidt M, Noonan J, Shug AL. Protection of the ischaemic myocardium by L-propionylcarnitine: effects on the recovery of cardiac output after ischaemia and reperfusion, carnitine transport, and fatty acid oxidation. Cardiovasc Res. 1986;20:536-541.

[13] Mingorance, C., Rodriguez-rodriguez, R., Justo, M.L., Herrera, M.D., 2011. Pharmacological effects and clinical applications of 69, 279-290.

[14] Mombouli JV, Vanhoutte PM. Endothelial dysfunction: from physiology to therapy. J Mol Cell Cardiol. 1999;31:61-74

[15] Félétou M, Vanhoutte PM. Endothelial dysfunction: a multifaceted disorder (The Wiggers Award Lecture). Am J Physiol Heart Circ Physiol. 2006;291:H985–H1002.

[16] Victor VM, Rocha M, Solá E, Bañuls C, Garcia-Malpartida K, Hernández-Mijares A. Oxidative stress, endothelial dysfunction and atherosclerosis. Curr Pharm Des. 2009;15:2988-3002.

[17] Herrera MD, Bueno R, De Sotomayor MA, Pérez-Guerrero C, Vázquez CM, Marhuenda E. Endothelium-dependent vasorelaxation induced by L-carnitine in isolated aorta from normotensive and hypertensive rats. J Pharm Pharmacol. 2002;54:1423-1427.

[18] Ferrari R, Ceconi C, Cargnoni A, et al. The effect of propionylL-carnitine on the ischemic and reperfused intact

myocardium and on their derived mitochondria. Cardiovasc DrugsTher. 1991;5:57-65

[19] Schönekess BO, Allard MF, Lopaschuk GD. Propionyl L-carnitine improvement of hypertrophied heart function is accompanied by an increase in carbohydrate oxidation. Circ Res. 1995;77:726-734

[20] Mister M, Noris M, Szymczuk J, et al. Propionyl-L-carnitine prevents renal function deterioration due to ischemia/reperfusion. Kidney Int. 2002;61:1064-1078

[21] Azzollini N, Cugini D, Cassis P, et al. Propionyl-L-carnitine prevents early graft dysfunction in allogeneic rat kidney transplantation. Kidney Int. 2008;74:1420-1428

[22] Surai PF (2015) Silymarin as a natural antioxidant: An overview of the current evidence and perspectives. *Antioxidants*. MDPI AG 4(1): 204-247

[23] Martinez T, Fabrias G and Camps F (1990) Sex pheromone biosynthetic pathway in Spodoptera littoralis and its activation by a neurohormone. *Journal of Biological Chemistry*. Elsevier 265(3): 1381-1387.

[24] Tillman, J.A., Seybold, S.J., Jurenka, R.A., Blomquist, G.J., 1999. Insect pheromones — an overview of biosynthesis and endocrine regulation 29, 481-514.

[25] Baxi KN, Dorries KM, Eisthen HL. Is the vomeronasal system really specialized for detecting pheromones? Trends Neurosci 2006; 29: 1-7.

[26] Semwal, A., Kumar, R., Vir, U., Teotia, S., Singh, R., 2013. Pheromones and their role as aphrodisiacs : A review. J. Acute Dis. 2, 253-261.

[27] Abdelhalim, M.A.K., Haseeb, A., 2016. Pharmacological Properties of Melanin and its Function in Health MiniReview Pharmacological Properties of Melanin and its Function in Health.

[28] Felix CC, Hyde JS, Sarna T, Sealy RC. Interactions of melanin with metal ions. Electron spin resonance evidence for chelate complexes of metal ions with free radicals, J. Am. Chem. Soc. 1978; 100: 3922-3926

[29] Shcherb VV, Babitskaya VG, Kurchenko VPN, Ikonnikova, NV, Kukulyanskaya TA. Antioxidant properties of fungal melanin pigments. Appled biochemistry and microbiology. 2000; 36: 491-495.

[30] Wu Y, Shan L, Yang S, Ma A. Identification and antioxidant activity of melanin isolated from Hypoxylon archeri, a companion fungus of Tremella fuciformi . Journal of Basic microbiology. 2008; 48: 217-221.

[31] Pugh ND, Balachandran P, Lata H, Dayan FE, Joshi V, Bedir E, Makino T, Moraes R, Khan I, Pasco DS. Melanin: dietary mucosal immune modulator from Echinacea and Other botanical supplements. Int Immunopharmacol. 2005; 5:637-647

[32] Pugh ND, Tamta H, Balachandran P, Wu X, Howell J, Dayan FE, Pasco DS. The majority of in vitro macrophage activation exhibited by extracts of some immune enhancing botanicals is due to bacterial lipoproteins and lipopolysaccharides. Int Immunopharmacol. 2008; 8:1023-1032.

[33] Sava V, Galkin B, Hong MY, Yang PC, Huang GS. A novel melanin-like pigment derived from black tea leaves with immuno-stimulating activity. Food Research International 2001, b.; 34:337-343.

[34] ElObeid AS, Kamal-Eldin A, Abdelhalim MAK and Haseeb AM (2017) Pharmacological Properties of Melanin and its Function in Health.

Basic & Clinical Pharmacology & Toxicology. Blackwell Publishing Ltd 120(6): 515-522.

[35] Sava VM, Hung YC, Blagodarsky VA, Hong MY, Huang GS. The liver-protecting activity of melanin-like pigment derived from black tea. Food Research International. 2003; 36:505-511.

[36] Kamei H, Koide T, Kojima T, Hasegawa M, Umeda T. Suppression of growth of cultured malignant cells by allomelanins, plant-produced melanins. Cancer Biother Radiopharm. 1997. a.; 12: 47-49.

[37] Hong Y, Song B, Chen H-D and Gao X-H (2015) Melanocytes and Skin Immunity. *The Journal of Investigative Dermatology Symposium* 17: 37-39.

[38] El-Obeid A, Al-Harbi S, Al-Jomah N and Hassib A (2006) Herbal melanin modulates tumor necrosis factor alpha (TNF-α), interleukin 6 (IL-6) and vascular endothelial growth factor (VEGF) production. *Phytomedicine*. Phytomedicine 13(5): 324-333.

[39] Thomas J, Shentu TP and K. D (2012) Cholesterol: Biosynthesis, Functional Diversity, Homeostasis and Regulation by Natural Products. *Biochemistry*. InTech. Available at: www.intechopen.com (accessed 29/03/21).

[40] Ling WH and Jones PJH (1995) Dietary phytosterols: A review of metabolism, benefits and side effects. *Life Sciences*. Life Sci, 195-206.

[41] Hanson, J. R. (1972). Biosynthesis of Terpenoids Compounds. The Chemical Society, 1, 41-43.

[42] Chen TC, Fonseca COD, Schönthal AH. Preclinical development and clinical use of perillyl alcohol for chemoprevention and cancer therapy. Am J Cancer Res. 2015;5(5):1580-1593.

[43] Bi X, Han L, Qu T, et al. Anti-Inflammatory effects, SAR, and action mechanism of monoterpenoids from Radix Paeoniae alba on LPS-stimulated RAW 264.7 cells. Molecules. 2017;22(5):715. doi:10.3390/molecules22050715

[44] Raut JS, Shinde RB, Chauhan NM, Karuppayil SM. Terpenoids of plant origin inhibit morphogenesis, adhesion, and biofilm formation by Candida albicans. Biofouling. 2013;29(1):87-96. doi: 201310.1080/08927014.2012.749398

[45] Gao S, Liu Z, Li H, Little PJ, Liu P, Xu S. Cardiovascular actions and therapeutic potential of tanshinone IIA. Atherosclerosis. 2012;220(1):3-10. doi:10.1016/j.atherosclerosis.2011.06.041)

[46] Shen, X., Liu, Y., Luo, X., & Yang, Z. (2019). Advances in Biosynthesis, Pharmacology, and Pharmacokinetics of Pinocembrin, a Promising Natural Small-Molecule Drug. MDPI, 1-14.

[47] Wu, C.X.; Du, G.H. Effects of pinocembrin on inflammation in rats with focal cerebral ischemia reperfusion injury. J. Shandong Med. College 2015, 37, 247-250.

[48] Said, M.M.; Azab, S.S.; Saeed, N.M.; El-Demerdash, E. Antifibrotic Mechanism of Pinocembrin: Impacton Oxidative Stress, Inflammation and TGF-β/Smad Inhibition in Rats. Ann. Hepatol. 2018, 17, 307-317.

[49] Celerino de Moraes Porto, I.C.; Chaves Cardoso de Almeida, D.; Vasconcelos Calheiros de Oliveira Costa, G.; Sampaio Donato, T.S.; Moreira Nunes, L.; Gomes do Nascimento, T.; Dos Santos Oliveira, J.M.; Batista da Silva, C.; Barbosa Dos Santos, N.; de Alencar E Silva Leite, M.L.; et al. Mechanical and aesthetics compatibility of Brazilian red propolis micellar nanocomposite as a cavity cleaning

agent. BMC Complement. Altern. Med. 2018, 18, 219.

[50] Liu, R.; Gao, M.; Yang, Z.H.; Du, G.H. Pinocembrin protects rat brain against oxidation and apoptosis induced by ischemia-reperfusion both in vivo and in vitro. Brain Res. 2008, 1216, 104-115.

[51] Rasul A, Millimouno FM, Ali Eltayb W, Ali M, Li J and Li X (2013) Pinocembrin: A novel natural compound with versatile pharmacological and biological activities. *BioMed Research International*. Hindawi Limited 2013. Available at: / pmc/articles/PMC3747598/ (accessed 29/03/21).

[52] Del Rio, D.; Rodriguez-Mateos, A.; Spencer, J.P.E.; Tognolini, M.; Borges, G.; Crozier, A. Dietary (poly) phenolics in human health: Structures, bioavailability, and evidence of protective effects against chronic diseases. Antioxid. Redox Signal. 2013, 18, 1818-1892.

[53] Gao, M.; Zhang, W.C.; Liu, Q.S.; Hu, J.J.; Liu, G.T.; Du, G.H. Pinocembrin prevents glutamate-induced apoptosis in SH-SY5Y neuronal cells via decrease of bax/bcl-2 ratio. Eur. J. Pharmacol. 2008, 91, 73-79.

[54] Meng F, Liu R, Gao M, Wang Y, Yu X, Xuan Z, Sun J, Yang F, Wu C and Du G (2011) Pinocembrin attenuates blood-brain barrier injury induced by global cerebral ischemia-reperfusion in rats. *Brain Research*. Elsevier 1391: 93-101.

[55] Mena P, Ascacio-Valdés JA, Gironés-Vilaplana A, Del Rio D, Moreno DA and García-Viguera C (2014) Assessment of pomegranate wine lees as a valuable source for the recovery of (poly)phenolic compounds. *Food Chemistry*. Food Chem 145: 327-334.

[56] Kallscheuer N.; Vogt, M.; Marienhagen, J. A novel synthetic pathway enables microbial production of polyphenols independent from the endogenous aromatic amino acid metabolism. ACS Synth. Biol. 2017, 6, 410-415.

[57] Chen, Z.J.; Wu, J.Q.; Ma, Y.; Wang, P.; Gu, Z.X.; Yang, R.Q. Advances in biosynthesis, regulation and bioactivity of phenolic acids in plant food raw materials. Food Sci. 2018, 39, 321-328.

[58] Schenck, C.A.; Maeda, H.A. Tyrosine biosynthesis, metabolism, and catabolism in plants. Phytochemistry 2018, 149, 82-102.

[59] Zhou, W.; Huang, Q.; Wu, X.; Zhou, Z.; Ding, M.; Shi, M.; Huang, F.; Li, S.; Wang, Y.; Kai, G. Comprehensive transcriptome profiling of Salvia miltiorrhiza for discovery of genes associated with the biosynthesis of tanshinones and phenolic acids. Sci. Rep. 2017, 7, 10554.

[60] Liu, F. Effect of PAP2 transcription factor on synthesis of salvianolic acids. Master's Thesis, Shaanxi Normal University, Xi'an, China, 2011

[61] Neta, M.; Moran, O.; Rinat, O.; Noga, S.P.; Biruk, A.; Aaron, F.; Gad, G.; Avichai, P.; David, W.; Michal, O.S. Phenylalanine and tyrosine levels are rate-limiting factors in production of health promoting metabolites in Vitis vinifera cv. Gamay Red cell suspension. Front. Plant Sci. 2015, 6, 538.

[62] Huang, Q.; Sun, M.; Yuan, T.; Wang, Y.; Shi, M.; Lu, S.; Tang, B.; Pan, J.; Wang, Y.; Kai, G. The AP2/ERF transcription factor SmERF1L1 regulates the biosynthesis of tanshinones and phenolic acids in Salvia miltiorrhiza. Food Chem. 2019, 274, 368-375.

[63] Ma, P.; Liu, J.; Zhang, C.; Liang, Z. Regulation of Water-Soluble Phenolic Acid Biosynthesis in Salvia miltiorrhiza Bunge. Appl. Biochem. Biotechnol. 2013, 170, 1253-1262.

[64] Jiang, R.W.; Lau, K.M.; Hon, P.M.; Mak, T.; Woo, K.S.; Fung, K.P. Chemistry and biological activities of caffeic acid derivatives from Salvia miltiorrhiza. Curr. Med. Chem. 2005, 12, 237-246

[65] Zhao, S.J.; Zhang, G.Y.; Liu, D.; Hu, Z.B. Advances in pharmacology and biosynthesis of water-soluble phenolic acids from Salvia miltiorrhiza. Chin. Tradit. Drug 2004, 3, 107-110.

[66] Razzaghi-Asl, N.; Garrido, J.; Khazraei, H.; Borges, F.; Firuzi, O. Antioxidant properties of hydroxycinnamicacids: A review of structure- activity relationships. Curr. Med. Chem. 2013, 20, 4436-4450.

[67] Zeng, F.; Chen, Y.B.; Xia, C.N.; Zhong, G.X. Advances in the synthesis and bioactivity of caffeic acid derivatives. Zhejiang Chem. Ind. 2012, 43, 24-28.

[68] Ma, X.H.; Ma, Y.; Tang, J.F.; He, Y.L.; Liu, Y.C.; Ma, X.J.; Shen, Y.; Cui, G.H.; Lin, H.X.; Rong, Q.X.;et al. The Biosynthetic Pathways of Tanshinones and Phenolic Acids in Salvia miltiorrhiza. Molecules 2015, 20, 16235-16254.

[69] Rong, Q.X.; et al. The Biosynthetic Pathways of Tanshinones and Phenolic Acids in Salvia miltiorrhiza. Molecules 2015, 20, 16235-16254.

[70] Xu, Z.; Luo, H.; Ji, A.; Zhang, X.; Song, J.; Chen, S. Global Identification of the Full-Length Transcripts and Alternative Splicing Related to Phenolic Acid Biosynthetic Genes in Salvia miltiorrhiza. Front. Plant Sci. 2016, 7, 100

[71] Sundaram, S.; Tripathi, A.; Gupta, D.K. Metabolic modeling of Rosmarinic acid biosynthetic pathway. Bioinformation 2010, 5, 168-172

[72] Di, P.; Zhang, L.; Chen, J.; Tan, H.; Xiao, Y.; Dong, X.; Zhou, X.; Chen, W. 13C tracer reveals phenolic acidsbiosynthesis in hairy root cultures of Salvia miltiorrhiza. ACS Chem. Biol. 2013, 8, 1537-1548

[73] Pandey, R.P.; Parajuli, P.; Koffas, M.A.G.; Sohng, J.K. Microbial production of natural and non-natural flavonoids: Pathway engineering, directed evolution and systems/ synthetic biology. Biotechnol. Adv. 2016, 34, 634-662.

[74] Mattera, R.; Benvenuto, M.; Giganti, M.G.; Tresoldi, I.; Pluchinotta, F.R.; Bergante, S.; Tettamanti, G.;Masuelli, L.; Manzari, V.; Modesti, A.; et al. Effects of Polyphenols on Oxidative Stress-Mediated Injury in Cardiomyocytes. Nutrients 2017, 9, 523.

[75] Eghbaliferiz, S.; Iranshahi, M. Prooxidant activity of polyphenols, flavonoids, anthocyanins and carotenoids: Updated review of mechanisms and catalyzing metals. Phytother. Res. 2016, 30, 1379-1391

[76] Hussain, T.; Tan, B.; Yin, Y.; Blachier, F.; Tossou, M.C.; Rahu, N. Oxidative stress and inflammation: What polyphenols can do for us? Oxid. Med. Cell. Longev. 2016, 2016, 7432797.

[77] Brglez Mojzer, E.; Knez Hrnˇciˇc, M.; Škerget, M.; Knez, Ž.; Bren, U. Polyphenols: Extraction methods, antioxidative action, bioavailability and anticarcinogenic effects. Molecules 2016, 21, 901.

[78] Wang, J., Xu, J., Gong, X., Yang, M., Zhang, C., & Li, M. (2019). Biosynthesis, Chemistry, and Pharmacology of Polyphenols from Chinese Salvia Species: A Review. 1-23.

[79] C.C.; Chang, Y.C.; Hu, W.L.; Hung, Y.C. Oxidative stress and Salvia miltiorrhiza in aging-associated cardiovascular diseases. Oxidative Med. Cell. Longev. 2016, 2016, 4797102.

[80] Adomakobonsu, A.G.; Chan, S.L.; Pratten, M.; Fry, J.R. Antioxidant activity of rosmarinic acid and its principal metabolites in chemical and cellular systems: Importance of physico-chemical characteristics. Toxicol. In Vitro 2017, 40, 248-255.

[81] Zhao, G.R.; Zhang, H.M.; Ye, T.X.; Xiang, Z.J.; Yuan, Y.J.; Guo, Z.X.; Zhao, L.B. Characterization of the radical scavenging and antioxidant activities of danshensu and salvianolic acid B. Food Chem. Toxicol. 2008, 46, 73-81

[82] Zhang, X.X.; Zhang, S.; Huang, X.Y.; Li, G.; Wang, Z.H.; Ma, C.Y. Anti-lipoperoxidation of Several Organic Polyphenol Acids in vitro. Coll. Life Sci. 2014, 26, 398-402

[83] Xue, L.; Wu, Z.; Ji, X.P.; Gao, X.Q.; Guo, Y.H. Effect and mechanism of salvianolic acid B on the myocardial ischemia-reperfusion injury in rats. Asian Pac. J. Trop. Med. 2014, 7, 280-284

[84] Zhang, W.; Song, J.K.; Zhang, X.; Zhou, Q.M.; He, G.R.; Xu, X.N.; Rong, Y.; Zhou, W.X.; Du, G.H. Salvianolic acid A attenuates ischemia reperfusion induced rat brain damage by protecting the blood brain barrier through MMP-9 inhibition and anti-inflammation. Chin. J. Nat. Med. 2018, 16, 184-193.

[85] Meng, Y. Danshen (Salvia miltiorrhiza) Prevents Myocardial Ischemia Reperfusion Injury by Regulate Nox2/ROS/JNK2/NF-κB pathway and Recude TRPC6/Ca2+ influx under High Glucose Condition. Anhui Med. Univ. 2016, 206.

[86] Liu, X.; Gao, Z.G.; Wu, Y.; Stevens, R.C.; Jacobson, K.A.; Zhao, S. Salvianolic acids from antithrombotic Traditional Chinese Medicine Danshen are antagonists of human P2Y 1 and P2Y 12 receptors. Sci. Rep. 2018, 8, 11380

[87] Lu, Y.; Li, Q.; Liu, Y.Y.; Sun, K.; Fan, J.Y.; Wang, C.S.; Han, J.Y. Inhibitory effect of caffeic acid on ADP-induced thrombus formation and platelet activation involves mitogen-activated protein kinases. Sci. Rep. 2015, 5, 13824.

[88] Yu, C.; Qi, D.; Lian, W.; Li, Q.Z.; Li, H.J.; Fan, H.Y. Effects of Danshensu on Platelet Aggregation and Thrombosis: In Vivo Arteriovenous Shunt and Venous Thrombosis Models in Rats. PLoS ONE 2014, 9, e110124.

[89] Huang M, Wang P, Xu S, Xu W, Xu W, Chu K and Lu J (2015) Biological activities of salvianolic acid B from *Salvia miltiorrhiza* on type 2 diabetes induced by high-fat diet and streptozotocin. *Pharmaceutical Biology*. Informa Healthcare 53(7): 1058-1065.

[90] Xu, X.; Hu, Y.; Zhai, X.; Liu, M.; Chen, Z.; Tian, X.F.; Zhang, F.; Gao, D.Y.; Ma, X.C.; Lv, L.; Yao, J.H. Salvianolic acid A preconditioning confers protection against concanavalin A-induced liver injury through SIRT1-mediated repression of p66shc in mice. Toxicol. Appl. Pharmacol. 2013, 273, 68-76

[91] Wang, Z.S.; Luo, P.; Dai, S.H.; Liu, Z.B.; Zheng, X.R.; & Chen, T. Salvianolic acid B induces apoptosis in human glioma U87 cells through p38-mediated ROS generation. Cell. Mol. Neurobiol. 2013

[92] Rasouli, H., Farzaei, M. H., & Khodarahmi, R. (2017). Polyphenols and their benefits : A review Polyphenols and their benefits : A review. International Journal of Food Properties, 00(00), 1-42.

[93] Warowicka, A., Nawrot, R., & Goździcka, A. (2020). Antiviral activity of berberine. Archives of Virology, 165(9), 1935-1945. https://doi.org/10.1007/s00705-020-04706-3.

[94] Khosrokhavar R, Ahmadiani AS (2010) Antihistaminic and anticholinergic activity of methanolic

extract of barberry fruit (Berberis vulgaris) in the Guinea-Pig Ileum. J Med Plants 13:99-105

[95] Akhter MH, Sabir M, Bhide NK (1979) Possible mechanism of antidiarrheal effect of berberine. Indian J Med Res 13:233-241.

[96] Rabbani GHBT, Knight J, Sanyal SC, Alam K (1987) Randomized controlled trial of berberine sulfate therapy for diarrhea due to enterotoxigenic Escherichia coli and Vibrio cholera. J Infect Dis 13:979-984.

[97] Amin AH, Subbaiah TV, Abbasi KM (1969) Berberine sulfate: antimidrobial activity, bioassay, and mode of action. Can J Microbiol 13:1067-1076.

[98] Sabir M (1971) Study of some pharmacological actions of berberine. Indian J Physiol Pharmacol 13:111-132.

[99] Ahmed T, Gilani AUH, Abdollahi M, Daglia M, Nabavi SF and Nabavi SM (2015) Berberine and neurodegeneration: A review of literature. *Pharmacological Reports*. Elsevier B.V., 970-979.

[100] Pang B, Zhao LH, Zhou Q, Zhao TY, Wang H, Gu CJ and Tong XL (2015) Application of berberine on treating type 2 diabetes mellitus. *International Journal of Endocrinology*. Hindawi Publishing Corporation. Available at: https://pubmed.ncbi.nlm.nih.gov/25861268/ (accessed 30/03/21).

[101] Zhao, L., Cang, Z., Sun, H., Nie, X., Wang, N., & Lu, Y. (2017). Berberine improves glucogenesis and lipid metabolism in nonalcoholic fatty liver disease. 1-8. https://doi.org/10.1186/s12902-017-0165-7.

[102] Zhu, X., Bian, H., & Gao, X. (2016). The Potential Mechanisms of Berberine in the. 1-12.

[103] Cicero, A. F. G., & Baggioni, A. (2016). Berberine and Its Role in Chronic Disease. 27-45.

[104] Kulkarni SK, Dhir A (2010) Berberine: a plant alkaloid with therapeutic potential for central nervous system disorders. Phytother Res 24(3):317-324

[105] Kulkarni SK, Dhir A (2008) On the mechanism of antidepressant-like action of berberine chloride. Eur J Pharmacol 589:163-172.

[106] Kong LD, Cheng CH, Tan RX (2001) Monoamine oxidase inhibitors from rhizoma of Coptis chinensis. Planta Med 67:74-76.

[107] Kumar A, Ekavali Chopra K et al (2015) Current knowledge and pharmacological profile of berberine: an update. Eur J Pharmacol 761:288-297.

[108] Comer, F., Tiwari, H. P., & Spenser, I. D. (1969). Biosynthesis of aristolochic acid1. Canadian Journal of Chemistry, 47(1), 481-487.

[109] Schutte, H. R., Orban, U., & Mothes, K. (1967). Biosynthesis of Aristolochic Acid. 1, 70-72.

[110] Spenser, I. D. (2015). Biosynthesis of aristolochic acid. February 2011. https://doi.org/10.1139/v69-070.

[111] Okhale, S., Egharevba, H., Ugbabe, G., & Igoli, J. O. (2019). Aristolochic acids in herbal medicine : Public health concerns for consumption and poor regulation of botanical products in Nigeria and West Africa. February. https://doi.org/10.5897/JMPR2018.6691

[112] American herbal product Association (AHPA) (2017). Botanical Identity References Compendium Aristolochic acid. Avialable at http://www.ahpa.org/Resources/TechnicalGuidance/ScientificAffairs/BotanicalAuthenticationProgram/

AHPABotanicalIdentity ReferencesCompendium.aspx

[113] Refik Gökmen M, Cosyns JP, Arlt VM, Stiborová M, Phillips DH, Schmeiser HH, Simmonds MSJ, Cook HT, Vanherweghem JL, Nortier JL and Lord GM (2013) The epidemiology, diagnosis, and management of aristolochic acid nephropathy: A narrative review. *Annals of Internal Medicine*, 469-477.

[114] Kupchan SM, Doskotch RW (1962). Tumor inhibitors I. Aristolochic acid, the active principle of Aristolochia indica. Journal of Medicinal Chemistry 5(3):657-659.

[115] Zhang H, Cifone MA, Murli H, Erexson GL, Mecchi MS, Lawlor TE (2004). Application of simplified in vitro screening tests to detect genotoxicity of aristolochic acid. Food and chemical Toxicology 42(12):2021-2028

[116] Li W, Chen Z, Liao Y, Liu H (2005a). Separation methods for toxic components in Traditional Chinese Medicines. Analytical Sciences 21(9):1019-1029.

[117] De Oliveira, M.R.; Peres, A.; Gama, C.S.; Bosco, S.M.D. Pinocembrin Provides Mitochondrial Protection by the Activation of the Erk1/2-Nrf2 Signaling Pathway in SH-SY5Y Neuroblastoma Cells Exposed to Paraquat. Mol. Neurobiol. 2017, 54, 6018-6031.

[118] Debelle, F. D., Vanherweghem, J. L., & Nortier, J. L. (2008). Aristolochic acid nephropathy: A worldwide problem. Kidney International, 74(2), 158-169.

[119] Barnes, J. L., Zubair, M., John, K., Poirier, M. C., & Martin, F. L. (2018). Carcinogens and DNA damage. 0, 1213-1224.

[120] Anger EE, Yu F and Li J (2020) Aristolochic Acid-Induced Nephrotoxicity: Molecular Mechanisms and Potential Protective Approaches. *International Journal of Molecular Sciences*. MDPI AG 21(3): 1157.

[121] Chen, S., Fan, M., Tseng, C., Ho, Y., & Ã, K. H. (2007). Pharmacokinetics and nephrotoxicity of aristolochic acid in rabbits. 50, 180-188.

[122] Wu, G. G. G., Yuan, M., Shen, S., Ma, X. X. X., Fang, J., Zhu, L. L., Sun, L., Liu, Z. Z. Z. Z., He, X. X., Huang, D., Li, T. T., Li, C. C. C., Wu, J. C. J. J., Hu, X., Li, Z., Song, L. L., Qu, K., Zhang, H. H. H. H. H. H. P., Gao, P., ... McCall, C. E. (2014). Supporting Information. In Nature. https://doi.org/10.1016/j.cmet.2014.08.003.

[123] Karadas F, Erdoğan S, Kor D, Oto G and Uluman M (2016) The effects of different types of antioxidants (Se, vitamin E and carotenoids) in broiler diets on the growth performance, skin pigmentation and liver and plasma antioxidant concentrations. *Revista Brasileira de Ciencia Avicola* 18(1): 101-115.

[124] Manimala MRA and Murugesan R (2014) In vitro antioxidant and antimicrobial activity of carotenoid pigment extracted from Sporobolomyces sp. isolated from natural source. *Journal of Applied and Natural Science* 6(2): 649-653.

[125] Palozza P and Krinsky NI (1992) Astaxanthin and canthaxanthin are potent antioxidants in a membrane model. *Archives of Biochemistry and Biophysics*. Arch Biochem Biophys 297(2): 291-295.

[126] Rosa AP, Scher A, Sorbara JOB, Boemo LS, Forgiarini J and Londero A (2012) Effects of canthaxanthin on the productive and reproductive performance of broiler breeders. *Poultry Science*. Elsevier 91(3): 660-666.

Chapter 3

Plants' Bioactive Metabolites and Extraction Methods

Achouak Madani, Nassima Mazouni and Mohammed Nedjhioui

Abstract

Plants are an inexhaustible source of bioactive compounds that have been utilized by Men since antiquity. Plants' bioactive substances are called primary and secondary metabolites. Primary metabolites are dedicated to biological functions and survival. Secondary metabolites are luxury products synthesized mainly to defend against predators and to communicate with other plants. Medicinal plants have ever since been a growing source of interest for scientific research as well as food, chemical, and pharmaceutical for their diverse applications for their antioxidant, antibacterial, stimulant, and inhibiting properties. Obtaining secondary metabolite can be challenging as the inadequate choice of extraction methods could alter or destroy bioactive compounds such as thermolabile substances. Hence, a beforehand knowledge of secondary metabolites properties and a mastery of extraction methods are primordial for effective extraction. Extraction methods range from conventional extraction such as solvent extraction to modern techniques like enzymatic assisted extraction EAE and microwave-assisted extraction MAE, which are more efficient methods and more environmentally friendly as they require very little to no use of solvents.

Keywords: plants' metabolites, essential oils, lipids, alkaloids, phenolic compounds, extraction, green extraction

1. Introduction

Plants produce a wide range of bioactive molecules that serve many purposes. Those that are destined for biological functions and surviving are called primary metabolites. On the other hand, those compounds that are luxury products are called secondary metabolites. They serve as extra nutritionals and may have adverse effects on other living organisms. Secondary metabolites have been used since antiquity in traditional medicine due to their benefits on human health. They are potent antioxidants, anticancer, antidiabetic, anti-inflammatory, stimulants... Their use has witnessed a dramatic increase in pharmaceutical, chemical, and food industries for the advantages they present. However, obtaining these compounds is a challenging process as they represent a small portion of plants raw material.

Diverse extraction methods, from conventional methods such as solvent extraction to modern techniques like enzymatic assisted extraction EAE, have been used to improve extraction yield in terms of quality and quantity as well as environmental safety. This chapter discusses the main active substances categories, their role for plants and men as well as the different extraction methods used to obtain them.

2. Essential oils

2.1 Definition

Essential oils (EO) are complex mixtures of volatile, lipophilic, and liquid compounds, extracted from different parts of a plant by physical processes. These interesting natural products have the characteristic odor of the plant.

According to ISO norm (ISO/D159235) "An essential oil is a product made by distillation with either water or steam or by mechanical processing of citrus rinds or by dry distillation of natural materiel. Following the distillation, the essential oil is physically separated from the water phase."

Essential oils are highly concentrated substances and therefore rarely used neat [1]' they evaporate slowly when exposed to air at room temperature, and because of this, they are sometimes referred to as volatile oils [1].

EO's quality depends on many factors:

- Origin and parts of the plant used.
- Harvest way and extraction technique.
- Purification process.

2.2 Essential oils composition

The EOs are composed of terpenoids in large proportion, phenylpropanoids, and other compounds of diverse origin such as that resulting from the degradation of fatty acids.

2.2.1 Terpenoids

The Terpenoids represent the major part of essential oil compounds; they are defined as substances composed of isoprene (2-methylbutadiene) units [2]; they are synthesized by acetate via the mevalonic acid pathway.

The Terpenoids are classified according to the number of C5 isoprene units that they contain; the classes are:

2.2.1.1 Hemiterpenes C5

Hemiterpenes are the simplest type of terpenoids; they consist of a single unit of isoprene. There is evidence that these compounds may assist in plant defense by repelling herbivores or by attracting predators and parasites of herbivores [3].

2.2.1.2 Monoterpenes C10

Monoterpenes are compounds made up of two units of isoprene, and of chemical formula $C_{10}H_{16}$, among the classes included in this group of terpenoids: iridoids and pyrethrins; they have anti-inflammatory pharmacological properties.

2.2.1.3 Sesquiterpenes C15

This class of terpenes has a crude formula $C_{15}H_{24}$ (three units of isoprene), these molecules are found in linear or cyclic form; they have several applications but are rarely found in EOs because of their low volatility.

2.2.1.4 Diterpenes C20

Diterpenes consist of a group of chemically heterogeneous compounds, their crude formula is C20H32 (four units of isoprene). Because of their higher boiling points, they are not considered to be essential oils; instead, they are classically considered to be resins, the material that remains after steam distillation of a plant extract [3].

Diterpenes can be classified as linear, bicyclic, tricyclic, tetracyclic, pentacyclic, or macrocyclic Diterpenes depending on their skeletal core [4].

2.2.1.5 Sesterterpenes C25

Sesterterpenes consist of 5 units of isoprene (25 carbon backbone). They exist in a wide variety of forms, including linear, monocyclic, bicyclic, tricyclic, tetracyclic, and macrocyclic frameworks [4].

2.2.1.6 Triterpenes C30

The triterpenes are made up of six isoprenes, they have relatively complex cyclic structures.

There are also tetraterpenes which consist of 8 units of isoprene (C40H64); and when the number of isoprene units exceeds 8 units, terpenes are called polyterpenes ($C > 40$).

2.2.2 Phenylpropanoids

The phenylpropanoids are a family of organic compounds with an aromatic ring and a three-carbon propene tail [5]; they are synthesized via shikimic acid pathway (biosynthesis).

2.3 Essential oils properties

The different properties of OEs are related to the species and their chemical composition.

2.3.1 Physicochemical properties

Essential oils are characterized by several physicochemical properties:

- The EOs are known by their volatile, odorous, and flammable character.
- They are colorless or have a yellowish color in the liquid state and at ambient temperature.
- They have a low polarity and a low solubility in water however they are soluble in alcohol and most of the organic solvents.
- They have a lower density than water (0.8 to 1.08).

2.3.2 Biological properties

Essential oils have various biological properties due to the variety of their chemical composition, they are using as:

- Antimicrobial;
- Antioxidant;
- Anti-inflammatory;
- Antiseptic

3. Lipids

3.1 Definition

Unlike the various families of basic molecules in the living world characterized by chemical structures, lipids are defined by their solubility not on the basis of their chemical structure [6].

Lipids are among the primary metabolites, they consist of a heterogeneous group of compounds characterized by their insolubility in water, on the other hand they are soluble in non-polar organic solvents such as: chloroform and alcohols.

Lipids play mainly important roles in plants, as signaling and energy storage compounds [7].

3.2 Classification

Lipids mainly include fatty acids, simple lipids and complex lipids.

3.2.1 Fatty acids

The fatty acids are carboxylic acids R-COOH; radical R is an aliphatic chain of hydrocarbon type of variable length which gives the molecule its hydrophobic character (fatty).

They are generally mono carboxylic, and characterized by a linear chain with an even number of carbons [8].

3.2.1.1 Saturated fatty acids

A fatty acid consists of a hydrocarbon chain, more or less long, strongly apolar, and a polar carboxyl group. The general formula of fatty acids is CH3 - (CH2) n – COOH [8]. The most abundant fatty acids are: palmitic acid and stearic acid (**Figure 1**).

3.2.1.2 Unsaturated fatty acids

The structure of unsaturated fatty acids includes one or more double bonds; the presence of these double bonds gives them specific physicochemical properties (**Figure 2**) [8].

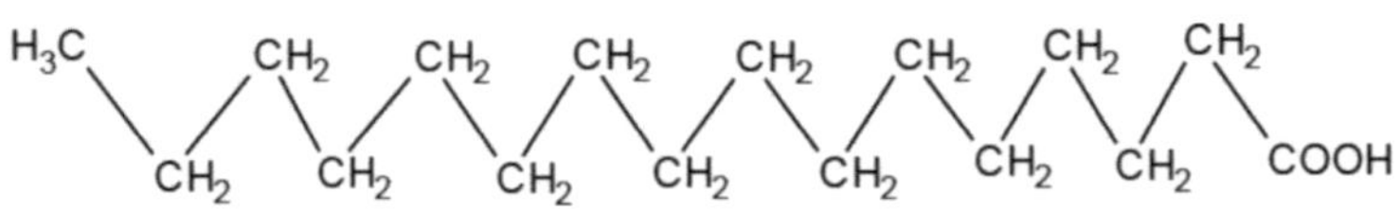

Figure 1.
Palmitic acid.

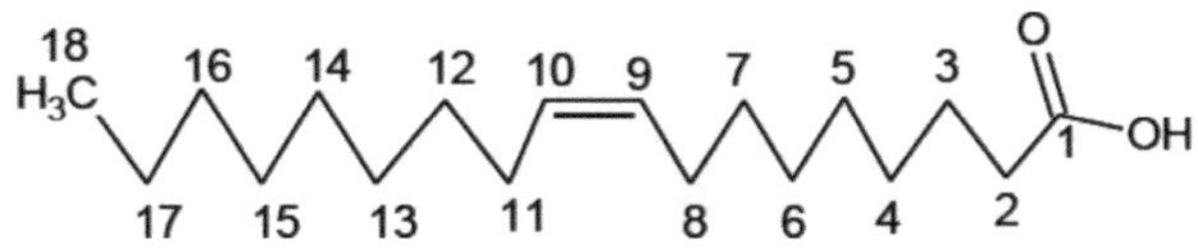

Figure 2.
Oleic acid.

3.2.1.3 Atypical fatty acids

Atypical fatty acids are characterized by an odd number of carbons; they are present in animal fats or in microbial lipids [8].

3.2.2 Simple lipids

Simple lipids or homolipids are ternary bodies (C, H, O), they are fatty acid esters, depending on the alcohol we distinguish the following classes [8]:

- **Acylglycerols:** or glycerides are esters of glycerol,
- **Cerides:** long chain alcohol esters (fatty alcohol),
- **Sterids:** sterol esters (polycyclic alcohol).

3.2.3 Complex lipids

Complex lipids are hetero lipids which contain phosphate, sulfate or carbohydrate groups. They are classified according to the molecule that fixes fatty acids [8].

3.2.3.1 Glycerophospholipids

These are the most numerous and most represented lipids which are constructed from the skeleton of a glycerol mono ester (**Figure 3**) [8].

3.2.3.2 Glyceroglycolipids

The alcohols of the C1 and C2 carbons of glycerol are esterified by fatty acids and the C3 carbon alcohol, unlike the glycerolipids, is not esterified, but it is linked to an ose by a glycosidic bond (**Figure 4**) [8].

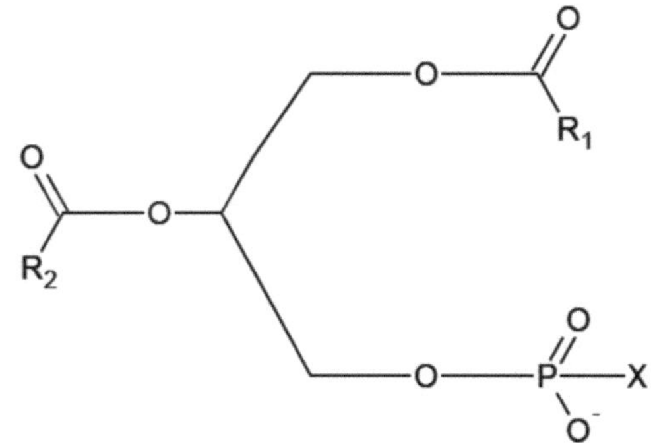

Figure 3.
Glycerophospholipids.

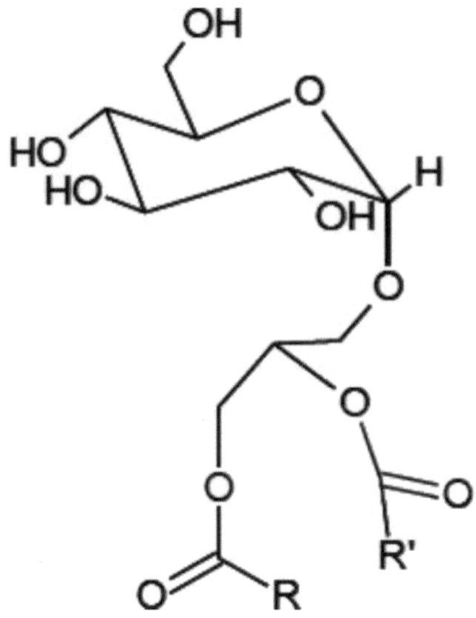

Figure 4.
Glyceroglycolipids.

3.2.3.3 Sphingolipids

The skeleton from which these lipids are made is not glycerol but a long-chain carbon diol amine of the sphingoid type; the binding of a fatty acid to the amine group gives a ceramide which is the lipid precursor molecule of this group [8].

4. Alkaloids

Alkaloids are natural compounds with low molecular weight [9] that are characterized by the presence of nitrogen atoms. They are mostly found in plants as secondary metabolites and some other animals and fungi [10].

They are derived from amino acids containing one or more heterocyclic nitrogen atoms [11]. Alkaloids have played a huge role in traditional medicine in various civilizations since antiquity due to their potential therapeutic properties. Toxic alkaloids were similarly used in poisonous arrows [12]. However, it was until 1804 that they were isolated and characterized by Friedrich Sertürner [13] and Derosne [9].

4.1 Classification

Due to their diverse structures, finding a common classification was complicated [14]. Some studies considered a classification based on the ring systems: indolizidine- and quinolizidine-based systems and quinoline-, quinazoline-, and acridone-based systems (**Figure 5**) [11]. Yet, this classification mode causes confusion as some alkaloids can be categorized in more than one class [15]. Henceforth, a new classification was adopted [9]:

True alkaloids: The nitrogen atom, originating from the precursor amino acid, is part of the alkaloid heterocyclic ring.

Protoalkaloids: The nitrogen atom, originating from the precursor amino acid, is not part of the alkaloid heterocyclic ring.

Pseudoalkaloids: They are not originated from amino acids, including steroids-like, purine-like and terpene-like alkaloids. Each class is divided to subclasses according to the precursor amino acid.

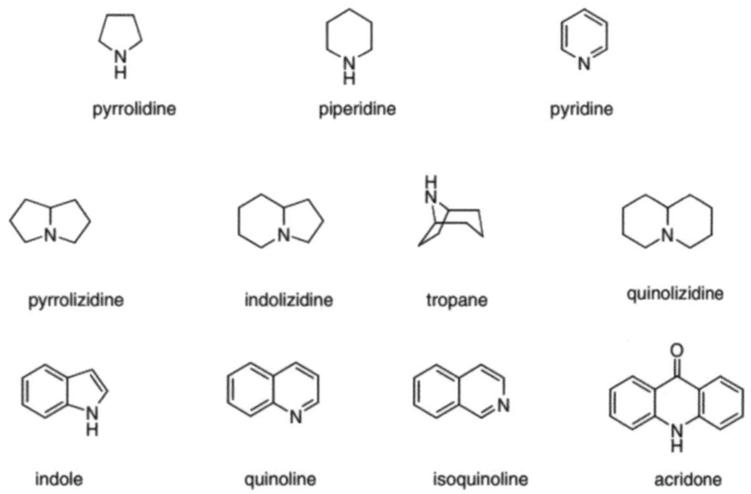

Figure 5.
Alkaloids classes based on ring systems [11].

4.2 Therapeutic usefulness

Alkaloids have a wide range of potential therapeutic activities. Morphine has narcotic effect, cocaine is a potent central nervous system stimulant, taxol as an effective anticancer chemical, colchicine as an anti-inflammatory [11], tubocurarine is a poisonous alkaloid used to relax muscles prior to surgeries [16], vinblastine, one of the antitumor alkaloids, and quinine being an effective anti-malarial [17].

4.3 Role for plants

Despite the various advantages they present for Man, their role in the plant is not well known. Some studies suggest that they help the process of seed formation [17] and protect the plant against herbivores and unwanted plants [18] whereas some others propose they are waste products [17].

4.4 Extraction methods

There exist as many extraction methods as the diversity of alkaloids structures [15]. They are extracted exploiting their solubility in organic solvents and water for their salts [19]. Base or acidic extractions followed by purification are widely used to obtain alkaloids.

Base extraction consists of the use of alkaline solvents including 1,2-dichloroethane, chloroform, diethyl ether or benzene. Base Alkaloids are then converted to salts by the addition of weak acids and are washed with water to crystallize. The purification method is repeated until the desired purity is reached.

Acidic extraction, on the other hand, processes the finely grounded raw plant material with weak acidic solutions mentioning acetic acid in water, methanol or ethanol… followed by the addition of basic solutions to convert alkaloids to basic forms to allow their extraction with organic solvents. The same purification steps are followed [14].

Different alkaloids are separated based either on the difference of solubility or boiling temperature [19].

5. Phenolic compounds

Phenolics are organic compounds containing one aromatic ring to which is attached one or more hydroxyl groups [11]. Phenolic compounds represent one of the most extensive groups of plants' secondary metabolites [20]. They are contained in a great share in the daily diet due to their countless benefits for human health and well- being. Henceforth, the consumption of phenolic-rich fruits and vegetables, such as berries and spinach, is increasingly recommended [20].

They are synthesized either through the shikimic acid or the acetic acid pathways. Phenylpropanoids are the result compounds of the shikimic acid pathway while simple phenols result from the acetic acid pathway acid. However, both pathways result in the formation of flavonoids, abundant phenolic compounds [21].

In addition to the presence of aromatic rings and hydroxyl substitutions, phenolic compounds are less likely to be found in the free form, rather, they are most frequently bound to other complex compounds as sugars to reduce their toxicity [21].

5.1 Classification

Due to the wide variety of phenolic compounds ranging from simple to polymeric, there exist multiple classifications:

- Based on the carbon chain:

From simple phenols to complicated lignins, 16 different classes can be distinguished. The main classes are shown in **Figure 6**.

- Based on solubility:

Depending on the form phenolics are found in plans (either free from any bound to other compounds or attached to cell wall polysaccharides or proteins), their solubility varies. Therefore, two categories can be defined:

- Soluble phenolics: Free phenolics found in the soluble fraction of the cell such as phenols, flavonoids, and tannins.

- Insoluble phenolics: Bound to other more complicated molecules to orm stable insoluble compounds. They majorly include condensed tannins and phenolic acids.

- Based on distribution:

Phenolic compounds are found in nature in different proportions. They thus can be divided to shortly distributed (simple phenols, pyrocatechol, hydroquinone, resorcinol ...), widely distributed (flavonoids and their derivatives, coumarins, and phenolic acids) and polymers (lignins and tannins) [21].

5.2 Main classes

5.2.1 Simple phenols

They are monomers constituted of various substitutions and represent the forming blocs of polymeric phenolic and acid compounds that make up the plant tissue. Some of them (p-hydroxybenzoic acid, protocatechuic acid, vanillic, syringic,

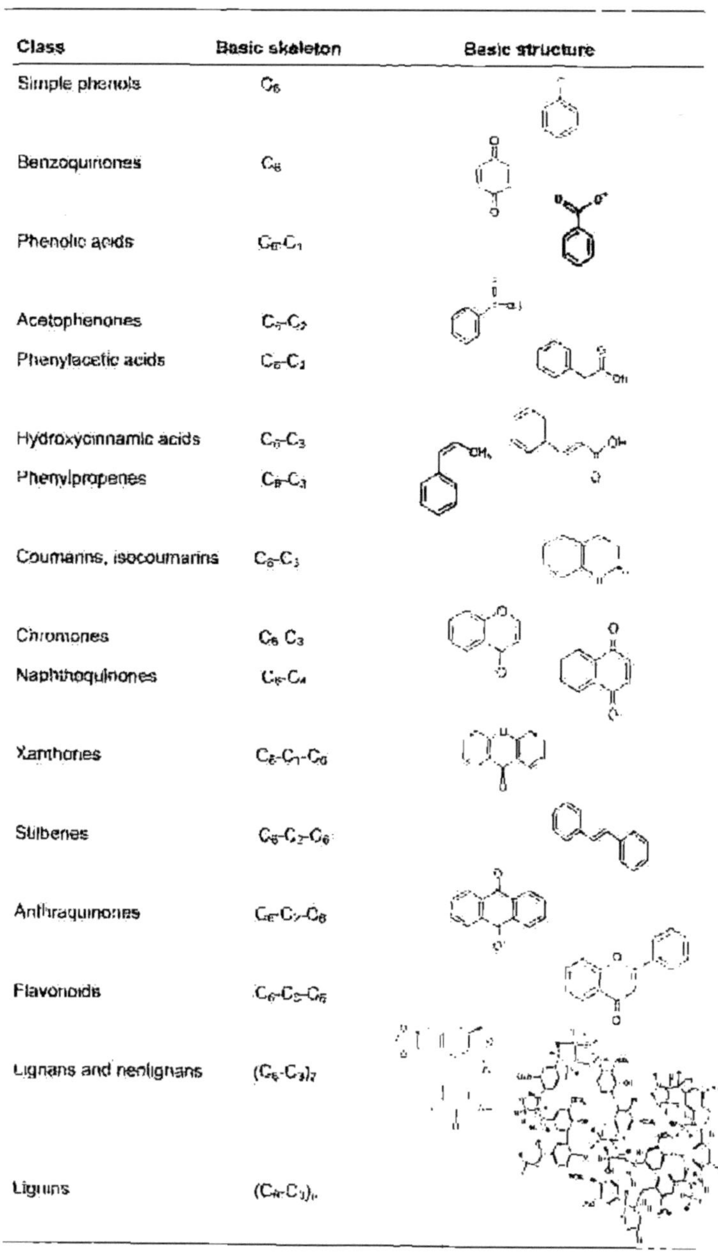

Class	Basic skeleton	Basic structure
Simple phenols	C_6	
Benzoquinones	C_6	
Phenolic acids	C_6-C_1	
Acetophenones	C_6-C_2	
Phenylacetic acids	C_6-C_2	
Hydroxycinnamic acids	C_6-C_3	
Phenylpropenes	C_6-C_3	
Coumarins, isocoumarins	C_6-C_3	
Chromones	C_6 C_3	
Naphthoquinones	C_6-C_4	
Xanthones	C_6-C_1-C_6	
Stilbenes	C_6-C_2-C_6	
Anthraquinones	C_6-C_2-C_6	
Flavonoids	C_6-C_3-C_6	
Lignans and neolignans	$(C_6$-$C_3)_2$	
Lignins	$(C_6$-$C_3)_n$	

Figure 6.
Classes of phenolic compounds [26].

salicylic, and gallic acids) can be obtained by the hydrolysis of the plant tissue whereas some other free simple phenols do not require the destruction the cell wall polymers (**Figure 7**) [11].

5.2.2 Flavonoids

They are compounds containing two aromatic rings attached by a bridge of triple carbon atoms (C6-C3-C6). The bridge often contains double carbon bonds and hetero- elements, which forms another mid-cycle. Therefore, flavonoids can be classified into 13 subgroups among which flavanols, flavones, isoflavones, anthocyanidins or anthocyanins, and flavanones (**Figure 8**) [21].

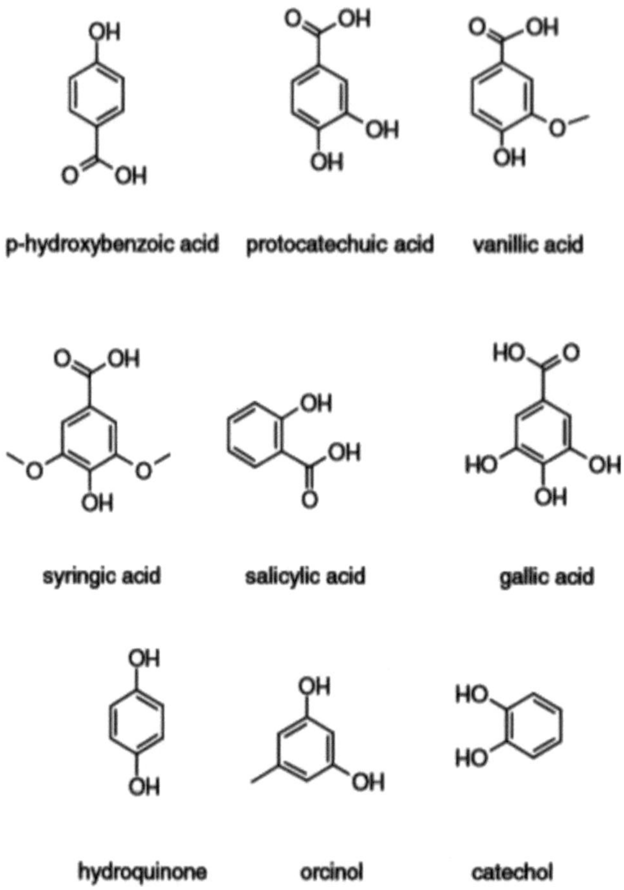

Figure 7.
Simple phenols [11].

5.2.3 Tannins

They are compounds with a medium to high molecular weight that comprise two major classes: hydrolysable and condensed tannins. Hydrolysable tannins are glucose or a polyhydric alcohol centered compounds esterified with gallic acid or hexahydroxydiphenic acid to form gallotannin and ellagitannins, respectively. They are hydrolyzed by enzymatic, acidic or base treatment. On the other hand, condensed tannins, also known as non-hydrolyzable tannins or recently proanthocyanidins, are polymers of catechin and leucoanthocyanidin. Their antioxidant activity is dependent on their degree of polymerization. High molecular weight tannins have been demonstrated to be fifteen to thirty times more active than simple phenols (**Figure 9**) [21].

5.2.4 Phenolic acids

As their name implies, phenolic acids are characterized by the presence of a phenolic ring and a carboxylic group. They comprise two categories:

- Benzoic acids:

They are the simplest phenolic acids in nature. They contain seven carbon atoms C6-C1 with hydroxyl or methoxyl roots.

DOI: http://dx.doi.org/10.5772/intechopen.96698

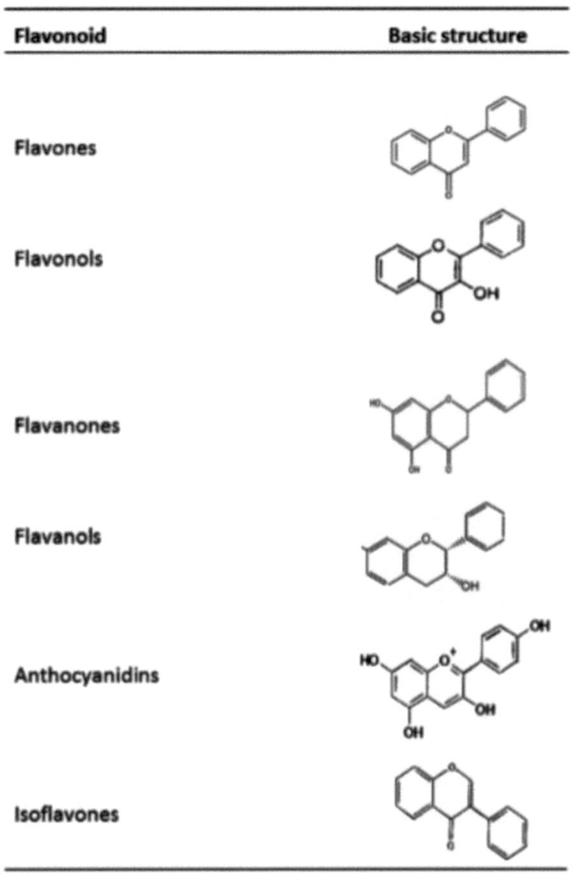

Figure 8.
Main classes of flavonoids [21].

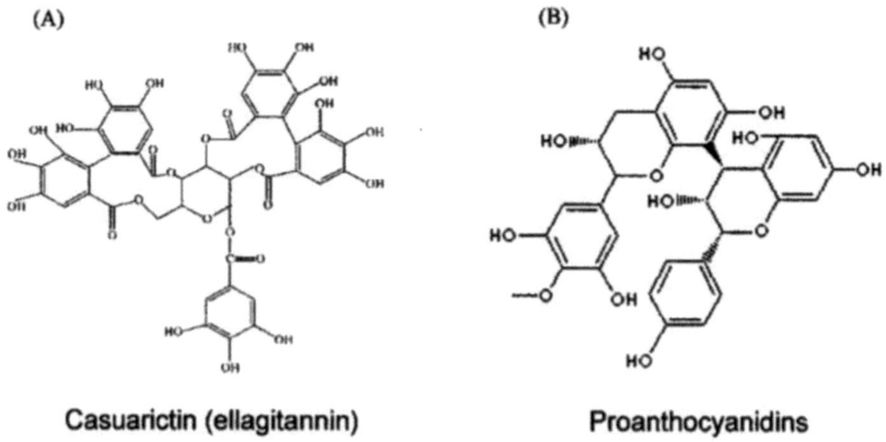

Figure 9.
Chemical structures of (A) hydrolysable tannins and (B) condensed tannins [21].

- Cinnamic acids:

They contain nine carbon atoms C6-C3 with hydroxyl or methoxyl roots.

Their nomenclature varies according to the nature of roots attached to the main ring [21].

5.3 Therapeutic usefulness

Phenolic compounds hold a variety of potential therapeutic properties ranging from antioxidant activity, anticancer [22], bacteriostatic, liver-protecting, anti-infection, cholesterol-lowering, immunity enhancement properties [23], cardio-protective and vasodilatory influences [20]. They have increasingly been part of the human diet for centuries for their benefits through the consumption of fruits and vegetables [21, 24].

5.4 Role for plants

Phenolics stand of great importance to plants. Not only are they responsible for the protection of the plants against exterior hazards [20], but they are also crucial to physiology and cellular metabolism. They play a key role in sensorial traits such as the plants' color and aroma, germination of seeds and reproduction [21].

6. Extraction methods

6.1 Hydrodistillation

6.1.1 Principle of the method

Hydrodistillation is a simple extraction techniquc which consists in putting in a flask proportional quantity of distilled water and the plant, then heating until boiling; the rising vapor condenses using a refrigerant to recover the distillate.

6.1.2 Mounting

See **Figure 10**.

6.2 Solvent extraction

This method consists in bringing the plant material into contact with the appropriate solvent; this protocol is carried out cold or hot. Among the most used solvents for the extraction of natural products from plant elements: petroleum ether, methanol, ethanol, and hexane. ne of the glassware used for this extraction technique is soxhlet (**Figure 11**).

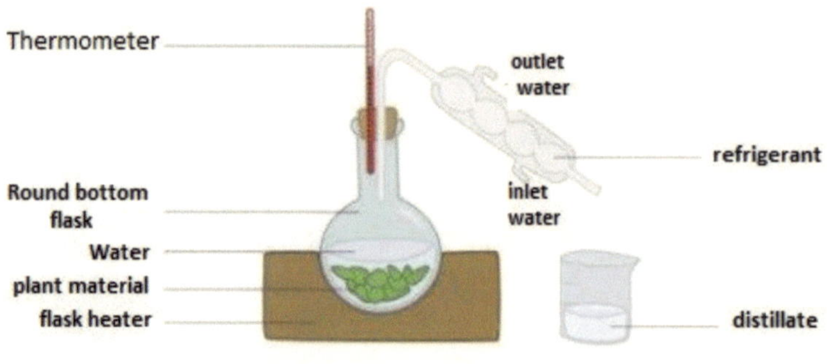

Figure 10.
Installation of Hydrodistillation.

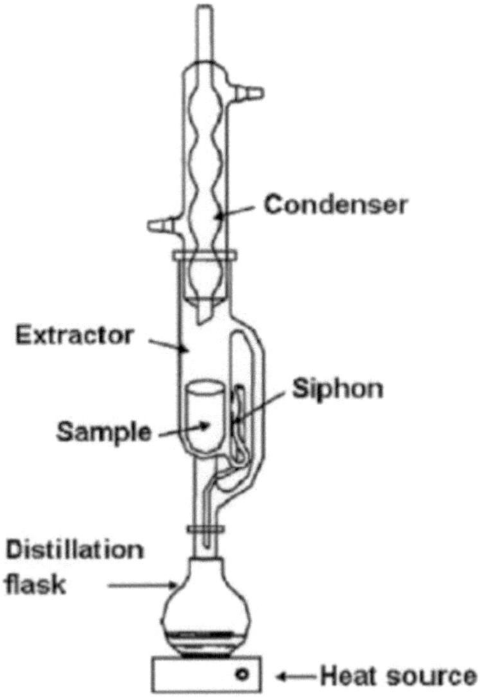

Figure 11.
Soxhlet.

6.3 Enzyme assisted extraction

Enzyme Assisted Extraction is an extraction technique that consists of the destruction of the source material cell wall using specific enzymes to liberate bio-active compounds [25]. Not only does this method augment the extraction yield compared to other conventional techniques, but it also is considered environmentally friendly as it does not require the use of toxic solvents [26, 27].

The most used enzymes are cellulases, hemicellulases, pectinases, and other hydrolytic enzymes [27]. These have the potential of catalyzing the hydrolysis of plant cell wall components such as polysaccharides and proteins. **Table 1** shows some of the useful plants metabolites and enzymes used for their extraction [26].

6.3.1 Plant cell walls

Due to their high insolubility and complex structure, plant cell walls are the major barrier to extracting bioactive compounds. Cellulose, the main component of cell walls, is a carbohydrate polymer that is characterized by low solubility and hydrogen- bonded crystalline fibers, which render its degradation greatly arduous. Components of the cell wall intervene in the process of extraction. Therefore, the nature of the cell wall matrix, the nature of the desired compounds, and their location are key factors controlling the extraction yield [25].

6.3.2 Principle and mechanism of action

Enzymatic assisted extraction is mainly based on the selectivity and ability of enzymes to intrude the matrix of the cell wall through interaction with the cell wall complex components. Thus, the release of bio-actives in the bulk solution is

Product type	Product	Source	Enzyme used	Maximum yield (%)
Oils and carotenoids	Oil	Grape seed	Cellulase, protease, xylase and pectinase	17.5
	Carotenoids	Marigold flower	Viscozyme, Pectinex, neutrase, corolase and HT-proteolytic	97
	Volatile oil	Mandarin peel	Xylan-degrading enzymes	15
	Carotene	Carrot pomace	Pectinex Ultra SP-L	0.0064
	Lycopene	Tomato	Pancreatin	2.5-fold
		Tomato	Cellulase and pectinase	206
	Capsaicin	Chilli	Cellulase, hemicellulase and pectinase	n.d.[a]
	Colourant	Pitaya	Pectinolytic, hemicellulolytic and cellulolytic enzymes	83.5
	Anthocyanin	Grape skin	Pectinex BE3-L	n.d.[a]
Glycosides	Sugar	Grapefruit peel waste	Cellulase and pectinase	0.6377
	Oligosaccharide	Rice bran	Cellulase	39.9
	Inulin	Jerusalem artichoke	Inulinase	n.d.[a]
	Starch	Cassava	Pectinase enzyme	45.6
	Pectin	Pumpkin	Xylase, cellulose, β-glucosidase, endopolygalacturonase and pectinesterase	14.0
Others	Vanillin	Vanilla green pods	β-glucosidase and pectinase	14–21
	Flavonoid (naringin)	Kinnow peel	Recombinant rhamnosidase	n.d.[a]
	Phenolics	Citrus peel	Celluzyme MX	65.5
	Proteins	Lentils and white beans	Glucoamylases	50.3
	Polyphenols	Grape pomace	Pectinolytic	98.1
	Catechins	Tea beverage	Pepsin	80
	Lignans	Flax	Cellulase and glycosidase	40.75 mg/g
	Soluble fibre	Carrot pomace	Cellulase-rich crude preparation	77.3

[a]*Abbreviation: n.d., not defined.*

Table 1.
List of bioactive compounds of industrial importance obtained by enzyme- assisted extraction from plants [26].

enhanced. Enzymes bind to their specific substrates by conformational complementarity forming the enzyme-substrate complex and therefore, allowing the hydrolysis to occur. This process is a function of various parameters such as temperature, hydrogen potential, enzyme concentration, the particle size of the substrate, and time of extraction that directly influence the efficiency of EAE. Optimizing these factors implies ensuring a high yield extraction in terms of quality and quantity [25].

6.3.3 Optimum operating conditions

The choice of enzymes is the first parameter to study. It is dependent on the chemical structure of the targeted compounds, the structural complexity of the cell wall, and the nature of the raw material. A combination of different enzymes is possible. Optimum temperature and pH are then selected based on the enzymes chosen (**Table 2**) [27].

Particle size is also a determinant parameter. Small particles were observed to have a better contact between enzymes and substrate and thus lead to a better release of bio-actives.

A prior understanding of the composition of the raw material, the structure of the cell wall, and the nature of the destinated compounds are necessary for the determination of the optimum operating conditions as it facilitates the selection of enzymes and optimizing related parameters [27].

6.3.4 Advantages of EAE

Enzyme Aided Extraction is an advantageous technique and has served as an objective for countless recent studies as it remarkably improves the extraction yield, does not alternate the bio-actives properties, and selectively removes the unwanted components of raw material. Moreover, it does not require the use of toxic organic solvents nor does it harm the environment [26–28].

6.4 Steam distillation

Steam distillation is practically the oldest and most famous way of essential oils' extraction [29]. Steam distillation is a separation process for temperature sensitive

Enzyme used	Bioactive extracted	Source material	Conditions used
Cellulase	Polysaccharides	Garlic	Temperature 45 °C, pH 5.0, time 80 min
α-Amylase and glucoamylase	Oleoresin	Turmeric	—
Cellulase, papain, and pectinase	Polysaccharides	Alfalfa	Temperature 52.7 °C, pH 3.87, time 2.73 h
Cellulase, pectinase, and protease	Seed oil	Pumpkin	Temperature 44 °C, time 66 min
Alginate lyase	Fucoxanthin and lipids	*Undaria pinnatifida*	Temperature 37 °C, pH 6.2
α-Amylase	Polysaccharides	*Panax ginseng*	—
Pectinase and cellulase	Carotenoids	Tomato waste	—
Lipase and phospholipase	Proteins	Olive pulp and stone	Temperature 30–40 °C, pH 7.0, time 15 min
Papain, protease, and trypsin	Fatty acids	*Strongylocentrotus nudus*	Temperature 40–55 °C, pH 7.8–8.5, time 180 min

Table 2.
Different enzymes and their optimum operating conditions [27].

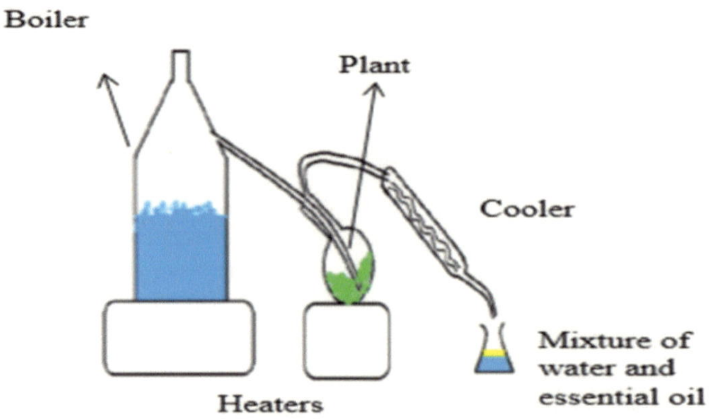

Figure 12.
Steam distillation.

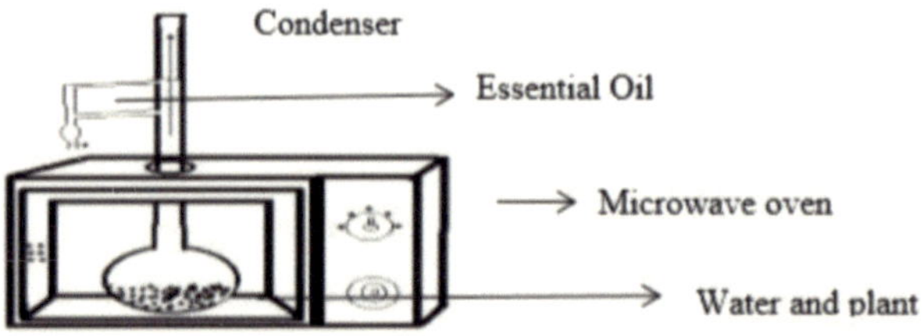

Figure 13.
Microwave assisted extraction.

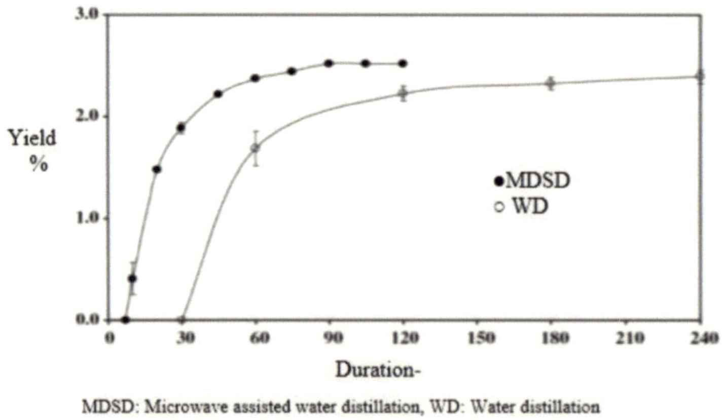

Figure 14.
Comparing water distillation and microwave assisted water distillation of obtaining thyme essential oil.

materials like oils which are insoluble in water and may decompose at their boiling point [30]. It consists of spraying with a certain amount of moisture on the plant material placed on the grid in a similar system to water distillation and allow steam transfer of the essential oils to occur [29]. The plant raw material is placed on a performed plate and the steam, generated by a separate boiler, volatilizes and carries essential oils as it passes through the plant material [29]. The vapors are then condensed in water cooling system only to be collected at the end of the cooler in a separate appropriate collector. Steam distillation is advantageous as it allows the vaporization of essential oils without reaching their boiling point [30]. Moreover, the boiler being kept in a separate chamber from the plant chamber keeps the ambient temperature at which the material to be distilled is located is kept below 100° C and hence it diminishes and prevents alterations due to heat effect. However, steam distillation is avoided for the reason that it is time and energy consuming and that, in case of high vapor flow rates, degradation of volatile compounds that can occur (**Figure 12**) [29].

6.5 Microwave assisted extraction

Microwave assisted extraction MAE is modern technique used in extraction. The first patent for MAE was in 1995for the extraction of a natural product using microwaves by Pare (**Figure 13**) [31].

The target for heating in dried plant material is the minute microscopic traces of moisture that occurs in plant cells. The heating up of this moisture inside the plant cell due to microwave effect, results in evaporation and generates tremendous pressure on the cell wall. The cell wall is pushed from inside due to the pressure and the cell wall ruptures. Thus, the exudation of active constituents from the ruptured cells occurs, hence increasing the yield of phytoconstituents [32]. MAE allows the boiling point to be reached earlier then supported water distillation as shown in **Figure 14**. It has also been determined that the amount and quality of essential oil obtained by microwave-assisted water distillation in 30 minutes is equivalent to the amount and quality of essential oil obtained in 4 hours and thirty minutes by water distillation [29]. Not only does MAE save time and energy, but it also is considered environment friendly as it uses very little or no solvents [31, 32].

7. Conclusion

The rich composition of plants (EOs, primary and secondary metabolites) offers several advantages in the phytotherapy field, therefore the use for different therapeutic purposes.

The extraction method of bioactive molecules depends on several factors and also depends on the plant properties.

Acknowledgements

Praise be to Allah, who guided us and helped us to complete this project, which would not have ended had it not been for God guidance.

We also address many thanks and appreciations to the professors Mr.Othmane Benkortbi, Mr.Mabrouk Hamadach, Mr.Yahia Belaissa, Mrs.Lefnaoui and Mrs. Yahoum for their tremendous generosity in sharing information, providing contacts and even some materials that we lacked. We are grateful!

We would like, as well, to thank Mr.Nesreddine Terkmane, a local herbs expert who helped, corrected and oriented us countless times.

Our acknowledgment is extended to the funding partners.

Conflict of interest

The authors declare no conflict of interest.

Notes/thanks/other declarations

We address the best most sincere thanks and appreciations to our families.

I, Achouak, am forever grateful for my parents Leila Gehlouz and Abderahmen, my lighted soul sister Melek and my two brothers Ismail and Abdeldjalil for their ever-lasting love and care, their unconditional support and their magical way of always turning hardships into something fun. Thank you!

I, Nassima, am pleased to extend sincere thanks and gratitude to my parents, a source of motivation and encouragement throughout my academic career, my sister and my twin soul Karima, who has always supported me, and my dear brothers Ahmed and Youcef.

Author details

Achouak Madani*, Nassima Mazouni and Mohammed Nedjhioui
University Dr Yahia Fares Medea, Medea, Algeria

*Address all correspondence to: achwakmadani08@gmail.com

References

[1] Patel, K: Phytochemical and microbiological studies of some Fijian plant extracts, University of the South Pacific 2005

[2] Baser, K H C; Buchbauer, G Handbook of essential oils: science, technology, and applications; CRC press, 2015.

[3] Cseke, L J; Kirakosyan, A; Kaufman, P B; Warber, S; Duke, J A; Brielmann, H L Natural products from plants; CRC press, 2016.

[4] Ludwiczuk, A; Skalicka-Woźniak, K; Georgiev, M In Pharmacognosy; Elsevier: 2017, p 233.

[5] Kawaguchi, H; Ogino, C; Kondo, A: Microbial conversion of biomass into bio-based polymers, Bioresource technology 2017, 245, 1664. Published Online: 2017

[6] McClements, D J Food emulsions: principles, practices, and techniques; CRC press, 2015.

[7] Suh, M C; Hahne, G; Liu, J R; Stewart, C N; Springer: 2015.

[8] ABDELRAZAG, H: Etude phytochimique et activité biologique de la plante :Limoniastrum guyonianum, KASDI MERBAH OUARGLA UNIVERSITY.

[9] De Almeida, A; De Faria, F; Dunder, R; Manzo, L; Manzo, B; Regina, A; Souza-Brito, M; Luiz-Ferreira, A: Recent Trends in Pharmacological Activity of Alkaloids in Animal Colitis: Potential Use for Inflammatory Bowel Disease, Evidence- based Complementary and Alternative Medicine 2017, 2017. DOI: 10.1155/2017/8528210. Published Online: 2017

[10] Kabera, J: Plant Secondary Metabolites: Biosynthesis, Classification, Function and Pharmacological Classification, Function and Pharmacological Properties, Journal of Pharmacy and Pharmacology 2 (2014) 2014, 2, 377. Published Online: 2014

[11] Hufford, C D: Natural Products from Plants By Peter B. Kaufman, Leland J. Cseke, Sara Warber (University of Michigan); James A. Duke (Retired USDA); and Harry L. Brielmann (Industrial Researcher). CRC Press, Boca Raton, FL. 1998. xiii + 343 pp. 15.5 × 23 cm. $129.95. ISBN 0-8493-3134-X, Journal of Natural Products 2000, 63, 736.DOI: 10.1021/np990729p. Published Online: 2000

[12] von Linné, C In Alkaloids - Secrets of Life; Aniszewski, T., Ed.; Elsevier: Amsterdam, 2007, p 141.

[13] Krishnamurti, C; Rao, S C: The isolation of morphine by Serturner, Indian J Anaesth 2016, 60, 861.DOI: 10.4103/0019-5049.193696. Published Online: 2016

[14] Edagha, I A; Davies, K G; Akpan, B C; Mbadugha, C C; Udoiso, W U: Ethanolic Extract of *Emilia sonchifolia* Leaves Possess Erythropoietic and Hepatoprotective Effect in Mice Infected with Plasmodium Berghei Berghei, Open Access Macedonian Journal of Medical Sciences 2014, 2, 11.DOI: 10.3889/oamjms.2014.002. Published Online: 2014

[15] Daly, J W: Alkaloids. Nature's Curse or Blessing By Manfred Hesse. Verlag Helvetica Chimica Acta, Zürich, Switzerland, and Wiley-VCH Weinheim, Germany. 2002. xii + 413 pp. 16 × 18 cm. $135. ISBN 3-906390-24-1, Journal of Natural Products 2003, 66, 901.DOI: 10.1021/np030705p. Published Online: 2003

[16] Kurek, J In Alkaloids-Their Importance in Nature and Human Life; IntechOpen: 2019.

[17] In Encyclopædia Britannica; Adam Augustyn, P. B., Brian Duignan, Alison Eldridge, Erik Gregersen, Amy McKenna, Melissa Petruzzello, John P. Rafferty, Michael Ray, Kara Rogers, Amy Tikkanen, Jeff Wallenfeldt, Adam Zeidan, and Alicja Zelazko., Ed.; Encyclopædia Britannica, inc.: 2018.

[18] Latif, S; Chiapusio, G; Weston, L A In Advances in Botanical Research; Becard, G., Ed.; Academic Press: 2017; Vol. 82, p 19.

[19] Gonçalves Paterson Fox, E; Russ Solis, D; Delazari dos Santos, L; Aparecido dos Santos Pinto, J R; Ribeiro da Silva Menegasso, A; Cardoso Maciel Costa Silva, R; Sergio Palma, M; Correa Bueno, O; de Alcântara Machado, E: A simple, rapid method for the extraction of whole fire ant venom (Insecta: Formicidae: Solenopsis), Toxicon 2013, 65, 5.DOI: https://doi.org/10.1016/j.toxicon.2012.12.009. Published Online: 2013

[20] Balasundram, N; Sundram, K; Samman, S: Phenolic compounds in plants and agri-industrial by-products: Antioxidant activity, occurrence, and potential uses, Food Chemistry 2006, 99, 191.DOI: https://doi.org/10.1016/j.foodchem.2005.07.042. Published Online: 2006

[21] Giada, M 2013, p 87.

[22] Abdelkader, B; Belkacem, N; Amrani, I 2017.

[23] Tian, S; Sun, Y; Chen, Z; Yang, Y; Wang, Y: Functional Properties of Polyphenols in Grains and Effects of Physicochemical Processing on Polyphenols, Journal of Food Quality 2019, 2019, 1.DOI: 10.1155/2019/2793973. Published Online: 2019

[24] Cheynier, V: Phenolic compounds: from plants to foods, Phytochemistry Reviews 2012, 11, 153.DOI: 10.1007/s11101-012-9242-8. Published Online: 2012

[25] Domínguez, H In Functional Ingredients from Algae for Foods and Nutraceuticals; Domínguez, H., Ed.; Woodhead Publishing: 2013, p 1.

[26] Puri, M; Sharma, D; Barrow, C: Enzyme-assisted extraction of bioactives from plants, Trends in biotechnology 2011, 30, 37.DOI: 10.1016/j.tibtech.2011.06.014. Published Online: 2011

[27] Sandesh J. Marathe, S B J, Sandip B. Bankar and Rekha S. Singhal.

[28] Ademola Monsur Hammed , I J, Azura Amid , Zahangir Alam , Tawakalit Tope Asiyanbi-H & Nazaruddin Ramli: Enzymatic Hydrolysis of Plants and Algae for Extraction of Bioactive Compounds, Food Reviews International 2013. DOI: 10.1080/87559129.2013.818012. Published Online: 2013

[29] Akdağ, A; Ozturk, E: Distillation Methods of Essential Oils, 2019. Published Online: 2019

[30] Božović, M; Navarra, A; Garzoli, S; Pepi, F; Ragno, R: Esential oils extraction: a 24-hour steam distillation systematic methodology, Natural Product Research 2017.DOI: 10.1080/14786419.2017.1309534. Published Online: 2017

[31] Destandau, E; Michel, T; Elfakir, C 2013, p 113.

[32] Jaiswal, P T a Y: An Overview of Microwave Assisted Extraction and its Applications in Herbal Drug Research, Research Journal of Medicinal Plants 2011, 5, 21.DOI: 10.3923/rjmp.2011.21.31. Published Online: 2011

Chapter 4

Biosynthesis of Natural Products

Stella O. Bruce and Felix A. Onyegbule

Abstract

Natural products are in the form of primary and secondary metabolites and are isolated chemical compounds or substances from living organisms. Terpenes, Phenolic compounds, and Nitrogen-containing compounds are secondary metabolites. The biosyntheses of secondary metabolites are derived from primary metabolism pathways, which consist of a tricarboxylic acid cycle (TCA), methylerythritol phosphate pathway (MEP), mevalonic and shikimic acid pathway. This chapter provides an overview of the diversity of secondary metabolites in plants, their multiple biological functions, and multi-faceted cultural history.

Keywords: natural products, biosynthesis, metabolites, phytochemicals, terpenes, phenolic compounds, nitrogen-containing compounds

1. Introduction

Natural products are lead compounds, which are frequently produced by plants and microbes as their secondary metabolites, and securing large quantities of such compounds for industrial and clinical applications has been a persisting problem [1]. Natural products (chemical compounds or substances) are isolated from living organisms [2]. Biogenesis belief that complex living things come only from other living things and also the production of new living organisms or organelles using reproduction. Chemistry of natural product is produced by the pathway of primary or secondary metabolism [3]. Metabolism is used to describe all chemical reactions which include maintaining the living state of cells of an organism [4]. Metabolism can be in form of catabolism or anabolism. Metabolites are a product of metabolism and restricted to small molecules [5].

Plants produce natural products with highly diverse structures; these products are called "secondary metabolites" in contrast to the "primary metabolites", which are essential for plant reproduction, and growth. The leaf, stem, root, or bark of the plant has plant secondary metabolites that have been produced, for example Alkaloids, Tannins, Flavonoids, and Phenolic compounds [6]. Most food, spices and herbs are indigenous plants has these secondary metabolites [7]. Plant secondary metabolites are exclusively produced by more than 30,000 different plants. They serve as defense compounds against pathogens and herbivores, as flower pigments that attract pollinators. Natural products have a strong impact on human culture and are used throughout human history as pigments, condiments, and pharmaceuticals [8].

This chapter therefore provides an overview of the biosynthesis of natural products, their multiple biological functions and multi-faceted cultural history.

2. Classification of plants secondary metabolite

Plants secondary metabolites can be classfied into three groups namely

- Terpenes
- Phenolic compounds
- Nitrogen-containing compounds [8]

2.1 Terpenes

Terpenoids constitutes the largest class of secondary products; they comprise of more than 40,000 different structures and are the largest natural products in plants [9, 10]. Terpenes consists of five-carbon isoprene units, and classified into hemiterpenes (C_5), monoterpenes (C_{10}), sesquiterpenes (C_{15}), diterpenes (C_{20}), triterpenes (C_{30}), tetraterpenes (C_{40}), and polyterpenes [11, 12]. Terpenoids originate from two different biosynthetic routes: plastid-located deoxyxylulose phosphate (DXP) pathway (also called methylerythritol phosphate or MEP pathway) and the cytosolic mevalonic acid (MVA) pathway (**Figure 1**) [13–15].

2.1.1 Hemiterpenes

This is a volatile compound synthesized from DMAPP, isoprene is the most abundant true hemiterpcnc from plants (**Figure 2**). The species that synthesize isoprene are found among ferns, mosses, angiosperms, and gymnosperms. The emission and production of isoprene are distributed very widely in the plant kingdom. Isoprene is emitted into the atmosphere and protects leaves to survive short periods of high temperature. Moreover, it increases the plant's tolerance towards ozone and reactive oxygen species [16]. Hemiterpenes may also act as signaling molecules. The highly volatile hemiterpene methacrolein are emitted in the leaves of sagebrush (*Artemisia tridentata*) (**Figure 2**) in addition to other volatile compounds like hexenal, monoterpenes, and methyl jasmonate when the plant is damaged, this

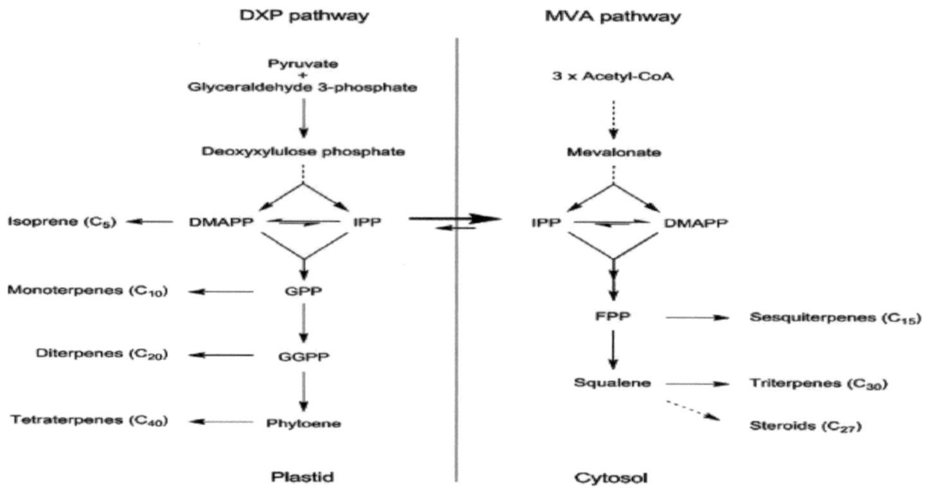

Figure 1.
Schematic overview of terpene biosynthesis in plants.

Figure 2.
Hemiterpenes.

is perceived by plants and enables them to react faster to a possible attack. A plant that is prepared in this manner is less likely to be damaged by herbivores [17]. The C-5 units derived from DMAPP are found in natural products of mixed biosynthetic origin, e.g., hop bitter acids, prenylated flavonoids, and hyperforin.

2.1.2 Monoterpenes

Monoterpenes originate from one molecule DMAPP and one molecule IPP that are joined in most cases head-to-tail, yielding all- trans geranyl diphosphate (GPP) (**Figure 3**). Several plant families, e.g., the Lamiaceae and Asteraceae, have glandular trichomes with secretory cells that produce terpenes and secrete them into a shared subcuticular storage cavity [18]. Conifers accumulate a complex mixture of mono-, sesqui-, and diterpenes, oleoresin, in resin blisters or ducts, which are covered by a layer of epithelial cells that secrete and synthesize the terpenes into the lumen [19]. As in the case of the conifers, many other plants accumulate monoterpenes in mixtures containing the larger sesqui- and diterpenes, rather than monoterpenes alone. The physiological function of monoterpenes is defense, the attraction of pollinators, and plant–plant communication [20]. The plant-insect interactions role of terpenes has been well-studied in conifers and the bark beetle. The oleoresin is secreted from the ducts or produced newly upon tissue damage by the beetle [21]. Ingested monoterpenes are converted by the beetles to pheromones that either attract more beetles or serve as anti-aggregation signals. Besides, conifer monoterpenes take part in tritrophic interactions and attract insect predators that feed on bark beetles [19]. Most aromatherapy, insecticides, perfumes and pharmaceutical products are made from monoterpenes [18]. The essential oil of corn mint (*Mentha arvensis* var. piperascens) produce more than 7000 tons of menthol every year either by total synthesis or from the steam-distilled. The cooling sensation stimulated by menthol is caused by the excitation of cation channels that serve as thermal receptors [22]. Two monoterpenes with promising anticancer effects are perrillyl alcohol and (+)-(R)- limonene [23], these two compounds

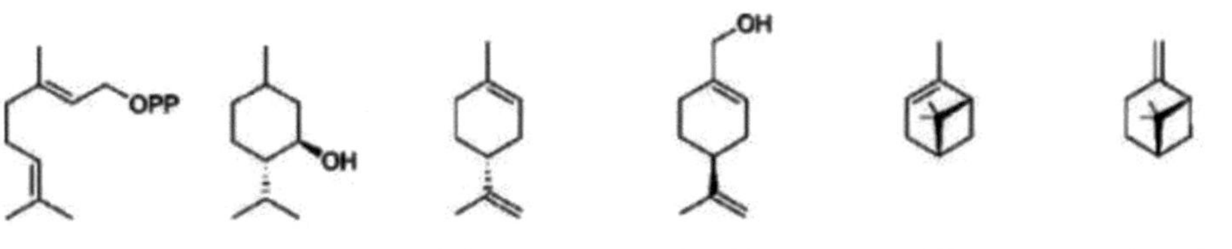

Figure 3.
Mono- and bicyclic monoterpenes derived from geranyl diphosphate (GPP).

suppress translation of 3-hydroxy-3-methylglutaryl-CoA (HMG-CoA) reductase, an enzyme of the MVA pathway and induce apoptosis [24]. The 3-hydroxy-3-methylglutaryl-CoA (HMG-CoA) reductase enzyme is a promising target for anti-tumor compounds because tumor cells have elevated HMG-CoA reductase levels and many proteins involved in cell growth are phenylated. The decreased terpene biosynthesis in humans leads to suppression of HMG-CoA reductase [25]. The secoiridoids yields from the cleavage of the cyclopentane ring of the iridoid skeleton, which are monoterpene indole alkaloids and the biosynthetic building units of the Ipecac alkaloids [25]. Many iridoids have an intense bitter-taste and therefore act as feeding deterrents (**Figure 4**) [26].

2.1.3 Sesquiterpenes

In general, sesquiterpenes are less volatile than monoterpenes; they contain three isoprene units and are formed by condensation of DMAPP with two molecules IPP, the central C_{15} intermediate farnesyl diphosphate (FPP) can be folded into mono-, bi- or tricyclic systems [9]. Initially, it was assumed that all sesquiterpenes are produced via the cytosolic MVA pathway. Recent studies, however, revealed that certain sesquiterpenes originate from isoprene units provided by the DXP pathway [12, 13] or by both biosynthetic routes [27]. This can be explained by the transport of isoprenoid precursors from the plastids to the cytosol [28]. Abscisic acid is a sesquiterpene phytohormone that is induced by drought and promotes stomatal closure and seed dormancy. Other sesquiterpenes take part in tritrophic plant-herbivore-parasite interactions [13]. The sesquiterpenes (E)-b-farnesene and the (E)-a-bergamotene attract the parasitic wasp *Cotesia marginiventris*, in maize infested with lepidopteran larvae [29]. Maize roots release (E)-b-caryophyllene (**Figure 5**) upon an attack of larvae of the beetle *Diabrotica virgifera* to attract the parasitic nematode *Heterorhabditis megidis* [30]. Many sesquiterpenes (sesquiterpene lactones) contain a pentacyclic lactone group, these compounds occur abundantly in the family Asteraceae, because of their bitter taste sesquiterpene lactones presumably serve as feeding deterrents of herbivores [31]. Pharmacologically active sesquiterpene lactones often show anti-inflammatory effects due to inhibition of the transcription factor NF- kB that mediates immunological responses and inflammation [32]. One of the most popular medicinal plants, chamomile (*Matricaria recutita*) is a sesquiterpenes with such activities. Antimigraine action of some sesquiterpene lactones, e.g., parthenolide from feverfew (*Tanacetum parthenium*), is mediated by inhibition of platelet aggregation and serotonin secretion [9]. The reason for the cytotoxicity

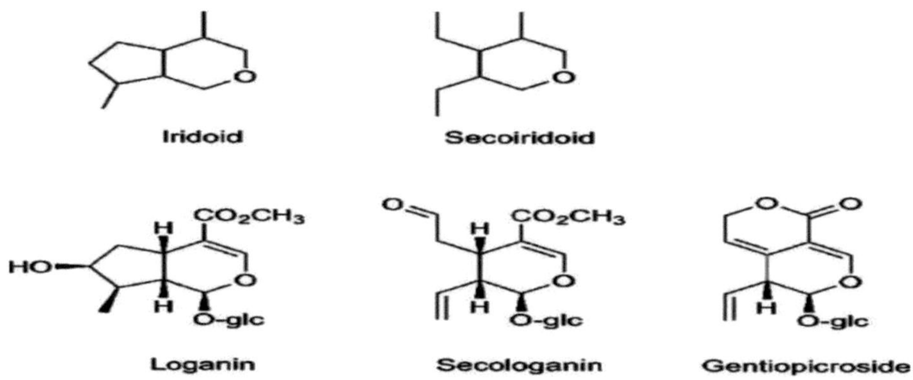

Figure 4.
Iridoids.

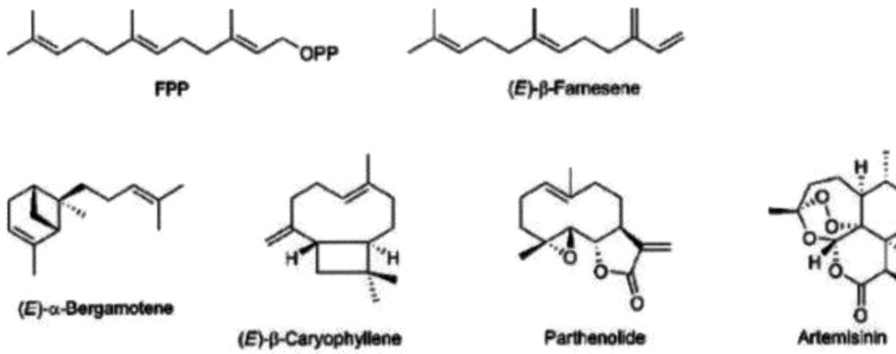

Figure 5.
Linear and cyclic sesquiterpenes.

and allergenicity of sesquiterpene lactones with a, b-unsaturated lactone is the alkylation of proteins. Artemisinin is a novel promising agent against malaria. Structurally, it is a tetracyclic sesquiterpene with a six-membered lactone ring and an unusual 1, 2, 4-trioxane ring (**Figure 5**). The president of North Vietnam, to the Chinese government (Ho Chi Minh) discovered artemisinin for a cure against malaria to support his troops in the malaria-infested jungles during the American/ Vietnamese war [33]. The ether extract from *A. annua*, and artemisinin (qing-haosu) revealed the antimalarial activity and the mode of action of artemisinin are still being investigated. Most likely, it interferes with sarco-endoplasmic reticulum calcium ATPase (SERCA) of *Plasmodium falciparum*, but other mechanisms, for example, alkylation of biological macromolecules or the production of reactive oxygen species [34]. The structural feature required for antimalarial activity is the peroxide bridge. Artemether and artesunate (two semisynthetic analogs), in efficiency comparison to artemisinin were developed and are now used as first-line therapy in the treatment of malaria, in combination with other antimalarial drugs like the lumefantrine and quinine analogs mefloquine. This combination tends to prevent resistance to Plasmodium. The artemisinin and its analogs success has triggered by extraction of the sesquiterpene from the plant because *A. annua* contains only 0.01–1.5% of artemisinin [35]. Therefore, an powerful and affordable drug for the people in malaria-endemic areas are necessary, either by breeding of *A. annua* plants with elevated artemisinin levels or biotechnological production of the artemisinin precursor artemisinic acid by cloning the biosynthetic genes from *A. annua* [35] and engineering the pathway into the bacterium *Escherichia coli* or yeast [36, 37].

2.1.4 Diterpenes

Diterpenes originate from the Plasmid DXP pathway and are synthesized from DMAPP and three molecules IPP yielding the C_{20} metabolite geranyl geranyl diphosphate (GGPP). GGPP is a smaller terpene; it can undergo rearrangements and cyclization to many different structures and also a precursor of the lipophilic phytyl side chain of chlorophyll and plastoquinone. Gibberellins are tetracyclic diterpenes that act as phytohormones and promote shoot elongation, flowering and seed germination [38]. Diterpenes like abietic- and levopimaric acid (**Figure 6**) are constituents of conifer oleoresin and function as a defense against herbivores and pathogens. After mono- and sesquiterpenes (turpentine) are removed from oleoresin by distillation, the solid diterpene fraction (rosin) is called colophonium. The mono and sesquiterpene containing distillate are used as oil of turpentine for the

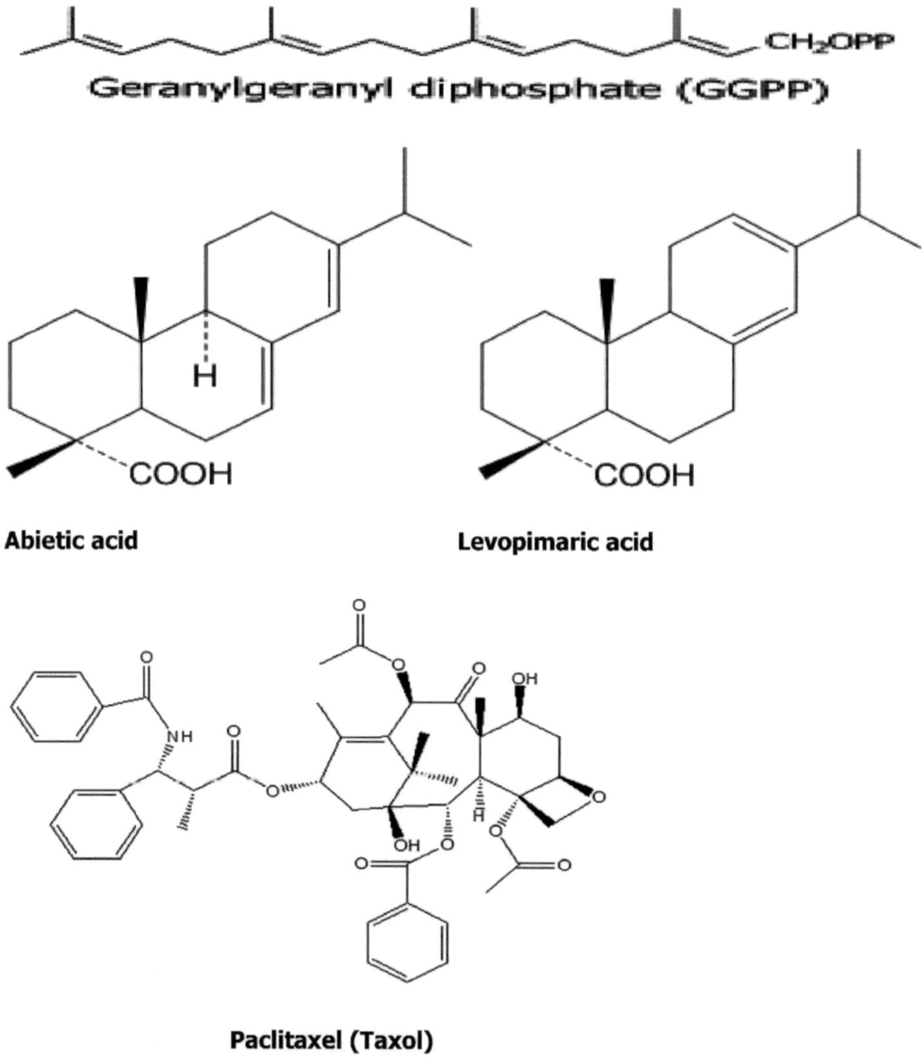

Figure 6.
Diterpenes.

thinning of paints and varnishes [39]. Paclitaxel binds to microtubules, stabilizes them against depolymerisation and thus blocks cell proliferation [40]. Paclitaxel is used in the therapy of cancers (breast, ovarian, lung, head and neck and Kaposi's sarcoma). In the bark of *T. brevifolia* (0.01–0.02%), paclitaxel occurs only in relatively low amounts and the trees grow slowly, other sources had to be found to supply enough of the diterpene for industrial production. Paclitaxel is obtained either by semisynthesis from baccatin III and 10-deacetylbaccatin III or from tissue cultures of various Taxus species, which can be extracted from the leaves and twigs of the common yew (*T. baccata*), a tree that grows much faster than *T. brevifolia* (**Figure 6**).

2.1.5 Triterpenes and steroids

Triterpenes are synthesized from two molecules of FPP that are joined by tail-to-tail condensation to squalene via the MVA pathway. Various structures, mostly tetra- or pentacyclic yields from cyclization of its metabolite 2, 3-oxidosqualene followed by rearrangements and methyl shifts. The precursor of plant steroids is 2, 3-Oxidosqualene (**Figure 7**). In this case, it is cyclized to the triterpene cycloartenol, which is then converted to the C-27 compound cholesterol with the loss of three methyl groups. In both triterpenes and steroids the oxygen of 2, 3-oxidosqualene is usually retained as hydroxy group at C-3. Phytosterols are lipophilic and are readily incorporated into the micelles involved in fat digestion. Esters of phytosterols are therefore used as cholesterol-lowering food additives [9]. A group of plant hormones (Brassinosteroids) is derived from campesterol. They regulate various biological processes, e.g., stem elongation, leaf expansion, seed germination, and xylem differentiation [38, 41].

2.1.6 Saponins

Monocots preferably accumulate steroidal saponins, which are abundant in the Agavaceae, Dioscoraeceae, and Yuccaceae. Triterpenoid saponins contain the lupane skeleton, tetracyclic dammarane backbone as aglycone and the pentacyclic a-amyrin (ursane), b-amyrin (oleanane). This aglycone is linked with one to three carbohydrate chains containing up to six sugar molecules or uronic acids [9, 42]. In the triterpene backbone, the first sugar chain is attached to the hydroxy group at C-3. When two or more carbohydrate chains are present, they are connected with carboxy or hydroxy groups at C-30 or C-28. Spirostanols and furostanols are two groups of steroid saponins. A tetrahydrofuran ring in furostanols is formed from the side chain of cholesterol, and the hydroxy group at C-26 is glycosylated. Upon cleavage of this sugar moiety, a second oxygen-containing heterocycle is formed, thus yielding a spirostanol (**Figure 8**). As in the case of the triterpene saponins, steroidal saponins carry a sugar chain at the C-3 hydroxy group [41, 43–47].

2.1.7 Tetraterpenes

Tetraterpenes are synthesized from two molecules GGPP by tail-to-tail addition and comprise only one group of compounds, the carotenoids. The tetraterpene chain is cyclized to a six-membered ring at either one or both ends. Carotenoids with hydroxy or epoxy functions are classified as xanthophylls [9]. The important physiological functions of carotenoids in plants, is that it act as accessory pigments of chlorophyll, since they are part of the light-harvesting complex. Besides, they quench triplet oxygen and singlet oxygen in case of excess light energy and thus protect the plant from photo-oxidative damage. As pigments of flowers and fruits,

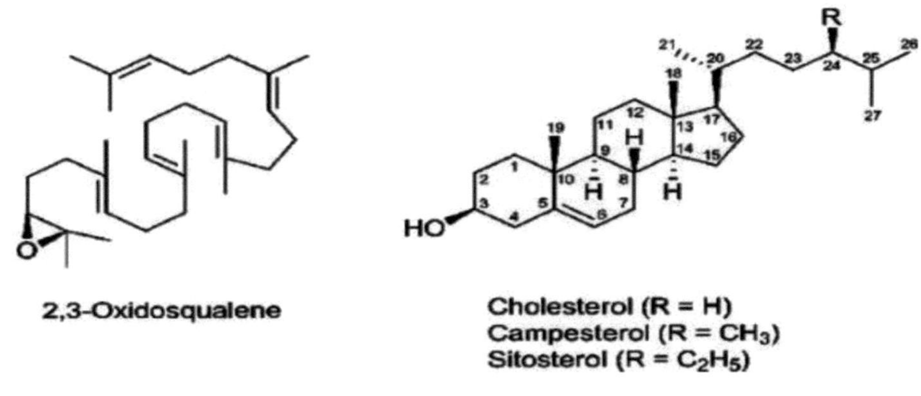

Figure 7.
Sterols derived from 2, 3-oxidosqualene.

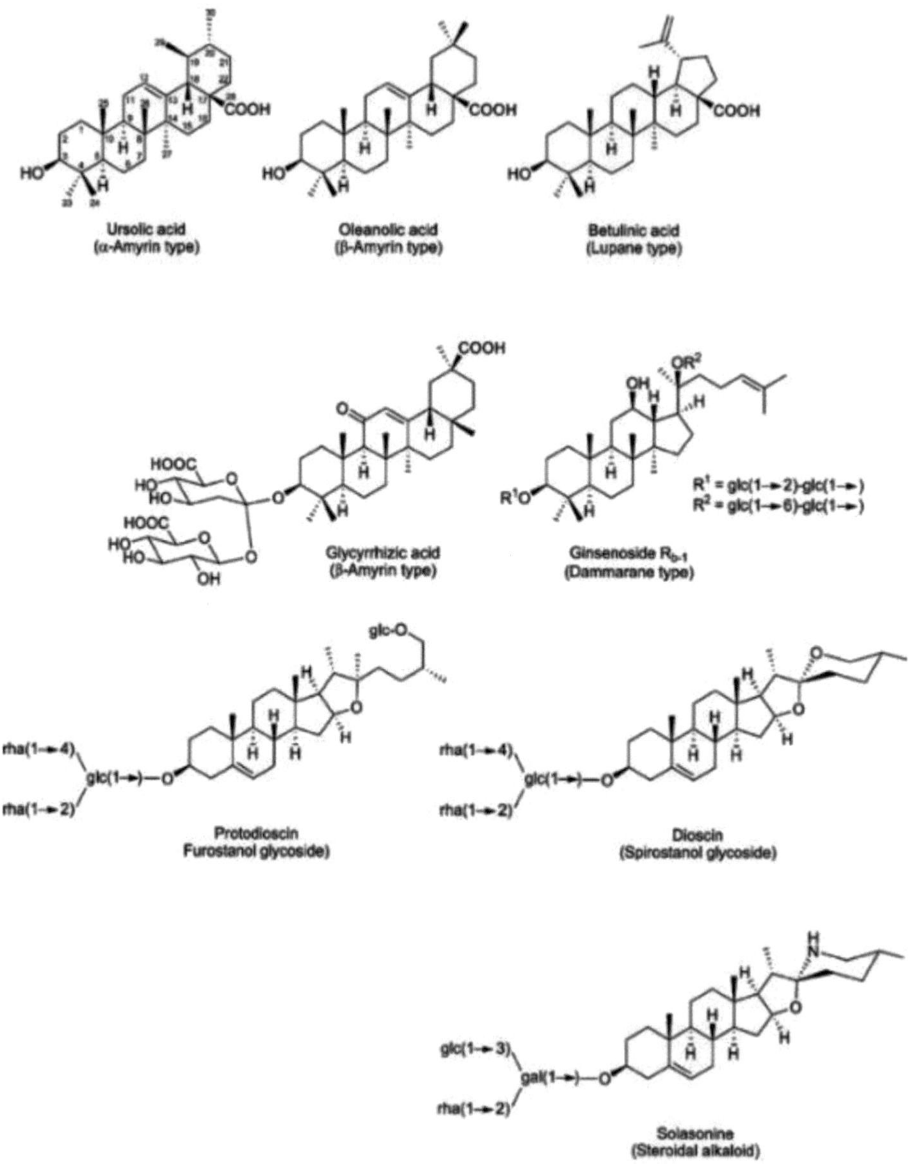

Figure 8.
Triterpene and steroid saponins.

carotenoids attract pollinators and seed dispersers [48]. Carotenoids (a-carotene, b-carotene) are essential for human health (**Figure 9**), b-cryptoxanthine is precursors of vitamin A. They serve as the pigment of the light receptors of the human eyes, and converted in the liver to vitamin A. To overcome vitamin A deficiency in areas with malnutrition, transgenic rice termed, golden rice "was developed that expresses high levels of carotenoid biosynthetic enzymes in the endosperm and accumulates elevated levels of carotenoids [49–51].

2.2 Phenolic compounds

Phenolic compounds (phenolic acids and polyphenols) are derivatives of the shikimic, pentose phosphate, and phenylpropanoid pathways in plants [52].

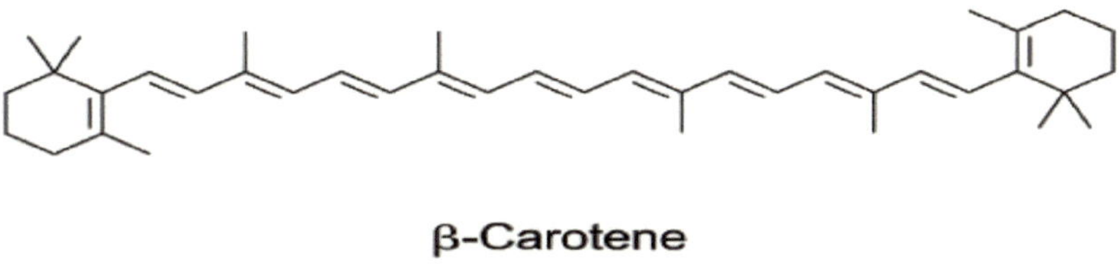

Figure 9.
β-Carotene.

Polyphenols are aromatic ring which contains a phenyl group and a hydroxyl functional group [53]. Plant phenolic compounds are lignin, flavonoids, carotenoids, tannins, and phytoalexins; they are responsible for antioxidant, antiaging, antiproliferation and anti-inflammatory activities. Vegetables, fruits and beverages are major sources of phenolics [54, 55]. Tannins significantly reduce the growth of many herbivores when added to their diets because they are generally toxic. Tannins can be seen in fruits like apples, blackberries, tea and red wine [56]. Tannins are mainly constituent of woody plants especially heartwood. Some derivatives of tannin include Gallic acid [56].

2.2.1 Phenol derivatives, especially flavonoids

The biosynthetic pathways are derived from the shikimate pathway (**Figure 10**), which is shared by indoles, and by several alkaloids and betalains. The phenylalanine

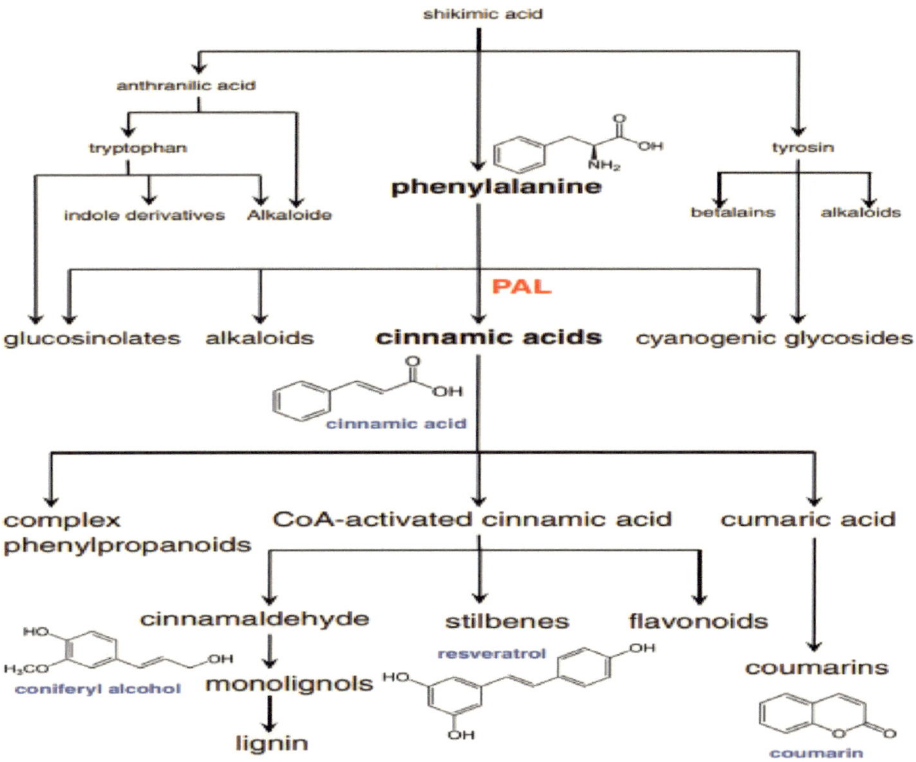

Figure 10.
Schematic overview for the biosynthetic pathways of selected phenols from phenylalanine as a precursor (bold). A key enzyme, phenylalanine ammonia-lyase (PAL), is shown (red). Some example structures are depicted (blue) [57].

is the precursor for the cinnamic acid derivatives and flavonoids, and it is converted by an enzyme, phenylalanine ammonia-lyase (PAL) to cinnamic acid. Rosmarinic acid has high antioxidative potential and also good aromatic qualities. The cinnamic acid derivatives serve as precursors for polymers (lignin), which is synthesized via cinnamaldehydes and monolignols. Much information is also available from maize and a legume, the latter also contains isoflavonoids (**Figure 11**). Other mutants in the pathway of, for example, the next enzyme encoding chalcone isomerase (which is responsible for the synthesis of naringenin), also show this phenotype, and consequently, the mutations were numbered consecutively, starting with "1." Mutations in the transcription factors that control the synthesis of flavonoids have similar phenotypes [57].

2.3 Nitrogen-containing compounds

Alkaloids are heterocyclic nitrogen compounds biosynthesized from amino acids. Alkaloids represent one of the biggest groups of natural products, with currently more than 12,000 known structures. In addition to alkaloids, benzoxazinoids, glucosinolates, and cyanogenic glucosides will be represented. Like alkaloids, these

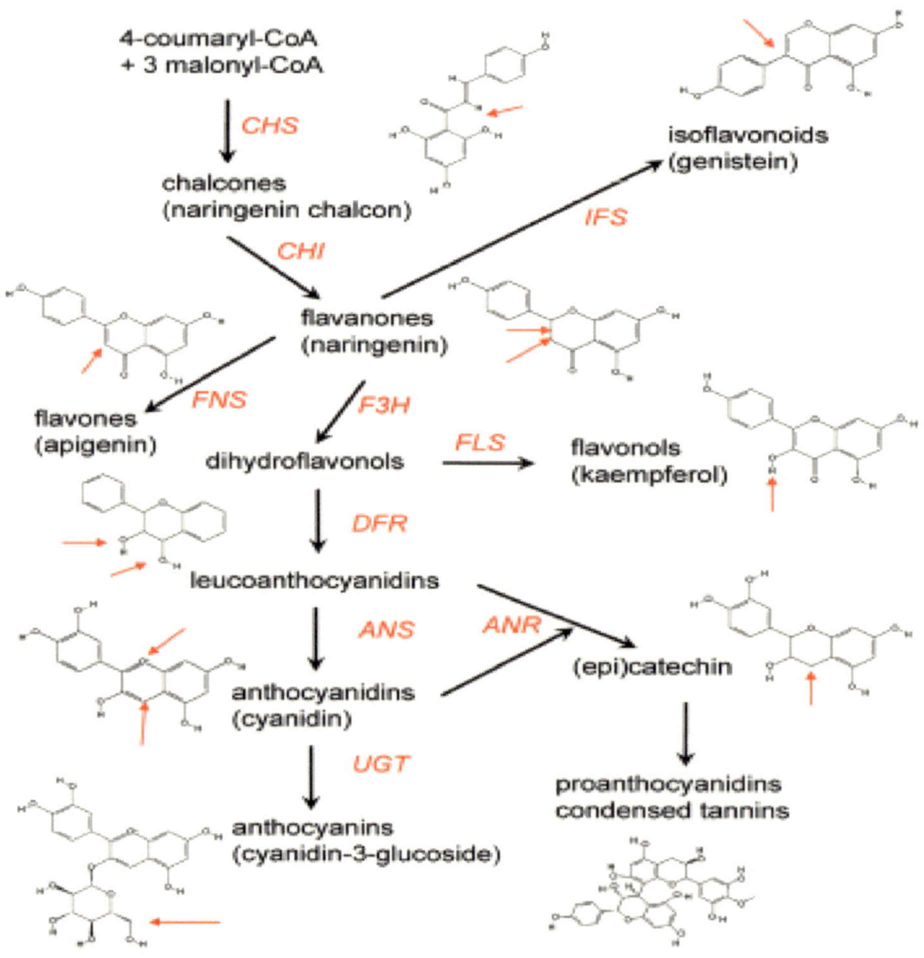

Figure 11.
The main pathways for flavonoid synthesis derived from different plant species. CHS: chalcone synthase; CHI: chalcone isomerase; IFS: isoflavonoid synthase; FNS: flavone synthase; F3H: flavanone-3- hydroxylase; FLS: flavonol synthase; DFR: dihdroflavonol reductase; ANS: anthocyanidin synthase; UGT: glycosyltransferase; ANR: anthocyanidin reductase [57].

metabolites contain nitrogen and are derived from amino acids. Morphine is an alkaloid isolated in 1805 from opium poppy *Papaver somniferum* [58, 59]. The role of alkaloids in the plant has been a subject of speculation for at least 100 years. Most alkaloids are now believed to function as a defense against especially mammals, because of the general toxicity and deterrence capacity [60]. One group of alkaloids, the pyrolizidine alkaloid illustrates how herbivores can become adapted to tolerate plant defensive substances and even use them in their defense [60].

2.3.1 Alkaloids

Alkaloid was introduced by a German Chemist, Carl F.W Meissnerin in 1815. Alkaloids are Alkali-like and derived from the word Alkali. They are a group of naturally occurring organic compounds which are basic, contain one or more nitrogen atoms normally of Heterocyclic nature. They also possess specific physiological actions on the human and animal bodies and are abundant in higher plants (Angiosperm). Major types of alkaloids and their examples are represented in **Table 1**. Families rich in alkaloids are- Apocynaceae, Rubiaceae, Solanaceae, Papaveraceae, Berberidaceae, etc. Alkaloids are present in many parts of the plant- Aerial part (Ephedra – Ephedrine), Entire plant (Vinca- Vincristine, Vinblastine), Leaves (Tea- Caffeine), Root (Rauwolfia- Reserpine), Bark (Cinchona- Quinine), Seed (Nuxvomica), Fruit (Black pepper- Piperine), Latex (Opium- Morphine, Codeine). Pharmacological uses include; Anagelsic, Antimalarial, Antispasmodic, Hypertension, Mental disorder, Anticancer etc. Alkaloids occur mainly in plants as Salts of organic acid (oxalic acid, citric acid, acetic acid, maleic acid, tartaric acid, fumaric, benzoic, etc). Functions in plants include; protective against insects and herbivores (bitterness and toxicity), a product of detoxification (a waste product) in a certain case, a reservoir for protein synthesis, and a source of nitrogen in case of deficiency. Many precursors are involved in various pathways, such as aromatic amino acids (tryptophan, tyrosine and phenylalanine), and also aspartate, glutamine, lysine, glycine and valine (**Figure 12**). Besides, the nonproteinogenic amino acid ornithine is an important precursor for various alkaloids.

Type	Plant source	Example	Uses
Pyrrolidine	Leaves of *Peruvian coca* shrub	Hygrine	Stimulants, Depressant
Tropine	*Atropa belladonna*	Atropine, Cocaine	Antidote of poison
Piperidine	Bark of bomegranate, Oil of hemlock. *Conium maculatum*	Coniine	Poison (paralyzes of motor neuron)
Pyramidine-pyridine	Tobacco leaf *Nicotina tabacum*	Nicotine	Respiratory stimulation
Quinoline	Cinchona tree	Quinine	Treatment of malaria
Isoquinoline	*Papaver somniferum* Seed of nuxvomica Strychnos	Codeine, morphine	Treatment of cough and Analgesic
Indole	*Claviceps purpurea*	Strychnine, Reserpine, Psilocybin	Treatment of hypertension, uterine atonia, postpartum bleeding, hallucination
Pyridine-piperidine	*Anabasis aphylla*	Anabasine	Antimicrobial, antioxidant

Table 1.
Major types of alkaloids [53].

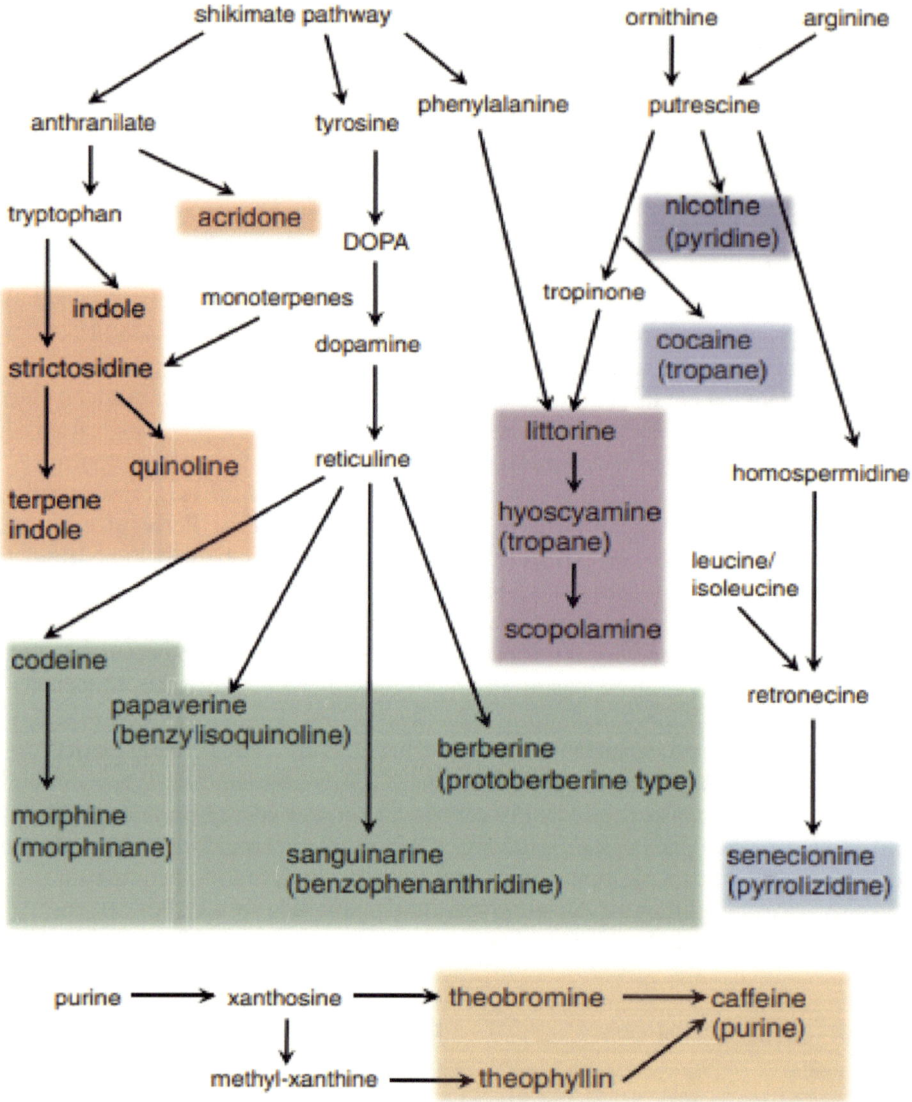

Figure 12.
Overview of the biosynthesis of selected alkaloids. Phenylalanine together with ornithine is needed for the synthesis of the second group of tropane alkaloids (violet). Caffeine and related substances are derived from purine (brown). The class of compounds is given in brackets [57].

For several alkaloids, two different precursors are needed for the biosynthetic pathways. In the case of terpene indole alkaloids (**Figure 12**), it is not only tryptophan that is involved as a precursor for the indole moiety, but also monoterpenes for the synthesis of side chains. Another example is the biosynthesis of the tropane alkaloids hyoscyamine and scopolamine, where ornithine and phenylalanine are required for the different parts of the molecule (**Figure 12**) [57].

2.3.2 Benzoxazinones

Benzoxazinones is a class of natural products known as cyclic hydroxamic acid, found in wheat, rye and maize in the family of Gramineae [61]. They act as plant resistance to insects and microbes. At present, it is still being investigated whether the pathway developed only once or several times independently after the divergence of monocots and dicots [62, 63]. Besides, they serve as feeding deterrents and reduce the vitality of pests. In particular, these metabolites confer resistance to one of the major corn pests, the European corn borer (*Ostrinia nubialis*) [64]. The mode of action of benzoxazinones can be explained by the modification of amino and thiol groups of biomolecules. The aldehyde function of the tautomeric open-ring form can react as an electrophile with NH_2 groups and form Schiff bases [65]. The structural prerequisite for this oxidation is an electron-donating substitution at C-7 of the benzoxazinone skeleton (**Figure 13**) [66]. Benzoxazinoids that have been bio-activated by N -acetylation may act as alkylating agents towards nucleic acids and proteins. Due to their toxicity, benzoxazinones can also function as allelochemicals and are therefore discussed as natural herbicides [67].

2.3.3 Glucosinolates

Glucosinolates are b -thioglucosides of (Z) - N - hydroximinosulfate esters (**Figure 14**). They share the first steps of cyanogenic glucoside biosynthesis. About 120 different structures of glucosinolates are known [68]. The glucosinolates are hydrolyzed by myrosinase (if the plant tissue is damaged), a thioglucosidase is spatially separated in the undamaged tissue [69] (**Figure 14**). The main product of the "mustard bomb" consisting of glucosinolates and myrosinase is isothiocyanates. These compounds are also responsible for many of the biological effects of glucosinolates, e.g., antibacterial, antifungal, nematicidal, and feeding deterrent activities [68]. The formation of hydrolysis products distinct from thiocyanates depends on the structure of the glucosinolates, pH, and the presence or absence of Fe^{2+} ions or specifier proteins [70]. Hydrolysis of b -hydroxyalknyl glucosinolates yields oxazolidine-2-thiones that can cause goiter by inhibiting the incorporation of iodine into thyroid hormones. To make the protein-rich seed cake that remains after the extraction of the oil suitable as animal foodstuff, Grape plants with low levels of glucosinolates have been developed by breeding efforts [68]. Sulforaphane enhances the excretion of cancerogenic compounds by inducing glutathione-S-transferase, UDP-glucuronosyl transferase, and NADPH quinone oxidoreductase (phase II detoxification enzymes) [70, 71]. The glucosinolates act as feeding deterrents, and many insect herbivores feed on plants containing these natural products. The detoxification of glucosinolates is known from two insect species which has two very different mechanisms [72]. The cabbage white butterfly (*Pieris rapae*) contains a specified protein that transforms glucosinolates in the presence of myrosinase to nontoxic nitriles that are excreted with the feces [73]. This requires either an endogenous myrosinase that is spatially separated from the glucosinolates in the insects or myrosinases from the gut microflora of their enemies [69].

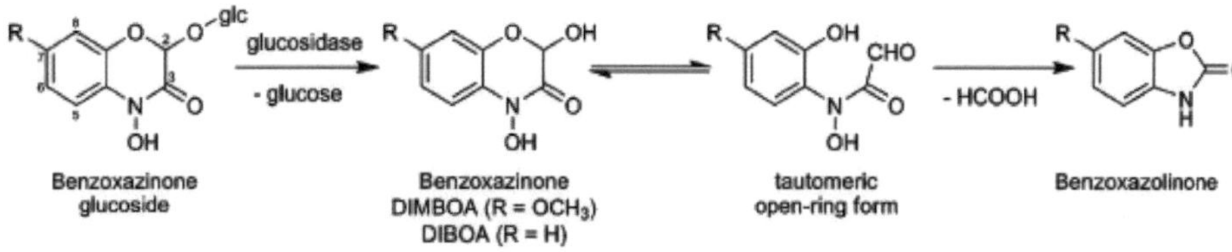

Figure 13.
Enzymatic and chemical degradation of benzoxazines with hydroxamic acid function [61].

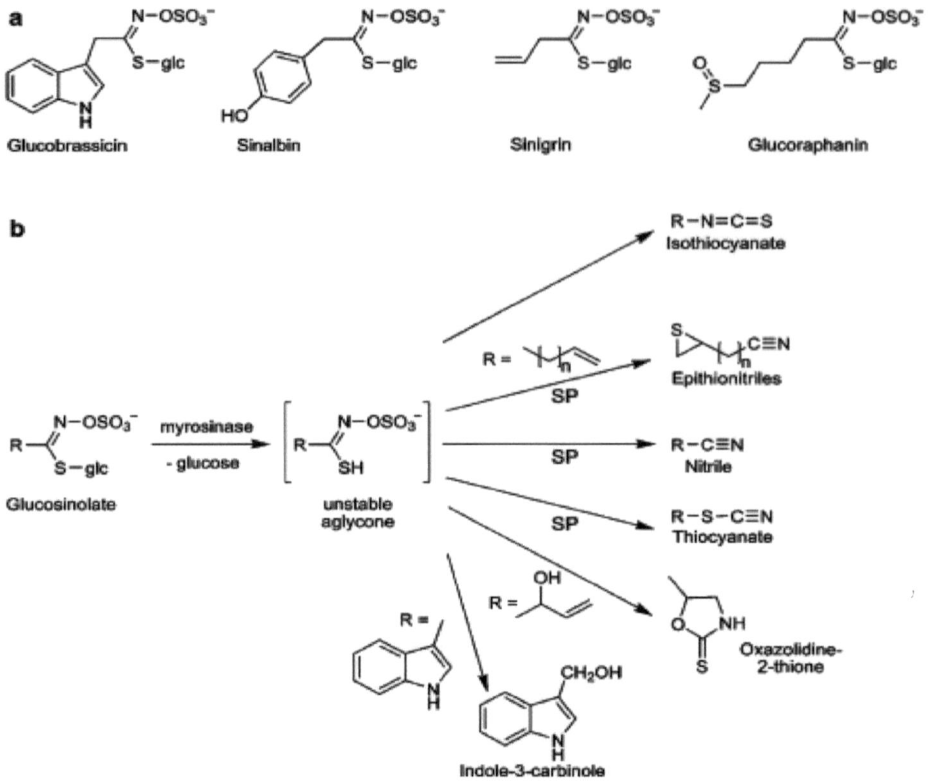

Figure 14.
Exemplary structures of glucosinolates (a) and hydrolysis of glucosinolates by myrosinase and rearrangement to various products (b) Isothiocyanates are the predominant degradation products.

2.3.4 Cyanogenic glycosides

Cyanogenic glucosides are b-glucosides of a–hydroxy nitriles (syn. cyanohydrins), which are derived from the five proteinogenic amino acids phenylalanine, tyrosine, valine, isoleucine, leucine, and the non-proteinogenic amino acid cyclopentenyl-glycine. About 2500 different plant species including ferns, gymnosperms, and angiosperms produce cyanogenic glycosides [74, 75]. Despite their widespread occurrence, these natural products are found predominantly in the families Araceae, Asteraceae, Euphorbiaceae, Fabaceae, Passifloraceae, Poaceae, and Rosaceae [9, 76]. Some of the most abundant molecules are amygdalin (Rosaceae), linamarin and lotaustralin (Fabaceae), and the epimers dhurrin and taxiphyllin in the genus Sorghum [75]. The b-glucosidic bond can also be hydrolyzed by intestinal bacteria in the gut of herbivores. The hydrogen cyanide toxicity can be explained by its affinity to metal ions. Cyanide ions complex iron (III) in the active site of cytochrome oxidase thus inhibits the respiratory chain [77, 78]. Cyanogenic glucosides act as feeding deterrents, by transferring all genes required for the formation of the cyanogenic glucoside dhurrin from *Sorghum bicolor* into Arabidopsis, proved that cyanogenic glucosides play a role in plant defense [79, 80]. Several herbivores, especially insects, can feed on plants containing these natural products, despite the toxicity of the cyanogenic glucosides, and the toxic compounds may act as phagostimulants. Cyanogenic glucosides act as defense compounds for some species of beetles, centipedes, and millipedes, but particularly many moths and butterflies (**Figure 15**). The compounds are either taken up by feeding on cyanogenic plants or synthesized by endogenous enzymes [77, 78]. It has been postulated that cyanogenic

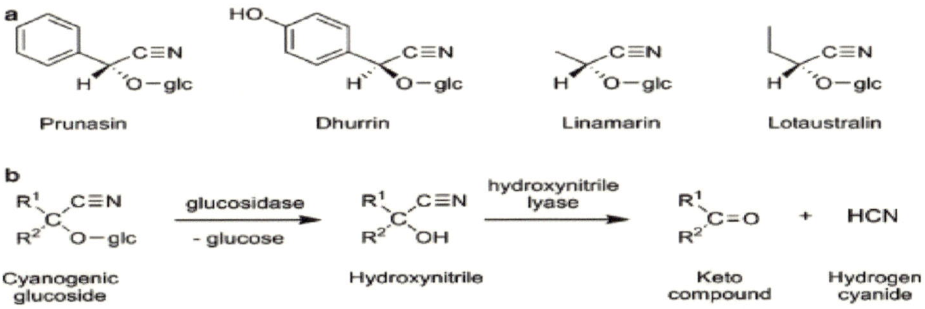

Figure 15.
Representative structures of cyanogenic glucosides (a) and degradation of cyanogenic glucosides with concomitant release of toxic hydrogen cyanide (b).

glucosides also serve as storage compounds for reduced nitrogen and sugar [81, 82]. These treatments often results in loss of protein, minerals, and vitamins. Various approaches to produce transgenic cassava with reduced content of cyanogenic glucosides in roots are currently underway [76, 80, 83, 84].

Author details

Stella O. Bruce[1]* and Felix A. Onyegbule[2]

1 Pharmacognosy and Traditional Medicine, Pharmaceutical Sciences, Nnamdi Azikiwe University, Awka, Anambra State, Nigeria

2 Department of Pharmaceutical & Medicinal Chemistry, Faculty of Pharmaceutical Sciences, Nnamdi Azikiwe University, Awka, Anambra State, Nigeria

*Address all correspondence to: stellaobruce@yahoo.com

References

[1] Smanski M.J, Zhou H, Claesen J, Shen B, Fischbach M.A, Voigt C.A (2016). Microbiol. Nat. Rev. 14, 135-149.

[2] Bhat S.V., Nagasampagi B. A. and Sivakumar M (2005). Chemistry of natural product. Berlin, New York. Springer, 6(6) 840.

[3] Finar I.L (2006). Stereochemistry and the chemistry of natural Product (5thedn.), Pearson Education Ltd, Patparganj Delhi, India Vol.2. 942p.

[4] Wink M (1998). Plant Breeding: Importance of plant secondary metabolite for protection against pathogen and herbivores Theoretical and Applied Genetics 75: pp225-233.

[5] Taiz L. and Zeiger E (2005). Plant physiology. Third Edition. Sinauer Association Inc., California, U.S. A. 690p.

[6] Hill A.F (1952). Economic Botany. A textbook of useful plant and plant Products (2nd edn.). MC-Graw-Hill Book Company Inc., New York, U.S. A. 743p.

[7] Okwu D.E (1999). Flavorings properties of species on cassava African Journal of Root Tuber Crops 3(2): pp19-21.

[8] Nwokeji P.A, Enodiana O.I, Ezenweani S.R, Osaro-Itota O, Akatah H.A (2016). The Chemistry of natural product: plant secondary metabolites. International Journal Of Technology Enhancements And Emerging Engineering Research,4(8): pp2347-4289.

[9] Dewick P.M (2002) Medicinal natural products: a biosynthetic approach, 2nd edn. Wiley, Chichester

[10] Laule O, Fürholz A, Chang H.S, Zhu T, Wang X, Heifetz P.B, Gruissem W, Lange M (2003). Crosstalk between cytosolic and plastidial pathways of isoprenoid biosynthesis in *Arabidopsis thaliana*. Proc Natl Acad Sci USA. 100: 6866-6871

[11] Schuhr C.A, Radykewicz T, Sagner S, Latzel C, Zenk M.H, Arigoni D, Bacher A, Rohdich F, Eisenreich W (2003). Quantitative assessment of crosstalk between the two isoprenoid biosynthesis pathways in plants by NMR spectroscopy. Phytochemistry Rev. 2, 3-16

[12] Dudareva N, Andersson S, Orlova I, Gatto N, Reichelt M, Rhodes D, Boland W, Gershenzon J (2005). The nonmevalonate pathway supports both monoterpene and sesquiterpene formation in snapdragon flowers. Proc Natl Acad Sci USA. 102, 933-938

[13] Piel J, Donath J, Bandemer K, Boland W (1998). Mevalonate-independent biosynthesis of terpenoid volatiles in plants: induced and constitutive emission of volatiles. Angew Chem Int Ed. 37, 2478-2481

[14] Gershenzon J and Croteau R (2012). Terpenoid in Herbivores: Their interaction with Secondary Plant Metabolite: the Chemical participant. Academic Press San Diego, California, USA. Vol.1 (2nd edn.), 409p

[15] Kessler A and Baldwin I.T (2001). Defensive function of herbivore Induced plant volatile emission in nature. Science. 29, pp2141-2144

[16] Sharkey T.D, Wiberley A.E, Donohue A.R (2008). Isoprene emission from plants: why and how. Ann Bot (Lond). 101, 5-18

[17] Baldwin I.T, Halitschke R, Paschold A, von Dahl C.C, Preston C.A (2006). Volatile signaling in plant-plant

interactions: "talking trees" in the genomics era. Science. 311, 812-815

[18] Croteau R.B, Davis E.M, Ringer K.L, Wildung M.R (2005). (−)-Menthol biosynthesis and molecular genetics. Naturwissenschaften. 92, 562-577

[19] Trapp S, Croteau R (2001). Defensive resin biosynthesis in conifers. Annu Rev Plant Physiol Plant Mol Biol 52, 689-724

[20] Mahmoud S.S, Croteau R.B (2002). Strategies fro transgenic manipulation of monoterpene biosynthesis in plants. Trends Plant Sci. 7, 366-373

[21] Phillips M.A and Croteau R.B (1999). Resin-based defenses in conifers. Trends Plant Sci 4, 184-190

[22] Jordt S.E, McKemy D.D, Julius D (2003). Lessons from peppers and peppermint: the molecular logic of thermosensation. Curr Opin Neurobiol. 13, 487-492

[23] Mo H and Elson C.E (2004). Studies of the isoprenoidmediated inhibition of mevalonate synthesis applied to cancer chemotherapy and chemoprevention. Exp Biol Med. 229, 567-585

[24] Dinda B, Debnath S, Harigaya Y (2007a). Naturally occurring iridoids. A review, Part 1. Chem Pharm Bull. 55, 159-222

[25] Dinda B, Debnath S, Harigaya Y (2007b). Naturally occurring iridoids. A review, Part 2. Chem Pharm Bull. 55, 689-728

[26] Seigler D.S (1998). Iridoid monoterpenes. In: Plant secondary metabolism. Kluwer, Dordrecht. 7, 61-71

[27] Adam K.P and Zapp J (1998). Biosynthesis of the isoprene units of chamomile sesquiterpenes. Phytochemistry. 48, 953-959

[28] Bick J.A and Lange B.M (2003). Metabolic cross talk between cytosolic and plastidial pathways of isoprenoid biosynthesis: Unidirectional transport of intermediates across the chloroplast envelope membrane. Arch Biochem Biophys. 415, 146-154

[29] Schnee C, Kollner T.G, Held M, Turlings T.C.J, Gershenzon J, Degenhardt J (2006). The products of a single maize sesquiterpene synthase form a volatile defense signal that attracts natural enemies of maize herbivores. Proc Natl Acad Sci USA. 103, 1129-1134

[30] Rasmann S, Kollner T.G, Degenhardt J, Hiltpold I, Toepfer S, Kuhlmann U, Gershenzon J, Turlings T.C.J (2005). Recruitment of entomopathogenic nematodes by insect-damaged maize. Nature. 434, 732-737

[31] Heinrich M, Robles M, West J.E, Ortiz de Montellano B.R, Rodriguez E (1998). Ethnopharmacology of *Mexican Asteraceae* (Compositae). Annu Rev Pharmacol Toxicol. 38, 539-565

[32] Lyß G, Knorr A, Schmidt T.J, Pahl H.L, Merfort I (1998). The anti-inflammatory sesquiterpene lactone helenalin inhibits the transcription factor NF- k B by directly targeting p65. J Biol Chem. 273, 33508-33516

[33] Hsu E (2006). Reflections on the 'discovery' of the antimalarial qinghao. Br J Clin Pharmacol. 61, 666-670

[34] White N.J (2008). Qinghaosu (Artemisinin): the price of success. Science. 320, 330-334

[35] Covello P.S, Teoh K.H, Polichuk D.R, Reed D.W, Nowak G (2007). Functional genomics and the biosynthesis of artemisinin. Phytochemistry. 68, 1864-1871

[36] Chang M.C.Y, Eachus R.A, Trieu W, Ro D.K, Keasling J.D (2007).

Engineering *Escherichia coli* for production of functionalized terpenoids using plant P450s. Nat Chem Biol. 3, 274-277

[37] Ro D.K, Paradise E.M, Quellet M, Fisher K.J, Newman K.L, Ndungu J.M, Ho K.A, Eachus R.A, Ham T.S, Kirby J, Chang M.C, Withers S.T, Shiba Y, Sarpong R, Keasling J.D (2006). Production of the antimalarial drug precursor artemisinic acid in engineered yeast. Nature. 440, 940-943

[38] Bishopp A, Mähönen A.P, Helariutta Y (2006). Signs of change: hormone receptors that regulate plant development. Development. 133, 1857-1869

[39] Wani M.C, Taylor H.L, Wall M.E, Coggin P, McPhail A.T (1971). Plant antitumor agents. VI. Isolation and structure of taxol, a novel antileukemic and antitumor agent from *Taxus brevifolia*. J Am Chem Soc. 93, 2325-2327

[40] Schiff P.B, Horwitz S.B (1980). Taxol stabilizes microtubules in mouse fibroblast cells. Proc Natl Acad Sci USA. 77, 1561-1565

[41] Bruce S.O., Onyegbule F.A., Ihekwereme C.P (2016). Evaluation of hepato-protective and anti-microbial activities of ethanol extracts and fractions of *Picralima nitida* seed and pod. Journal of Phytomedicine and Therapeutic, 1(2): 1-21.

[42] Hostettmann K.A, Marston A (1995). Saponins. Cambridge University Press, Cambridge. Pp548

[43] Güçlü-Üstündag Ö, Mazza G (2007). Saponins: properties, applications and processing. Crit Rev Food Sci Nutr. 47, 231-258

[44] Osbourn A (1996). Saponins and plant defence - a soap story. Trends Plant Sci. 1, 4-9

[45] Radad K, Gille G, Liu L, Rausch W.D (2006). Use of ginseng in medicine with emphasis on neuro degenerative disorders. J Pharmacol Sci. 100, 175-186

[46] Liby K.T, Yore M.M, Sporn M.B (2007). Triterpenoids and rexinoids as multifunctional agents for the prevention and treatment of cancer. Nat Rev Cancer. 7, 357-5369

[47] Li F, Goila-Gaur R, Salzwedel K, Kilgore N.R, Reddick M, Matallana C, Castillo A, Zoumplis D, Martin D.E, Orenstein J.M, Allaway G.P, Freed E.O, Wild C.T (2003). PA-457: a potent HIV inhibitor that disrupts core condensation by targeting a late step in Gag processing. Proc Natl Acad Sci USA. 100, 13555-13560

[48] Howitt C.A, Pogson B.J (2006). Carotenoid accumulation and function in seeds and non-green tissues. Plant Cell Environ. 29, 435-445

[49] Ye X, Al-Babili S, Klöti A, Zhang J, Lucca P, Beyer P, Potrykus I (2000). Engineering the provitamin A (beta-carotene) biosynthetic pathway into (carotenoid-free) rice endosperm. Science. 287, 303-305

[50] Riccioni G, Bucciarelli T, Mancini B, Corradi F, Di Ilio C, Mettei P.A, D'Orazio N (2007). Antioxidant vitamin supplementation in cardiovascular diseases. Ann Clin Lab Sci. 37, 89-95

[51] U.S. Preventive Services Task Force (2003). Routine vitamin supplementation to prevent cancer and cardiovascular disease: recommendations and rationale. Ann Intern Med. 139:51-55

[52] Randhir R., Lin Y.T., Shetty K (2004). Stimulation of phenolics, antioxidant and antimicrobial activities in dark germinated mung bean sprouts in response to peptide and

phytochemical elicitors. Process Biochem. 39:637-646.

[53] Taiz L and Zeiger E (2005). Plant physiology. Third Edition. Sinauer Association Inc., California, U .S. A. 690p

[54] Kondo T, Yoshida K, Kawai T, Tamura H and Goto T (2012). Structural basis of blue colour development in flower petal from *Commelina communis*. Nature. 358, pp515-518

[55] Li J. Lee O, Raba R and Last R.L (2003). Arabidopsis flavonoid mutant are hypersensitive to UV-B radiation. Plant Cell. 85, pp171-179

[56] Onyegbule F.A, Okoli O.G, Bruce S.O (2019). "*In vivo* Evaluation of the Antimalarial Activity of the Aqueous Ethanol Extract of *Monodora myristica* Seed in Albino Mice", International Journal of Science and Research (IJSR), Volume 8 Issue 6, 1530-1538

[57] Gutzeit H.O and Ludwig-Muller J (2014). Plant Natural Products: Synthesis, Biological Functions and Practical Applications, First Edition. Wiley-VCH Verlag GmbH & Co. KGaA. Pp1-80.

[58] Edeoga H.O, Okwu D.E and Mbaebie B.O (2009). Phytochemical constituent of some Nigeria medicinal plants. African Journal of Biotechnology. 4 (7): pp685-688

[59] Fessenden R.J and Fessenden J.S (1997). Organic Chemistry (2nd edn.). Willard Grant Press, Massachusetts, U. S. A. 57p

[60] Hartmann T (2013). Chemical ecology of pyrrolizidine alkaloids. Planta. 207, pp483-495

[61] Sicker D, Frey M, Gierl A (2000). Role of natural benzoxazinones in the survival strategy of plants. Int Rev Cytol. 198, 319-346

[62] Bravo H.R, Lazo W (1993). Antimicrobial activity of cereal hydroxamic acids and related compounds. Phytochemistry. 33, 569-591

[63] Bravo H.R, Lazo W (1996). Antialgal and antifungal activity of natural hydroxamic acids and related compounds. J Agric Food Chem. 44, 1569-1571

[64] Grombacher A.W, Russell W.A, Guthrie W.D (1989). Resistance to first-generation European corn borer (*Lepidoptera: Pyralidae*) and DIMBOA concentration in midwhorl leaves of the BS9 maize synthetic. J Kans Entomol Soc. 62, 103-107

[65] Pérez F.J, Niemeyer H.M (1989). Reaction of DIMBOA with amines. Phytochemistry. 28, 1831-1834

[66] Atkinson J, Morand P, Arnason J.T, Niemeyer H.M, Bravo H.R (1991). Analogs of the cyclic hydroxamic acid 2,4-dihydroxy-7-methoxy-2H-1,4-benzoxazin-3-one (DIMBOA): decomposition to benzoxazolinones and reaction with beta-mercaptoethanol. J Org Chem. 56, 1788-1800

[67] Hashimoto Y, Shudo K (1996). Chemistry of biologically active benzoxazinoids. Phytochemistry. 43, 551-559

[68] Fahey J.W, Zalcmann A.T, Talalay P (2001). The chemical diversity and distribution of glucosinolates and isothiocyanates among plants. Phytochemistry. 56:5-51

[69] Halkier B.A, Gershenzon J (2006). Biology and biochemistry of glucosinolates. Ann Rev Plant Biol. 57, 303-333

[70] Wittstock U, Burow M (2007). Tipping the scales - specifier proteins in glucosinolate hydrolysis. IUBMB Life. 59, 744-751

[71] Higdon J.V, Delage B, Williams D.E, Dashwood R.H (2007). Cruciferous vegetables and human cancer risk: epidemiologic evidence and mechanistic basis. Pharmacol Res. 55, 224-236

[72] Ratzka A, Vogel H, Kliebenstein D.J, MitchellOlds T, Kroymann J (2002). Disarming the mustard oil bomb. Proc Natl Acad Sci USA. 99, 11223-11228

[73] Wittstock U, Agerbirk N, Stauber E..J, Olsen C.E, Hippler M, Mitchell-Olds T, Gershenzon J, Vogel H (2004). Successful herbivore attack due to metabolic diversion of a plant chemical defense. Proc Natl Acad Sci USA. 101, 4859-48564

[74] Hegnauer R (1986). Chemotaxonomie der Pflanzen. Vol 7. Birkhäuser Verlag, Basel Volume 27, Issue 8, Pages 2423-2427

[75] Seigler D.S (1991). Cyanide and cyanogenic glucosides. In: Rosenthal GA, Berenbaum MR (eds) Herbivores: their interactions with secondary plant metabolites. Academic, San Diego, CA. 1, 35-77

[76] Bruce S.O, Onyegbule F.A, Ezugwu C.O (2019). Pharmacognostic, physicochemical and phytochemical evaluation of the leaves of *Fadogia cienkowskii* Schweinf (Rubiaceae). Journal of Pharmacognosy and Phytotherapy. Vol. 11(3), pp. 52-60

[77] Zagrobelny M, Bak S, Møller B.L (2008). Cyanogenesis in plants and arthropods. Phytochemistry. 69, 1457-1468

[78] Selmar D, Lieberei R, Biehl B (1988). Mobilization and utilization of cyanogenic glycosides: the linustatin pathway. Plant Physiol. 86, 711-716

[79] Jones D.A (1998). Why are so many food plants cyanogenic? Phytochemistry. 47, 155-162

[80] Okoye V.O, Bruce S.O and Onyegbule F.A (2020). Phytochemical screening and pharmacognostic properties of *Peuraria phaseoloides* leaves (roxb) benth. (fabaceae). International Journal of Public Health, Pharmacy and Pharmacology, 2020; 5(2)11-24.

[81] Sánchez-Pérez R, Jørgensen K, Olsen C.E, Dicenta F, Møller B.L (2008). Bitterness in almonds. Plant Physiol. 146, 1040-1052

[82] O'Brien GM, Taylor A.J, Poulter N.H (1991). Improved enzymatic assay for cyanogens in fresh and processed cassava. J Sci Food Agric. 56, 277-289

[83] Jørgensen K, Bak S, Busk P.K, Sørensen C, Olsen C.E, Puonti-Kaerlas J, Møller B.L (2005). Cassava plants with a depleted cyanogenic glucoside content in leaves and tubers. Distribution of cyanogenic glucosides, their site of synthesis and transport, and blockage of the biosynthesis by RNA interference technology. Plant Phys. 139, 363-374

[84] Siritunga D, Sayre R (2007). Transgenic approaches for cyanogen reduction in cassava. AOAC Int. 90, 1450-1455

Chapter 5

The Need to Use Microorganisms and Their Biosynthesized Bioactive Metabolites for Biological and Medical Activities

Zeinab Amanzadeh

Abstract

Some microorganisms (MOs) such as bacteria, fungi and aquatic creatures synthesize bioactive secondary metabolites synthesis that known as natural products. A series of landmark metabolomics studies by using mass spectrometry (MS) or nuclear magnetic resonance (NMR) analysis to identify potentially important microbial metabolites that derive from the intestine microbes. Vital roles for numerous microbial metabolic pathways in host physiology had been long established, such as in the synthesis of vitamin K and the synthesis of water-soluble B vitamins including biotin, folates, nicotinic acid, pyridoxine, riboflavin, cobalamin and panthotenic acid, the degradation of nutritional oxalates, and amendment of bile salts. These metabolites have biological and medical activities. The medical activities including antimicrobial, immunosuppressive, anticancer, and anti-inflammatory, antibiotic, antitumor, antioxidant activities, etc. Also biological activities derive from metabolites microbial transformation have effects on physiological processes such as gut and immune homeostasis, energy metabolism, vascular function, and neurological behavior such as short-chain fatty acids, trimethylamine N-oxide, tryptophan and tyrosine derivatives, and oxidized fatty acids. Using recombinant DNA technology synthesize a wide array of biopharmaceutical products, such as recombinant proteins, offering significant advances in treating a broad spectrum of medical. Such interventions will require modulating either bacterial species or the bacterial biosynthetic enzymes required to synthesis these metabolites.

Keywords: microorganism, metabolism, biosynthesis, metabolite, biological activities

1. Introduction

The biosynthesis process is multi-step for formation of organic compounds in a living microorganism by chemical energy (e.g. ATP). In biosynthesis, simple compounds or substrates by enzyme-catalyzed are modified, converted into other compounds, or joined together to form macromolecules [1]. Some important biological macromolecules include: proteins, which are composed of amino acid monomers joined via peptide bonds, and DNA molecules, which are composed of nucleotides joined via phosphodiester bonds [1].

Biosynthesis occurs due to a series of chemical reactions in which precursor compounds, catalytic enzymes, cofactors, and chemical energy are necessary for these reactions to take place. The biosynthetic processes responsible for the production can mentioned photosynthesis, lipogenesis, glycolysis, glyconeogenesis and Krebs cycle [2].

These biological processes result in the biosynthesis of intermediates which proceed towards manufacturing of secondary metabolites via alternate biosynthetic routes responsible for metabolite diversity in living organisms [3].

Metabolism is one of the biggest factors of inter-kingdom interactions along with the ones between microorganisms and their multicellular hosts. Ordinarily notion to fuel energy necessities and provide constructing blocks for biosynthetic pathways, metabolism is now liked for its position in imparting metabolites, small-molecule intermediates generated from metabolic techniques, to perform various regulatory features to mediate symbiotic relationships between microbes and their hosts [4].

Metabolite is divided into two main categories in living organisms: Primary and Secondary metabolite. Primary metabolites include biological molecules i.e., vitamins, amino acids, nucleosides, Organic acids, Acetone-butanol, Ethanol, Vitamins, fats, carbohydrates and proteins, essential for the survival and well-being of the organism and are produced to sustain cell growth. Secondary metabolites are compounds with varied and sophisticated chemical structures, produced by microorganisms after the rapid growth phase. These compounds are not essential for growth [5].

2. Microorganism metabolites

Microorganism such as bacteria and fungi are inhabitants of diverse habitats worldwide. Due to which, they have evolved to cope with adverse conditions [6]. The structurally diverse secondary metabolites produced by them possess biological activities such as antibiotic, antimicrobial, immunosuppressive, anticancer, and anti-inflammatory activities, many of which have been developed as treatments and have potential therapeutic applications for human diseases [6]. The produce secondary metabolites, also known as natural products. Aside from natural products, the recent development of recombinant DNA technology has sparked the development of a wide array of biopharmaceutical products, such as recombinant proteins, offering significant advances in treating a broad spectrum of medical illnesses and conditions [6, 7].

2.1 Characteristics of secondary metabolites

Secondary metabolites (SMs) are organic compounds with complex chemical structures and diverse physiological functions. Secondary metabolites include antibiotics, pigments, and other bioactive compounds (Bioactive word means Biologically Active). Many of these compounds have important agricultural and medical applications [8, 9].

Microorganisms are noted as a rich source of bioactive secondary metabolites and bioactive metabolites. Some of these bioactive metabolites, such as antibiotics, siderophores, immunosuppressants and degradative enzymes are also useful in medicine and biotechnology. These play a role in defense mechanisms against predators [8].

Many microorganisms synthesize secondary metabolite molecules that play essential ecological roles of their complex and heterogeneous microenvironments.

Commonly, the genes governing the biosynthesis of secondary metabolites are clustered collectively, and increasingly gene clusters accountable for the biosynthesis of secondary metabolites were located [10].

The provision of clusters has improved purposeful investigations of biosynthetic pathways of secondary metabolites. An intensive understanding of the enzymatic method is required for metabolic engineering to enhance manufacturing of secondary metabolites and for combinatorial biosynthesis to generate novel compounds or derivatives. Secondary metabolites are usually produced at some point of the desk bound section of growth in microorganisms [11].

2.1.1 Secondary metabolites have the subsequent traits

1) Secondary metabolites (SMs) may be produced only with the aid of a few microorganisms. 2) They will be inclined to be produced at the terminal of exponential growth or within the direction of substrate-restricted situations. 3) They're created from common metabolic intermediates but use specialized pathways encoded via a specific gene. Those products are not nessarary for the organism's very own growth, duplicate, and regular metabolism. 4) Secondary metabolites have uncommon chemical linkages, for instance, β-lactam rings, cyclic peptides, unsaturated linkage of polyacetylenes and polyenes, big macrolide rings, and so forth. 5) Increase situations, particularly the composition of the medium inside a fermentation machine, control the formation of secondary metabolites. 6) Those compositions are produced as a collection of carefully associated systems. 7) Secondary metabolic compositions can be overproduced [5].

2.2 Why secondary metabolites are produced by the organisms?

Secondary metabolites seem to act the organisms that produce them as (1) competitive tools used in opposition to different microorganisms, flora, bugs, and large animals; (2) sexual hormones; (3) agents of plant–microbe symbiosis and plant increase stimulation; (4) metallic transporting dealers; and (5) differentiation effectors [12]. Secondary metabolites have a first-rate impact at the fitness, nutrients, and economics of communities. Antibiotics are the most essential of the secondary metabolites. The alarming rise in emergence and occurrence of antibiotic resistance poses a primary danger to human healthcare. It is clean that novel antibiotics are urgently had to combat this trouble [12, 13].

Different secondary metabolites are insecticides, pesticides, pigments, xenobiotics, effectors of ecological competition and symbiosis, pheromones, enzyme inhibitors, immunomodulating factors, receptor antagonists and agonists, insecticides, antitumor agents, immunosuppressives, cholesterol-lowering factors, plant protectants, and growth promotants of animals and herbals. As a stop result, they have wonderful monetary significance [13].

3. Use tools to identify significance microbial metabolites

Intense interest in the intestine microbes over the past decade has led to understanding of diet–microbiota–host interactions suggests significant opportunities to create new therapeutic approaches, including selectively altering the microbial production of molecules to promote human health and prevent disease [14].

A sequence of landmark metabolomics research over the past decade have appreciably superior our know-how via the usage of mass spectrometry (MS) or nuclear magnetic resonance (NMR) evaluation to select out in all likelihood crucial microbial metabolites that derive from the gut microbes, which might be enriched or depleted in diseased humans, or that can be used to are expecting physiological response to meals or different interventions [13].

Researchers have established a number of metabolites which can play essential roles in human fitness and ailment, together with short-chain fatty acids (SCFAs) and long-chain fatty acid metabolites which inclusives conjugated linoleic acid and 10-hydroxy-cis-12-octadecenoate, trimethylamine and trimethylamine N-oxide, tyrosine and phenylalanine metabolites collectively with hippuric acid, phenylacetylglycine, phenyl sulfate, paracresyl sulfate, phenylpropionylglycine, cinnamoylglycine and equol sulfate and tryptophan metabolites together with indole, indole-three- propionate and indoxyl-sulfate [13, 15].

A number of the metabolites diagnosed by manner of these research result from the transformation of unique nutritional components via pick out species of microbes that express the important enzymes to behave on these additives. For that reason, the variable presence of microbes using those eating regimen-established metabolic pathways can be key to knowledge the variable host reaction to particular nutritional components and susceptibility to illnessess [13].

lots work stays to completely symbolize the physiological results of those and the many other microbial metabolites that can be essential in human health [16].

Accordingly, it appears there may be a vast want for cautiously controlled research to decide the physiological outcomes of each recognized microbial metabolite and its particular mechanisms of action [16]. Moreover, so that you can fully take advantage of the capacity of the gut microbiota for disease prevention, we need a much more expertise of ways dietary additives and host genetics affect the manufacturing of numerous metabolites. The gut microbiota for human health, the remarkable progress of the last decade suggests that such approaches have significant potential to revolutionize therapeutic approaches to human disease [17].

4. Biosynthesis of vitamins by probiotic bacteria

The connection among vertabrates and the microbial cells that reside of their gastrointestinal tracts relies on a complicated molecular, with microbial metabolites acting as essential mediators of this a complex molecular. Important roles for numerous microbial metabolic pathways in host body structure were lengthy mounted, along side in the production of a few vitamins, the degradation of dietary oxalates, and change of bile salts [13].

Vitamins are crucial micronutrients which may be frequently precursors to enzymes, which all living cells require to carry out biochemical reactions. Since human body cannot synthesize many vitamins, simply so they want to be externally received [18]. The use of vitamins-generating microorganisms can be a natural and marketable approach to the usage of pseudo-vitamins which may be chemically produced, and could permit for the producing of foodstuffs with better levels of vitamins that could lessen unwanted facet outcomes. Probiotic bacteria, further to commensal microorganism observed inside the human intestine, consisting of Lactobacillus and Bifidobacterium, can de novo synthesize and supply nutrients to human body [18].

Within the human body, groups of the intestine microbiota are capable of synthesize vitamin K and the production of water-soluble B vitamins including cobalamin (vitamin B12), folate (vitamin B9), pyridoxine (vitamin B6), riboflavin (vitamin B2), and thiamine (vitamin B1). All of these vitamins are essential for the body and serve as a co-factor for the specific enzymes [19].

5. Using of microorganisms as valuable resource for healthy food

Microorganisms are taken into consideration a treasured resource for novel wholesome food ingredients and biologically lively compounds. Microorganisms have increasingly been used to synthesis value-added products with numerous functions inside the agricultural, foods and pharmaceutical industries [20].

These value-added compounds can also embody enzymes, prebiotics, fatty acid, antioxidants, proteins, polysaccharides, organic acids, and biofuels. For this reason, microbial biosynthesis offers a renewable, environmentally benign route, sustainable feedstocks and economically appealing alternatives [1].

Furthermore, the recent advancement in analytical measurement, such as chromatography, with a particular reference to ultrahigh-performance liquid chromatography (UHPLC) coupled with mass spectrometry (MS), allowed the simultaneous analysis of various compounds, with rapid and accurate results [21]. Functional foods and natural-health products comprise quite a wide range of food ingredients, with various bioactive compounds responsible for their activity in disease prevention and/or health promotion [22].

Prebiotics serve via various mechanisms, which includes producing vitamins, interacting with host immune structures, stopping pathogen adhesion to host cells, and affecting the morphological shape of the intestine, all of which likely act via the modulation of intestinal microbiota. A broad sort of dietary compounds may satisfy those criteria. so far, the maximum promising dietary fibers with promising prebiotic capabilities are nondigestible oligosaccharides containing 3–9 sugar monomers [15].

5.1 Fructooligosaccharides

Dietary carbohydrates especially Fructooligosaccharides are notably emerging as an important prebiotic due to their hypocaloric, bifidogenic, and noncariogenic functions. The possible health benefits associated with the consumption of Fructooligosaccharides has led to their increased acceptance as food ingredients and alternative sweeteners used in diabetic formulations [15].

5.2 Omega-3 PUFAs

There are numerous benefits of long-chain omega-3 PUFAs, particularly eicosapentaenoic acid (EPA, C20:5) and docosahexaenoic acid (DHA, C22:6), for human health. PUFAs, which are involved in many vital biological activities, such as inflammatory, immune, and cancer processes. In fact, PUFAs form a unique class of food constituents with numerous functions; they are considered food and nutritional products with specific health-promoting activities, modulating the risk of certain diseases [15].

PUFAs are fatty acids (FAs) are found in particular in fish, vegetable oils, inclusive of sunflower, flaxseed, soybean, rapeseed, and marine belongings. In latest many years, there has been interest within the utilization of microorganisms as ability promising producers of determined on PUFAs. But, the growth in PUFA call for and the decline in fish shares have stepped forward the attention paid to microorganisms, for the reason that microorganisms can be cultivated underneath managed conditions with immoderate growth rates and that additionally they do not compete for land for unique meals manufacturing. The principle microbial resources of PUFAs are marine algae, fungi, and microorganism [15, 23].

6. Microbial sources in recombinant drug discovery

An expansion of organisms, like bacteria, fungi, and plant produce secondary metabolites, called natural products. Natural merchandise are prolific sources and a revelation for numerous clinical factors with extensively divergent chemical structures and natural factors functions, at the side of antimicrobial, immunosuppressive, anticancer and anti-inflammatory features, many of that have been developed as remedies and feature functionality recuperation programs for human sicknesses [6, 12].

These structurally and chemically diverse molecules act as a remarkable class of therapeutics to heal various ailments. Aside from natural products, the recent development of recombinant DNA technology has sparked the development of a wide array of biopharmaceutical products, such as recombinant proteins, offering significant advances in treating a broad spectrum of medical illnesses and conditions [24].

6.1 Natural product from fungal sources

Fungi were used for a long time by humankind for plenty functions, inclusive of food manufacturing (beer, wine, leavened bread, soy substances and so on), remedies, and in regular existence. Loads of years ago, fungi were used to deal with intestinal illness. Since the producing of penicillin, which turn out to be isolated from the fungus Penicillium notatum, fungi had been a rich source of many recuperation factors. Fungi are a highrich supply of biologically energetic secondary metabolites [25].

Many healing factors, which includes cyclosporine and mycophenolic acid (immunosuppressive interest), fusidic acid and griseofulvin (antimicrobial interest), and exclusive novel semisynthetic antifungal pharmacy, in conjunction with anidulafungin and caspafungin, have been originated from fungal metabolites [6].

One of the most vital mediciene are statins, invlusive mevastatin from Penicillium citrinum and lovastatin from Aspergillus terreus. Statins as the most important member of antilipidemic medicienes for the remedy of cardiovascular ailments, also are derived from microbial resources. Fungal metabolites are not only crucial for treatment but additionally for plant safety. For example, the producing of strobilurins from Strobilurus species, precipitated compounds for artificial fungicides, which includes trifloxystrobin [25].

6.2 Natural product from bacteria sources

Almost three-quarters of microbial-produced bioactive compositons are from actinomycete bacteria. Extra than 500 species of streptomycetes that are the most extensively identified organization, generating extensive range of biologically energetic compositions. They will be gram-positive aerobic filamentous (regularly soil) bacteria. They often produce spores and are characterized by manner of the producing of geosmin, a risky metabolite that provide them "earthy" scent. The spore germination technique relies upon at the environmental conditions [25]. In everyday conditions, the germination of streptomycete spores starts off evolved by arthrospore (substrate mycelium), however inside the case of nutrient depletion, the increase starts with aerial mycelium. In one-of-a-kind phrases, underneath favorable conditions, a fully matured mycelia is produced. Beneath drastic conditions, alternatively, the aerial mycelium is subdivided with the resource of septa, then into spores, which in turn can, underneath sure conditions, germinate into mycelium [25–27].

Actinomycetes are recounted to provide diverse sorts of antibiotics, in particular, peptides/glycopeptides, angucyclinone, tetracyclines, phenazines, macrolides, anthraquinones, polyenes, anthracyclines, β-lactams, piercidins, octaketides, benzoxazolophenanthridines, heptadecaglycosides, and lactones [6].

6.2.1 Bioactive activities of natural products of bacterial sources

The secondary metabolites produce in actinomycetes is greatly affected by various fermentation parameters, such as nutrients availability, pH, aeration, temperature, mineral salts, heavy metals, precursors, inducers, and inhibitors, which often vary from organism to organism [28]. Streptomycetes are a rich source of many bioactive compounds. Most antifungals derived from streptomycetes tend to be macrolide polyene, such as nystatin, produced by streptomyces. About two-thirds of bioactive compounds are produced by this group, and they have many clinical efficacies against different kinds of organisms, such as bacteria, fungi, and parasites [29].

Further, antitumor features, such as aclacinomycin A, actinomycin D, bleomycin, daunorubicin, mithramycin, mitomycin C, and nogalamycin (synthesized with the aid of Streptomyces glalilaeus, Streptomyces antibioticus, Streptoverticillium verticillium, Streptomyces paecetius, Streptomyces argillaceus, Streptomyces lavendulae, and Streptomyces nogalater, respectively). These medications can act on DNA by using altering its function via mechanisms, including intercalation, cross-linking, DNA strand fracture, or interacting with DNA non-intercalatively [29].

7. Microbial metabolites for medical and anticancer activities

The search for novel microbial metabolites has shown Mevinolin, a potent cholesterol-lowering agent was isolated from Aspergillus terreus. Aspercilin was isolated from Aspergillus alliaceus. Later on, benzodiazepines were derived from aspercilin and used for curing anxiety or insomnia [30].

7.1 Fungal metabolites

Norsolorinic acid, isolated from Aspergillus spp. was reported to cause apoptosis in breast cancer (MCF-7) and human bladder cancer (T-24) cells [30, 31].

Extracts of Penicillium steckii and Aspergillus sydowii induced cytotoxicity in human cervical carcinoma cell line (HeLa). Whereas, extract of *Alternaria alternata* showed cytotoxic activity against *Staphylococcus aureus*, *Escherichia coli* and HeLa cells.

Similarly, ethanolic extracts of Fomitopsis pinicola induce cytotoxicity in various cancer cell lines including human hepatoma, colorectal, lung and breast cancer cells along with synergistic effects with cisplatin in vivo [6, 29].

Aspergillus parasiticus, a type of fungal endophytic, isolated from *Sequoia sempervirens* was reported to be a producer of sequoiatones A and B which showed moderate anticancer potential with the highest activity against breast cancer cell lines.

Torreyanic acid, isolated from endophytic fungi of *Torreya taxifolia* tree exhibited apoptotic activity in protein kinase C sensitive cancer cells [9]. Endophytic fungi, Taxomyces andreanae and Nodulisporium sylyiforme have been reported to produce taxol (an anticancer drug previously isolated from the pacific yew tree) [9].

Along with anticancer potential, this drug also exhibited antifungal activity against Pythium, Phytophthora and Aphanomyces spp. A novel polykedite,

5-hydroxyramulosin isolated from an endophytic fungus of Cinnamomum mollissimum showed both antifungal activities against Aspergillus niger and anticancer activity against murine leukemia cells [9].

7.2 Bacterial metabolites

Two bacterial strains, *Escherichia coli* and Bacillus subtilis are genetically more variable from each other than humans are from corns [9].

Nisin, a bacteriocin has been used as bio-preservative. Rapamycin, an antifungal agent was isolated from soil inhabiting Actinomycetes. It is also used for inhibiting organ rejection in transplant patients. Amrubicin hydrochloride, an anticancer compound was isolated from Streptomyces peucetius in 2002. A new class of antibiotics called Pumalicidins A, B, C, D, E, F and G were obtained from the culture broth of *Bacillus pumilus* [9, 31].

Homologs of Bacillomycin D, isolated from Bacillus subtillis (B38) were reported to have anti-oxidative and DNA protective activities. Additionally, Surfactin, produced by Bacillus subtilis CYS191, was reported to induce apoptosis in human breast cancer cells (MCF-7) by causing oxidative stress [9, 31]. Streptomyces hygroscopicus was identified as a producer of antifungal prenylated indole, galbonolides A and B, elaiophylin and its derivatives and herbimycins. Pterocidin, produced by endophytic Streptomyces hygroscopicus showed cytotoxic effects in human lung, ovarian, glioblastoma and melanoma cells [9].

There was an analogue of signal peptide 27 that produced by using *Streptococcus pneumoniae* has proven cytotoxicity in opposition to leukemia, gastric and breast cancer via cell permeabilization and induction of caspase-unbiased apoptosis [9]. Also any other anticancer peptide Entap (Enterococcal anti-proliferative peptide) remoted from Enterococcus spp. make induction of autophagous apoptosis and inhibits proliferation in numerous cancers [9].

8. Biosynthesis of nanoparticles by microorganisms

The improvement of eco-friendly technology in substances synthesis is of sizable significance to amplify their natural activities. In recent times, a selection of inorganic nanoparticles (NP) with properly-defined chemical shape, duration, and compunds had been synthesized by using of special microorganisms, and their capabilities in lots of cutting-edge methodological areas had been analyzed [32]. Elements that constitutive nanoparticles having one or greater dimensions of the order of 100 nm or a good deal less have attracted extremely good attention due to their uncommon and fascinating features, and applications tremendous over their bulk contrary numbers [32].

There are huge types of bodily, chemical, natural, and hybrid techniques available to synthesize awesome kinds of nanoparticles. Despite the fact that bodily and chemical techniques are extra famous inside the synthesis of nanoparticles, use of poisonous chemicals greatly limits their biomedical functions, especially in scientific fields [32].

Consequently, improvement of reliable, non-toxic and eco-friendly techniques for synthesis of nanoparticles is of utmost significance to expand their biomedical applications. One of the options to gather this motive is to use microorganisms to produce nanoparticles. These elements synthesized with the aid of manner of a biogenic enzymatic system are a long way superior, in several techniques, to those elements synthesized via chemical techniques [33].

Nanoparticles are produced whilst the microorganisms take target ions from their surroundings after which flip the metal ions into the detail metal via enzymes generated thru the cellular functions. It is able to be categorized into intracellular and extracellular synthesis in step with the area where nanoparticles are organized [33].

The intracellular system inclusives transporting ions into the microbial cellular to shape nanoparticles internal using enzymes. The extracellular synthesis of nanoparticles includes taking the metallic ions at the surface of the cells and reduction of ions the use of enzymes [32, 33].

8.1 Types of NPs

NPs are formed in two synthesized: metal and oxide by organisms. Metal nanoparticles such as gold (Au), silver (Ag), platinum (Pt), mercury (Cu), cadmium (Cd), selenium (Se), mercury (Hg) and chromium (Cr) are synthesized by microorganisms. Metal nanoparticles are usually spherical, flat and cube shapes, Pyramidal and Irregular polygonal. Their size of diameters range from 2 to 180 nm. The most important nanopatiticles are gold and silver. Gold nanoparticles are made by *Sargassum wightii* and Rhodococcus sp., Shewanella oneidensis, Plectonemaboryanum, *Plectonema boryanum* andYarrowia lipolytica, silver nanoparticles by Trichodermaide vir, Phaenerochate, chrysosporium and Bacillus lichformeniis [32–34].

The most important oxide nanoparticles include TiO_2, $BaTiO_3$, Sb_2O_3, $BaTiO_3$, ZrO_2, Fe_3O_4 and Fe_2O_3. Their size of diameters nanoparticles range from 3 to 80 nm. The size of these nanoparticles ranges from 3 to 80 nm and forms Rectangular, rhombic, hexagonal, Cubo-octahedral, Nanopowders, Wormhole-like, Bullet-shaped, Tetragonal, Spherical and Pseudohexagonal/irregular or rhombohedral is synthesized. These are by some bacterial and fungal microarchanis like Shewanella oneidensis, Lactobacillus sp.oxysporum, oxysporum, *Saccharomyces cerevisiae* and yeast cells are synthesized [32–34].

8.2 Utility of nanoparticles

Treatment using nanoparticles (nanomedicine) is a attractive and growing discipline to study with super potentialities possibilities for the improvement of the analysis and treatment of human illnesses. Dispersed nanoparticles are usually employed in nanobiomedicine as fluorescent natural, drug and gene delivery entrepreneurs, and in programs which include biodetection of pathogens, tissue engineering, tumor destruction through heating (hyperthermia), MRI comparison enhancement, and phagokinetic research [33].

The biosynthesized nanoparticles have been used in a variety of applications including drug carriers for targeted delivery, cancer treatment, gene therapy and DNA analysis, antibacterial agents, biosensors, enhancing reaction rates, separation science, and magnetic resonance imaging (MRI) [33]. Some of the purposes of using nanoparticles include drug delivery, antibacterial agent, biosensor, reaction rate enhancement agent and magnetic separation and detection. Here, we provide some examples to illustrate these applications [34].

8.2.1 Drug delivery

Magnetic nanoparticles like Fe3O4 and Fe2O3 are known to be biocompatible. They have been actively investigated for targeted cancer treatment (magnetic hyperthermia), stem cell sorting and manipulation, guided drug delivery, gene therapy and DNA analysis, and MRI [33].

8.2.2 Antibacterial agent

Silver nanoparticles (AgNPs) had been biosynthesized the use of fungus Trichoderma viride. These nanoparticles have been evaluated for their improved antimicrobial functions with numerous antibiotics towards gram-positive and gram-negative bacteria. The antibacterial functions of ampicillin, canamycin, erythromycin, and chloramphenicol had been elevated within the presence of silver nanoparticles (AgNPs) towards test traces [35].

The best improving effect changed into observed for ampicillin in opposition to take a look at lines. The end result confirmed that the aggregate of antibiotics with AgNPs has higher antimicrobial effects and provided beneficial insight into the development of latest antimicrobial dealers [35].

8.2.3 Biosensor

Nanoparticles possess interesting electronic and optical properties and can be used in biosensor applications. They are synthesized from spherical selenium nanoparticles by the Bacillus subtilis with diameters ranging. These sensors exhibited good electrocatalytic activity towards the reduction of H2O2 due to the good adhesive ability and biocompatibility of Se nanomaterials [36]. These H2O2 biosensors had high sensitivity and affinity for H2O2 with a detection limit of 8×10^{-8} M. The selenium (Se) nanomaterials-modified electrode will probably be promising for a wide range of applications related to the detection of H2O2 in food, pharmaceutical, clinical, industrial and environmental analyses [36].

8.2.4 Reaction rate enhancement agent

Magnetic nanoparticles have been used to improve the microbiological reaction rates.

In fact, magnetic nanoparticles were utilized not only for their catalytic function but also for their good ability to disperse i.e. use of the coated microbial cells of Pseudomonas delafieldii with magnetic Fe3O4 nanoparticles to fulfill desulfurization of dibenzothiophene. The high surface energies of nanoparticles resulted in their strong adsorption on the cells [36].

8.2.5 Magnetic separation and detection

Magnetic elements conjugated with organic molecules, that rectangular degree enticing substances for building assay structures, are deliberate to be used as a biological label [37]. Aggressive luminescence catalyst immunoassays utilizing antibodies immobilized onto microorganism magnetic elements had been evolved for the fast and sensitive detection of tiny molecules, like xenobiotics, hormone and cytotoxic detergents. The employment of magnetic particles as a solid-phase adsorbent is similar temperament for polymer extraction techniques because of they will be actually manipulated through smooth software of a magnet [37].

9. Conclusions

Microorganisms had been present in the world 4 billion years ago and have been evolving and increasing into new surroundings ever since, current anywhere. Their presence has driven the development of latest ecosystems, a number of which allowed the evolution of extra complex organisms. Issues about the supply of healthy, secure meals for people have elevated research for the replacement of chemical substances with inexperienced biomaterials. Considered one of strategies of creating and processing renewable monomers and polymers are presently studied, thinking of their benefits and downsides. There is a need to supply a green substitute for foods, medicinal drug, and pharmaceutical packages. Microbial synthesis has attracted extremely good attention because of the ease of the process and transformation into critical primary and secondary metabolites. The function of natural merchandise in meals, medicine, remedy, and agriculture fields is extensively highlighted because of their chemical stability and biocompatibility. Microorganisms were employed in large-scale manufacturing of a diffusion of biochemicals, antibiotics, and inside the processing of meals and feeds.

It is increasingly possible to identify potential vitamin-producing strains and interpret the intertwined mechanisms for their biosynthesis, because of the expanding availability of genome sequences, which could be used to expand the vitamin-producing capacities of the human intestine. Probiotic bacteria, as well as commensal bacteria found in the human intestine, such as Lactobacillus and Bifidobacterium, can de novo synthesize and supply vitamins to human body.

The usage of vitamin-producing types supplied a brand new perspective at the precise makes use of probiotics. Many vitamin-producing bacteria overproduce B vitamins and K vitamin, that could allow them to organically enrich uncooked meals materials like soy, milk, meat, and greens with B vitamins, preventing the want for additives. Consequently, the meals enterprise ought to take benefit of those novel and efficient vitamin-producing types to feature nutritional value to fermented merchandise and economically viable. It's far an increasing number of feasible to identify capacity vitamin-producing types and interpret the intertwined mechanisms for their biosynthesis, due to the increasing availability of genome sequences, which will be used to increase the vitamin-generating capacities of the human intestine. Probiotic microorganism, in addition to commensal microorganism observed inside the human intestine, along with Lactobacillus and Bifidobacterium, can de novo synthesize and supply vitamins to human intestine. In human body, bacteria can produce vitamin K and most of the water-soluble B vitamins.

There were terrific trends in the field of microorganism synthesized nanoparticles and their functions over the past decade. Although an incredible deal work is wanted to enhance the overall performance of synthesis, the manipulate of particle duration and morphology. The biosynthesis of nanoparticles by using microbes is idea to be clean, safe, and environmentally appropriate "green chemistry" techniques. Biochemical, molecular and cellular mechanisms that mediate the synthesis of biological nanoparticles need to be studied in element to increase the rate of synthesis and to obtain favored duration and form of nanoparticles. In assessment with microorganism, fungi can produce big amounts of nanoparticles because they could secrete large quantities of proteins which directly translate to higher productivity of nanoparticles. The biogenic approach is further supported by way of the reality that most of the people of the bacteria inhabit ambient conditions of varying temperature, pH, and stress condition. With the present day development and the continued efforts in improving particle produce performance and exploring their biological and medical functions. We hope so, the implementation of these processes on a largescale and their commercial packages in medication and fitness care will take vicinity inside the coming years. It appears distinctly in all likelihood that future studies will become aware of many other disorder states in which intestine microbial metabolites are significantly enriched or depleted. It is far vital to keep in mind that by using the use of themselves such studies do no longer display causality.

Acronyms and abbreviations

NMR	nuclear magnetic resonance
MOs	microorganisms
SM	secondary metabolites
MS	mass spectrometry
NPs	nanoparticles
EPA	eicosapentaenoic acid
DHA	docosahexaenoic acid
ATP	adenosine triphosphatase
PUFAs	polyunsaturated fatty acids
DNA	deoxyribonucleic acid
MCF	cytotoxic cadmium chloride
FAs	fatty acids
UHPLC	ultrahigh-performance liquid chromatography
MRI	magnetic resonance imaging
Entap	enterococcal anti-proliferative peptide
Au	gold
Ag	silver
Pt	platinum
Cu	mercury
Cd	cadmium
Se	selenium
Cr	chromium
Hg	mercury
HeLa cell	Henrietta Lacks cell
AgNPs	silver nanoparticles
Fe3O4	magnetite
Fe2O3	maghemite
H2O2	Hydrogen peroxide

Author details

Zeinab Amanzadeh
Faculty of Natural Science, University of Tabriz, Tabriz, Iran

*Address all correspondence to: zeinab_amanzadeh@yahoo.com

References

[1] Abdel-Aziz S, Abo Elsoud M, H.Anise A: Handbook of **Food Biosynthesis** Microbial Biosynthesis: A Repertory of Vital Natural Products. 2017; 25 p. **DOI:**10.1016/B978-0-12-811372-1.00003-8

[2] Andryukov B, Mikhailov V, Besednova N: The biotechnological potential of secondary metabolites from marine macteria. FEBS Letters.2019; 7;176-192. DOI:10.3390/jmse7070176

[3] Mosunova O, Navarro-Muñoz J, Collemare J. The Biosynthesis of Fungal Secondary Metabolites: From Fundamentals to Biotechnological Applications. FEBS Letters. 2020; Encyclopedia of Mycology. DOI:10.1016/B978-0-12-809633-8.21072-8

[4] McCarville J,Chen G, Cuevas V, Troha K, Janelle S. Ayres: Microbiota Metabolites in Health and Disease. FEBS Letters. 2020; 38:147-70. DOI: 10.1146/annurev-immunol-071219-125715

[5] Nigam P, Singh A, , Lyngby, Denmark: Production of Secondary Metabolites – Fungi. FEBS Letters. 2014; 1319-1328

[6] Pham JV, Yilma MA, Feliz A, Majid MT, Maffetone N, Walker JR, Kim E, Cho HJ, Reynolds JM, Song MC, Park SR and Yoon YJ: A Review of the Microbial Production of Bioactive Natural Products and Biologics. FEBS Letters. 2019; DOI: 10.3389/fmicb.201901404

[7] Bachmann B, Van Lanen S, Baltz R. H: Microbial genome mining for accelerated natural products discovery: is a renaissance in the making J. Ind. Microbiol. Biotechnol. FEBS Letters .2014; 41:175-184. DOI: 10.1007/s10295-013-1389-9

[8] Guoqing T, HuaRong T: Biosynthesis and regulation of secondary metabolites in microorganisms. FEBS Letters.2013. 7: 581-583. DOI: 10.1007/s11427-013-4501-5

[9] Mushtaq S, Haider Abbasi B, Uzair B, Abbasi R: NATURAL PRODUCTS AS RESERVOIRS OF NOVEL THERAPEUTIC AGENTS. FEBS Letters. 2018; 17:420-451. DOI:10.17179/excli2018-1174

[10] Gutzeit H, Ludwig J: Handbook of Plant Natural Products Biosynthesis and Chemical Properties of Natural Substances in Plants. 1st ed. Wiley-VCH Verlag GmbH & Co. KGa. 2014; 1 p.

[11] Janardhan A, Praveen Kumar A, Viswanath B, Saigopal D, Narasimha G: Production of bioactive compounds by actinomycetes and their antioxidant properties. FEBS Letters.2014; DOI:10.1155/2014/217030

[12] Sanchez S, Demain A: Secondary Metabolites. FEBS Letters. 2011; 2: 155-167. DOI: 6/B978-0-444-64046-8.00012-4

[13] Zhang L, Davies S: Microbial metabolism of dietary components to bioactive metabolites: opportunities for new therapeutic Interventions. FEBS Letters. 2016; DOI: 10.1186/s13073-016-0296-x

[14] Dorrestei P, Mazmanian S, Knight R: Finding the Missing Links among Metabolites, Microbes, and the Host. FEBS Letters.2014; DOI: 10.1016/j.immuni.2014.05.015

[15] Sorour N, Tayel A, Abbas R, Abonama O: Handbook of Food Biosynthesis Microbial Biosynthesis of Health-Promoting Food Ingredients. 2017; 55 p. DOI:10.1016/B978-0-12-811372-1.00002-6

[16] Claus SP, Tsang TM, Wang Y, Cloarec O, Skordi E, Martin FP: Systemic multicompartmental effects of

the gut microbiome on mouse metabolic phenotypes. Mol Syst Biol. 2008;4:219. doi:10.1038/msb.2008.56.

[17] Marcobal A, Kashyap P, Nelson T, Aronov P, Donia M, Spormann A: A metabolomic view of how the human gut microbiota impacts the host metabolome using humanized and gnotobiotic mice. ISME J. 2013; 7:10. doi:10.1038/ismej.2013.89.

[18] LeBlanc JG, Laino JE, del Valle MJ, Vannini V, van Sinderen D, Taranto MP, de Valdez GF, de Giori GS, Sesma F. B-group vitamin production by lactic acid bacteria – current knowledge and potential applications. J Appl Microbiol. 2011;111:1297-1309. DOI: 10.1111/j.1365-2672.2011.05157.x

[19] Gu Q, Li P: Handbook of Intech Open. Biosynthesis of Vitamins by Probiotic Bacteria. FEBS Letters. 2016. DOI:10.5772/63117

[20] Verardo V , Gómez-Caravaca A.M, Tabanelli G: Bioactive Components in Fermented Foods and Food By-Products. FEBS Letters. 2020; DOI:10.3390/foods9020153

[21] Martinez K, Leone V, Chan E: Microbial Metabolites in Health and Disease: Navigating the Unknown in Search of Function. FEBS Letters. 2017;DOI: 10.1074/jbc.R116.752899

[22] Hernández-Almanza A, Muñiz-Márquez D, de la Rosa O, Navarro V, Martínez-Medina G, Rodríguez-Herrera R, Aguilar C: Handbook of **Food Biosynthesis**. *Microbial* Production of Bioactive Pigments, Oligosaccharides, and Peptides. FEBS Letters. 2017; **DOI:1016/ B978-0-12-811372-1.00004-X**

[23] Druart C, Neyrinck AM, Vlaeminck B, Fievez V, Cani PD, Delzenne NM: Role of the lower and upper intestine in the production and absorption of gut microbiota-derived PUFA metabolites. FEBS Letters. 2014;9(1), e87560. DOI: 10.1371/ journal.pone.0087560.

[24] Vrancken K , Anne J: Secretory production of recombinant proteins by Streptomyces. Future Microbiol. FEBS Letters .2009;4: 181-188. DOI: 10.2217/17460913.4.2.181

[25] Yun Lim F, Sanchez J, Wang C, Keller N: Toward Awakening Cryptic Secondary Metabolite Gene Clusters in Filamentous Fungi. FEBS Letters. 2012; 517: 303-324. DOI:10.1016/ B978-0-12-404634-4.00015-2

[26] Bumpus S, Evans B, Thomas P, Ntai I, Kelleher N: A proteomics approach to discovering natural products and their biosynthetic pathways. Nature Biotechnology. 2009; 27:951-956.DOI: 10.1038/nbt.1565

[27] Davies J: Specialized microbial metabolites: functions and Origins. FEBS Letters. 2013; 66: 361-364. DOI: 10.1038/ja.2013.61

[28] Nigam P, Singh A: Production of Secondary Metabolites – Fungi. FEBS Letters. 2014; pp 1319-1328. DOI: 10.1016/ B978-0-12-384730-0.00202-0

[29] Abdel-Razek A , El-Naggar M , Allam A, Morsy, Othman S: Microbial Natural Products in Drug Discovery. FEBS Letters.2020; 8: 470-489. DOI:10.3390/pr8040470

[30] Nataraj B, Ali S.A, Behare P, Yadav H: Postbiotics-parabiotics: the new horizons in microbial biotherapy and functional foods. FEBS Letters. 2020; 19:168-190. DOI: 10.1186/ s12934-020-01426-w

[31] Taghinasab M, Jabaji S: Cannabis microbiome and the role of Endophytes in modulating the production of secondary metabolites. FEBS Letters.2020; 8:355-371. DOI:10.3390/ microorganisms8030355

[32] Hulkoti N, Taranath T.C: Biosynthesis of nanoparticles using microbes. FEBS Letters. 2014; 121:474-483. DOI: 10.1016/j.colsurfb.2014.05.027

[33] Li X, Huizhong Xu, Chen Z-S, Chen G: Biosynthesis of nanoparticles by microorganisms and their applications. FEBS Letters.2011; doi:10.1155/2011/270974. DOI:10.1155/2011/270974

[34] Lef'evre C. T, Abreu F, Schmidt M: Moderately thermophilic magnetotactic bacteria from hot springs in Nevada. FEBS Letters.2010; 11: 3740-3743. DOI: 10.1128/AEM.03018-09

[35] Kalishwaralal K, Banumathi E, Pandian S. R: Silver nanoparticles inhibit VEGF induced cell proliferation and migration in bovine retinal endothelial cells. FEBS Letters. 2009; 73:51-57. DOI:10.1186/1477-3155-7-8

[36] Wang T, Yang L, Zhang B, Liu J: Extracellular biosynthesis and transformation of selenium nanoparticles and application in H2O2 biosensor. FEBS Letters.2010; 80: 94-102. DOI:10.1016/j.colsurfb.2010.05.041

[37] Xiang L, Wei J, Jianbo S, Guili W, Feng G, Ying L: Purified and sterilized magnetosomes from Magnetospirillum gryphiswaldense MSR-1 were not toxic to mouse fibroblasts in vitro. FEBS Letters. 2007; 45:75-81. DOI:10.1111/j.1472-765x.2007.02143.x

Chapter 6

Biosynthesis of Diverse Class Flavonoids *via* Shikimate and Phenylpropanoid Pathway

Mohd Rehan

Abstract

Flavonoids are natural products, which are useful in the protection of various types of human diseases. Several bioactive flavonoids as chalcones, flavonols, flavanol, flavones, flavanone, flavan, isoflavonoids, and proanthocyanidin, are found in parts as leaves, root, bark, stem, flowers, weed, fruits of plant species. Flavonoids are synthesized in higher plant species via the shikimate pathway, phenylpropanoid and polyketide pathway. The chalcones and flavanones are central intermediates of the pathway, which give several diverse classes of flavonoids. Central intermediates pathway (chalcones and flavanones pathway) depends on plants species and group of enzymes such as hydroxylases, reductases and isomerases to give different classes of flavonoids skeleton. The anthocyanins, isoflavonoids and condensed tannin (proanthocyanidins) are an important class of flavonoids, which synthesized by flavanones. Mostly, biosynthesis of flavonoids start from phenylpropanoid pathway. The phenylpropanoid pathway starts from shikimate pathway. The shikimate pathway starts from phosphoenol pyruvate and erythrose 4-phosphate.

Keywords: flavonoids, biosynthesis, shikimate pathway, phenylpropanoid pathway, tannins

1. Introduction

Flavonoids, are the largest class of secondary metabolites, having polyphenolic structure, which widely distributed in several parts as leaves, root, stem, bark, fruit, flower, weed, of diverse plant species [1]. The flavonoids play a key role to provide pigments in plant as dark blue and red color of berries, yellow and orange color of citrus fruits. These flavonoids play similar role as vitamins in the human body [2]. The flavonoids are constituted by 15 carbon atoms, which are arranged in C_6-C_3-C_6 backbone skeleton rings, in which ring A and ring B are linked by three carbon ring C [3]. The skeleton of ring represented in **Figure 1**.

On the basis of substitution pattern, flavonoids can be classified into major subgroups as chalcone, flavanone, dihydroflavonol, flavanol, flavones, isoflavone, flavonol, leucoanthocyanidin, proanthocyanidin (condensed tannins), anthocyanin [4]. The nature of these flavonoids depends on the basis of degree of hydroxylation, structural class, conjugations, substitutions and degree of polymerization [5]. Approximately, 9000 diverse type flavonoids have been reported and sure this number will be increased [6]. The diverse type flavonoids show diverse biological function as protection from UV radiation, apoptosis,

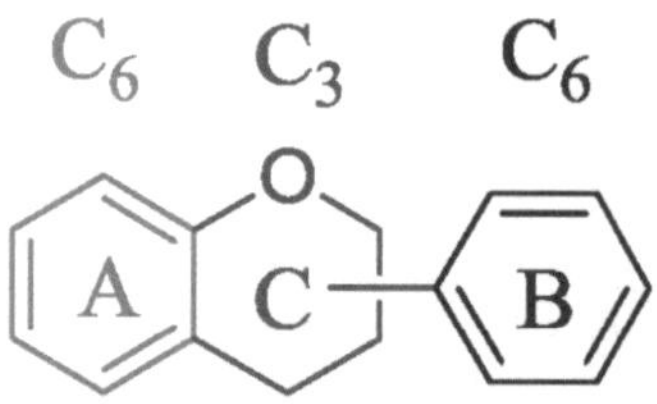

Figure 1.
Basic skeleton C_6-C_3-C_6 of ring A, B, and C in flavonoids.

treatment of psoriasis [7, 8]. The diverse class of flavonoids have been isolated from several plant species as quercetin and apigenin from *Cymbopogon citratus* [9], pinostrobin and cardamonin chalcone from rhizomes of *Boesenbergia rotunda* [10], 6-aldehydo-isoophiopogonanone A, 6-aldehydo-isoophiopogonanone B, methylophiopogonanone A and methylophiopogonanone B from fibrous roots of *Ophiopogon japonicus* [11]. The diverse type flavonoids were synthesized in plant species via shikimate and phenylpropanoid pathway. Several types enzyme as DAHP synthase, DHQ synthase, SA kinase, PAT, ADT, 4CL, CHS, CHI, F3H, DFR play key role in the biosynthesis of flavonoids [12, 13].

2. Shikimate pathway

Shikimate pathway plays high potential role in the biosynthesis of flavonoids. Several key enzymes are involved in this pathway for biosynthesis of shikimic acid. This pathway starts with the aldol condensation reaction of phosphoenol pyruvate (PEP) and D-erythrose 4-phosphate (E4P) to generate seven carbon keto acid, 3-deoxy-D-arabino-heptulosonate –7-phosphate (DAHP). This reaction catalyzes by 3-deoxy-D-arabino-heptulosonate –7-phosphate synthase (DAHPS) enzyme. The DAHPS is a highly potential enzyme of the shikimate pathway. Two DAHPS genes as DHS1 and DHS2 are found in *Arabidopsis thaliana* plants [14]. From literature, it is identified that DHS1 is more produced by infiltration or by physical wounding with pathogen in both tomato and Arabidopsis [15]. The DAHP is transformed to 3-dehydroquinic acid (DHQ) by intramolecular cyclization reaction in presence of DHQ synthase enzymes.

In most bacteria, DHQS is monofunctional and in some organism, it behaves multifunctional enzyme, which catalyze 2, 3, 4, and 5 steps of the shikimate pathway. The DHQS is a small part of larger AROM protein, which is pentafunctional peptide containing enzyme [16, 17]. The *Neurospora crassa* and *Aspergillus nidulans* DHQS enzyme found in nature as part of the AROM protein [18]. The DHQ converts into 3-dehydroshikimic acid (DHS) by losing a water molecule.

In the fourth step, DHS is transformed into shikimic acid by removing water molecule. The phosphorylation of shikimic acid is done by activating of shikimate kinase enzyme in the fifth step reaction. The shikimic acid with ATP is phosphorylated at the 5-OH group of shikimic acid converts into shikimic acid 3-phosphate (S3P). The shikimate kinase enzyme is not found in the human cell, but is an essential enzyme of many bacterial pathogens [19, 20]. The shikimic acid 3-phosphate converts into 3-enolpyruvyl shikimate –5-P (EPSP) by EPSP synthase enzymes.

The EPSPS is activating of shikimic acid 3-phosphate in the sixth step reaction of the shikimate pathway. According to intrinsic glyphosate sensitivity, it enzyme has been classified as a class I EPSP synthases and class II EPSP synthases [21, 22].

The class I EPSP synthases are found in plant and some bacteria as *Escherichia coli* and *Salmonella typhimurium*. The class II EPSP synthases is found several bacteria species as *Streptococcus pneumonia*, *Streptococcus aureus*. The EPSP converts into chorismic acid (CHA) by eliminating of the pi group at C-3.

The end product of shikimate pathway is chorismic acid, which found in plants, fungi, bacteria and some parasites [23]. The chorismate synthases (CS) is divided

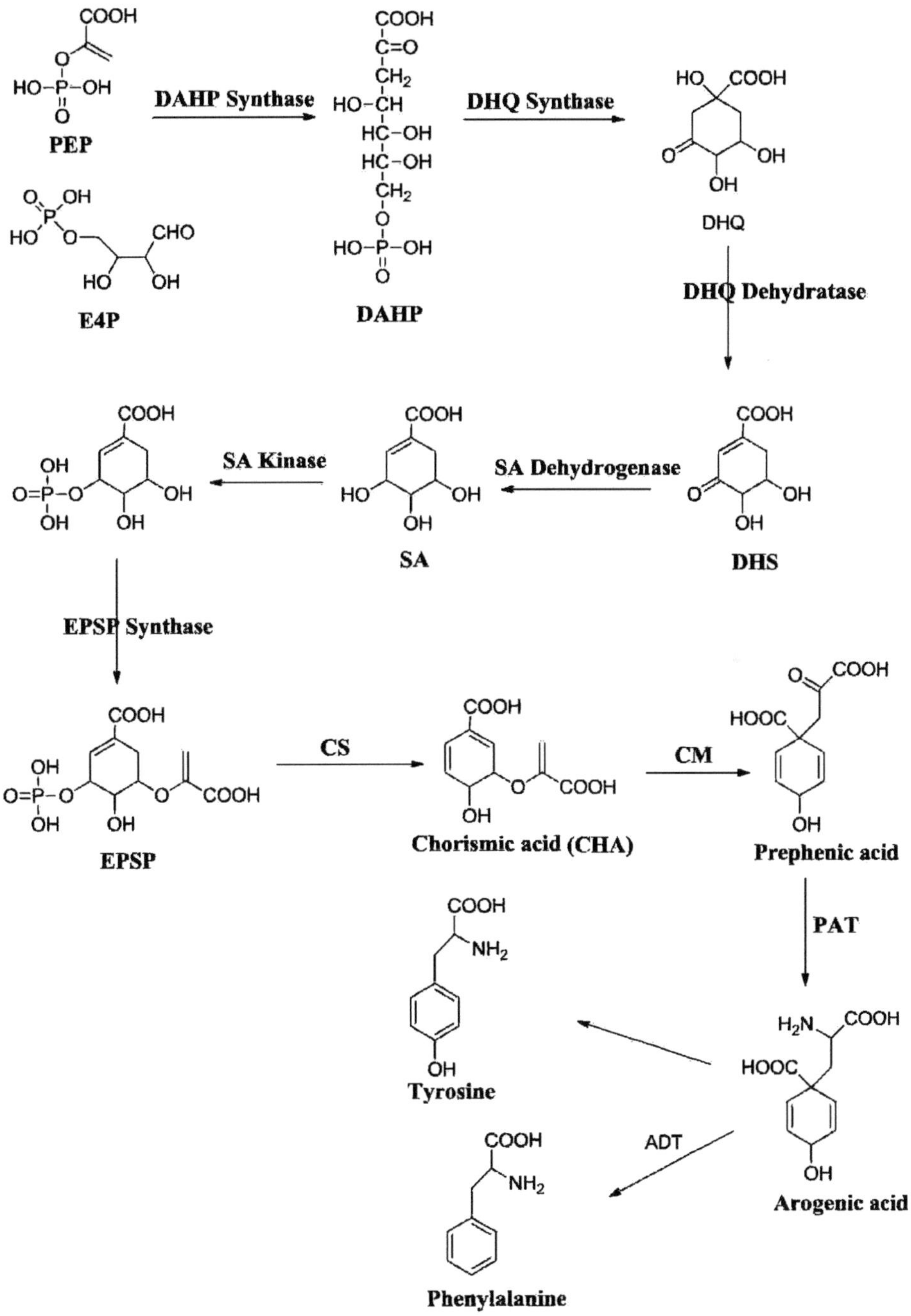

Figure 2.
Shikimate pathway in biosynthesis of flavonoids.

within one of two functional groups as fungal type bifunctional CS and plant, bacterial type monofunctional CS [24, 25].

The chorismate mutase (CM) is a first step enzyme of the tyrosine and phenylalanine biosynthesis. It activates of chorismic acid, which converts into prephenic acid by claisen rearrangement [26]. On the basis functional and structural, multiple form of this enzyme exists. Some monofunctional example from *Serratia rubidaea*, *Bacillus subtilis* [27], *Aspergillus nidulans* [28]. In presence of this enzyme, chorismic acid change into prephenic acid.

The prephenate aminotransferase (PAT) play a key role in phenylalanine biosynthesis. It catalyzes first step product (prephenic acid) into arogenic acid [29]. The arogenate dehydratase (ADT) is a last step enzyme of phenylalanine biosynthesis, which catalyzes of arogenic acid into amino acid phenylalanine [30]. In the arabidopsis genome, six ADT genes as ADT1-ADT6 are found, whereas ADT4 and ADT5 were dominant in roots and stems [31]. The shikimate pathway with enzyme activity is summarized in **Figure 2**.

3. Phenylpropanoid pathway

The shikimate pathway plays the main role in the biosynthesis of flavonoids, which provides amino acid phenylalanine. The phenylalanine ammonia lyase (PAL) is an enzyme of first step reaction in phenylpropanoid pathway. The presence of this

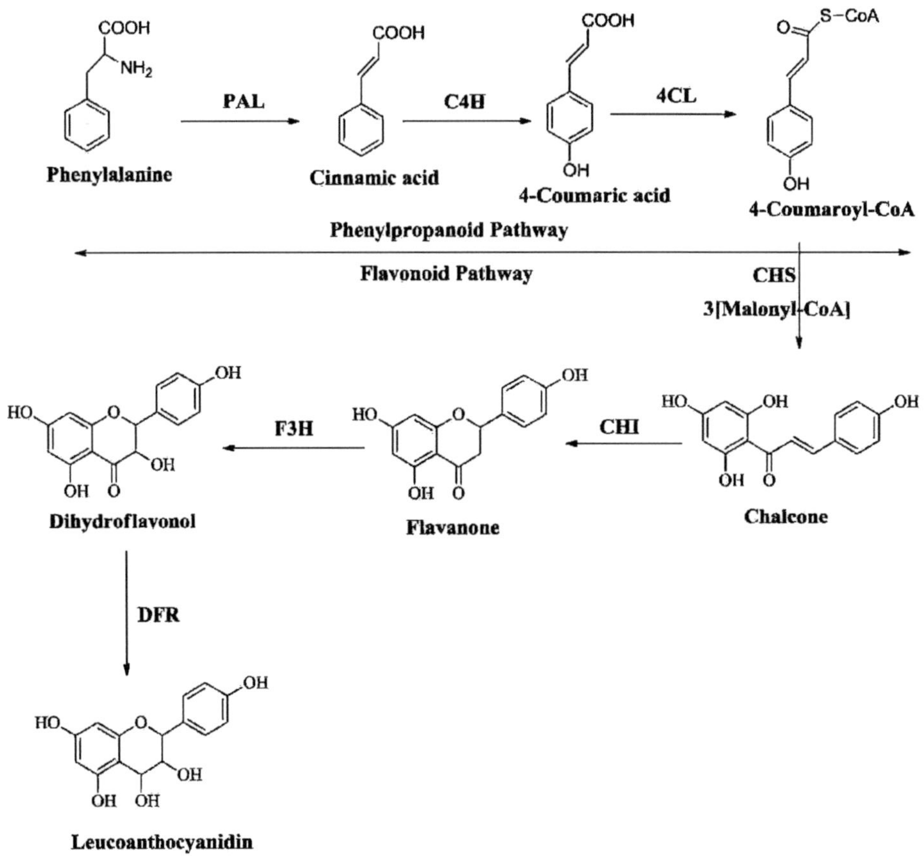

Figure 3.
Phenylpropanoid pathway in biosynthesis of flavonoids.

enzyme has been reported in different types of plant species [32] as certain fungi [33], few prokaryotic organisms, including *Streptomyces* [34, 35], algae, including *Dunaliella marina* [36] and detected in several species representing gymnosperms, ferns, lycopods, monocots, and dicots [37]. This enzyme converts phenylalanine into cinnamic acid and remove the ammonium ion.

The cinnamate −4-hydroxylase (C4H) plays a crucial role in conversion of trans-cinnamic acid in 4-coumaric acid. This acid, yielding 4-coumaroyl-CoA by catalyzing of 4-coumaroyl-CoA-ligase (4CL). The 4-coumaroyl-CoA-ligase (4CL) plays a pivotal role in phenylpropanoid biosynthesis pathway and produced coumarin skeleton. Mostly, a multiple isoform of 4CL are found in higher plants. These isoforms have distinct catalytic properties and expression profiles in plant tissue [38].

The initial step of flavonoids biosynthesis is the condensation reaction of one molecule 4-coumaroyl-CoA with three molecules of malonyl-CoA to yielding chalcone (2′,4′,6′,4-tetrahydroxy chalcone) by catalyzing the chalcone synthase (CHS) enzyme [39]. chalcone synthase (CHS) enzyme plays key role in the biosynthesis of flavonoids and isoflavonoids. The plant polyketide synthase is a big family called superfamily, CHS is a member of this family [40]. The chalcone isomerized into flavanone by activating of chalcone flavanone isomerase (CHI) enzyme. The flavanone is the intermediate pathway of flavonoids, which divided into many different flavonoids classes [41, 42]. The modification of flavanone into the basic skeleton of flavonoids, depends on the species and a group of enzymes as hydroxylases, reductases, isomerases [43]. The phenylpropanoid pathway in the biosynthesis of flavonoids summarized in **Figure 3**.

4. Flavonoids pathway

The shikimate and phenylpropanoid pathway play important role in biosynthesis of flavonoids. After this pathway flavonoids pathway starts, which produce various diverse type flavonoids in presence of several enzymes. The isoflavonoid synthase (IFS) is a main enzyme, which converts a flavanone into isoflavone. In soybean, two isoform of IFS genes as IFS-1 and IFS-2 are found, which play a crucial role in the isoflavones biosynthesis [44, 45]. The role of this enzyme summarized in **Figure 4**.

The flavonol synthase (F3H) is a key enzyme of the biosynthesis in the central flavonoid pathway. It plays a pivotal role in the conversion of flavanone into dihydroflavonol. It has been isolated from various plant species (more than 50 plants) [46, 47]. The flavonol synthase (FLS) is a highly activating enzyme, which converts of dihydroflavonol into flavonol. The first FLS gene was known from *P. hybrida* [48] and other FLS gene were known from various plant species as *A. thaliana* [49], *E. grandiflorum* [50] etc.

The dihydroflavonol reductase (DFR) is a essential enzyme, which catalyzes dihydroflavonol into leucoanthocyanidin and are precursors of anthocyanidins and proanthocyanidins [51]. The DFR genes have been cloned in several plant species as *Lotus japonicas* [52], *Ginkgo biloba* [53], *Brassica rapa* [54]. *The DFR can overexpression in apple and tobacco, which increase anthocyanin production* [55, 56].

The proanthocyanidins is known condensed tannins (polymers), which produced by condensation of flavan-3-ol monomeric units as epicatechin and catechin. It catalyzes in the presence of two enzymes as leucoanthocyanidin reductase (LAR) and anthocyanidin reductase (ANR). The LAR is the main enzyme of anthocyanin biosynthesis pathway, which converts leucoanthocyanidin into catechin, while ANR converts anthocyanidin into epicatechin [57–59]. *The CsLAR gene is found in tobacco, which*

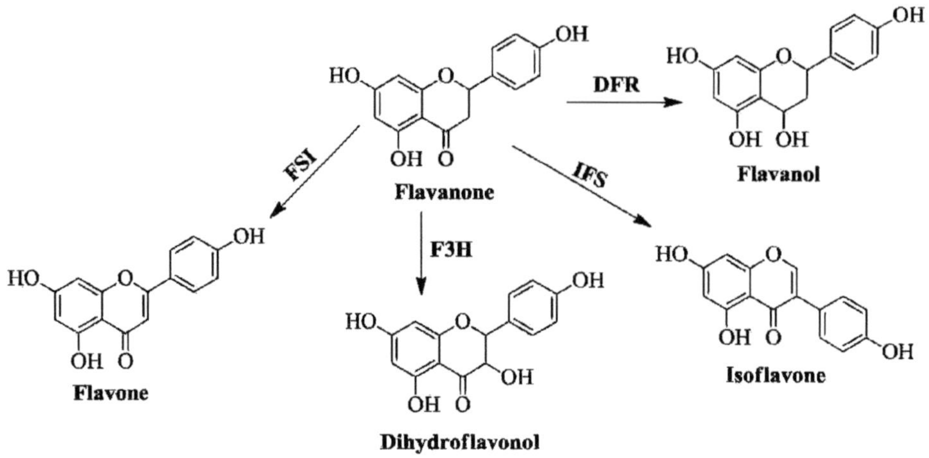

Figure 4.
The essential role of enzyme in flavonoids pathway.

accumulation of higher level of epicatechin than catechin while ANR in tea and grape-vine is involved in biosynthesis of mixture of catechin and epicatechin from anthocyanidin [60, 61]. *The proanthocyanidins have been reported from various plant species* [62, 63]. *The catalyzing properties of these enzymes are showed in* ***Figure 5****.*

4.1 Chalcones

Chalcone synthase plays potential role in the biosynthesis of flavonoids/isoflavonoids pathway. The CHS is a member of the polyketide synthase family, which play a key role flowering plant as providing floral pigment, insect repellents, UV Protectants and antibiotics [64]. The chalcones are called open chain

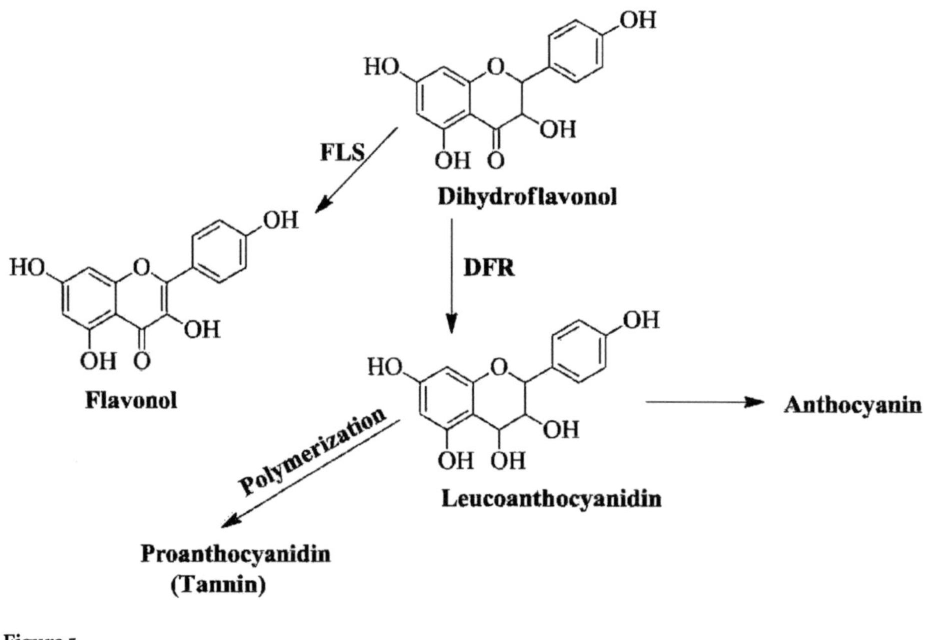

Figure 5.
Biosynthesis of tannins and anthocyanin in flavonoids pathway.

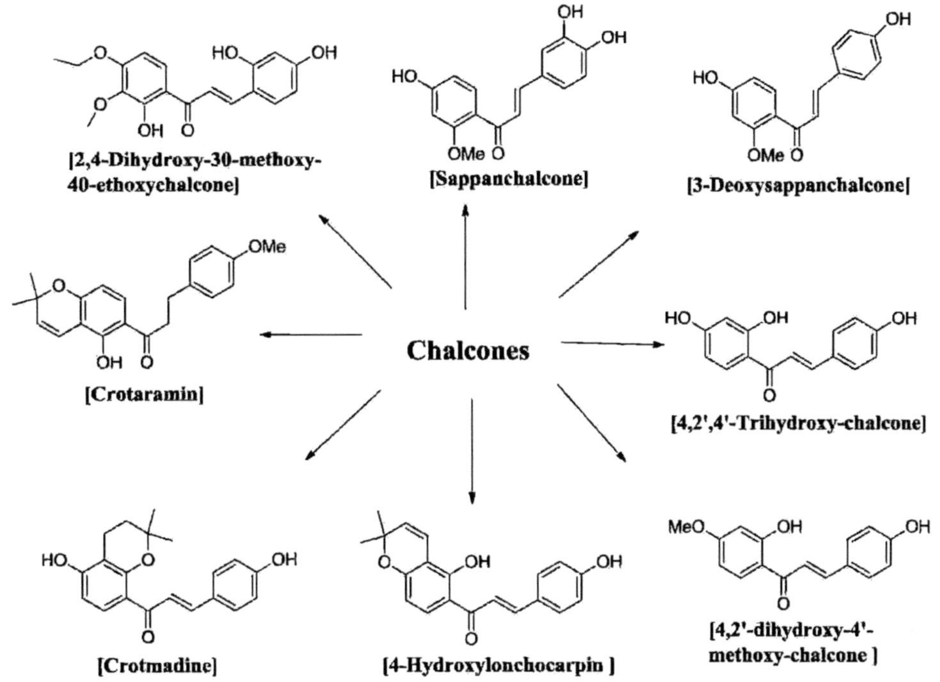

Figure 6.
Various types chalcones isolated from several plants.

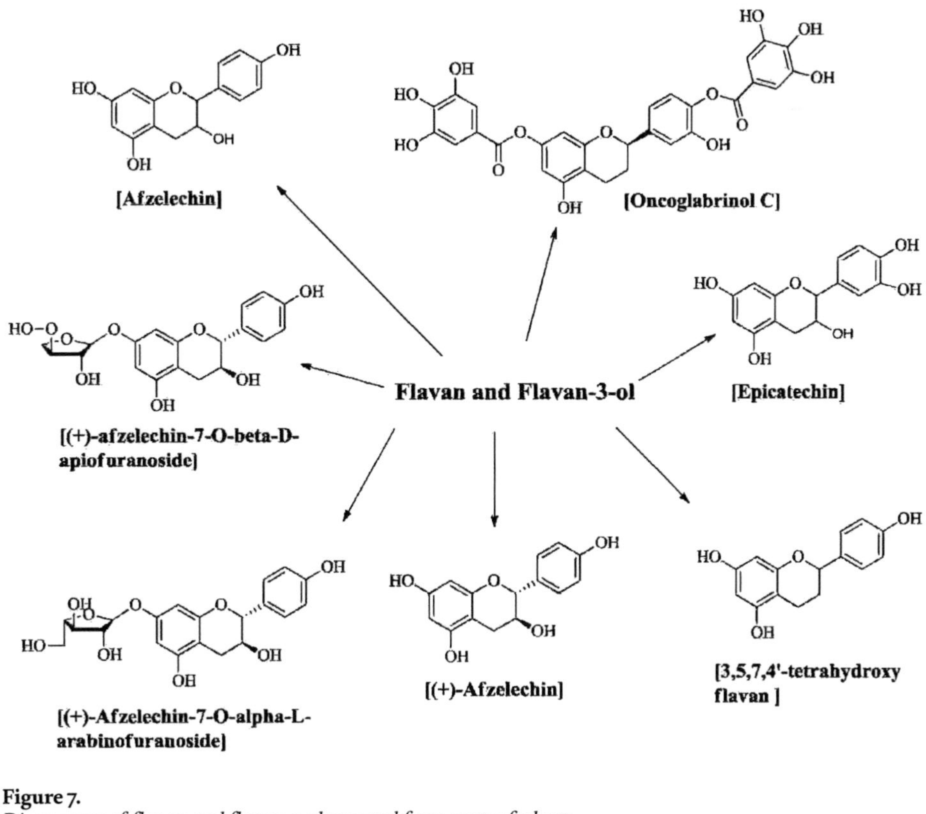

Figure 7.
Diverse type of flavan and flavan-3-ol reported from parts of plants.

flavonoids, which have 15 carbon structure and arranged in C_6-C_3-C_6 skeleton. The modification of chalcones can be done by methylation, condensation, and hydroxylation. These chalcones can be distributed in many parts of plants as fruits, seed, bark, stem, flowers [65].

Various diverse type chalcones have been reported from many plant species such as 2,4-dihydroxy-30-methoxy-40-ethoxychalcone from *Caragana pruinosa* [66], two chalcones, sappanchalcone and 3-deoxysappanchalcone from *Haematoxylum campechianum* [67], 4,2′,4′-trihydroxy-chalcone 4,2′-dihydroxy-4′- methoxy-chalcone, 4-hydroxylonchocarpin, crotmadine chalcones *Codonopsis cordifolioidea* root [68], and crotaramin chalcone from *Crotalaria ramosissima* plant [69]. These chalcones are showed in **Figure 6**.

4.2 Flavan and Flavan-3-ol

Many different flavan and flavan-3-ol are summarized in **Figure 7**, which have been isolated from many plants as afzelechin from steam bark of *Pinus halepensis* [70], oncoglabrinol C from *Oncocalyx glabratus* [71], epicatechin, and 3,5,7,4′-tetrahydroxy flavan from stem bark of *Embelia schimperi* [72], three flavan-3-ol derivatives as (+)-afzelechin, (+)-afzelechin-7-O-α-L-arabinofuranoside and (+)-afzelechin-7-O-β-D-apiofuranoside from *Polypodium vulgare* L. rhizomes [73].

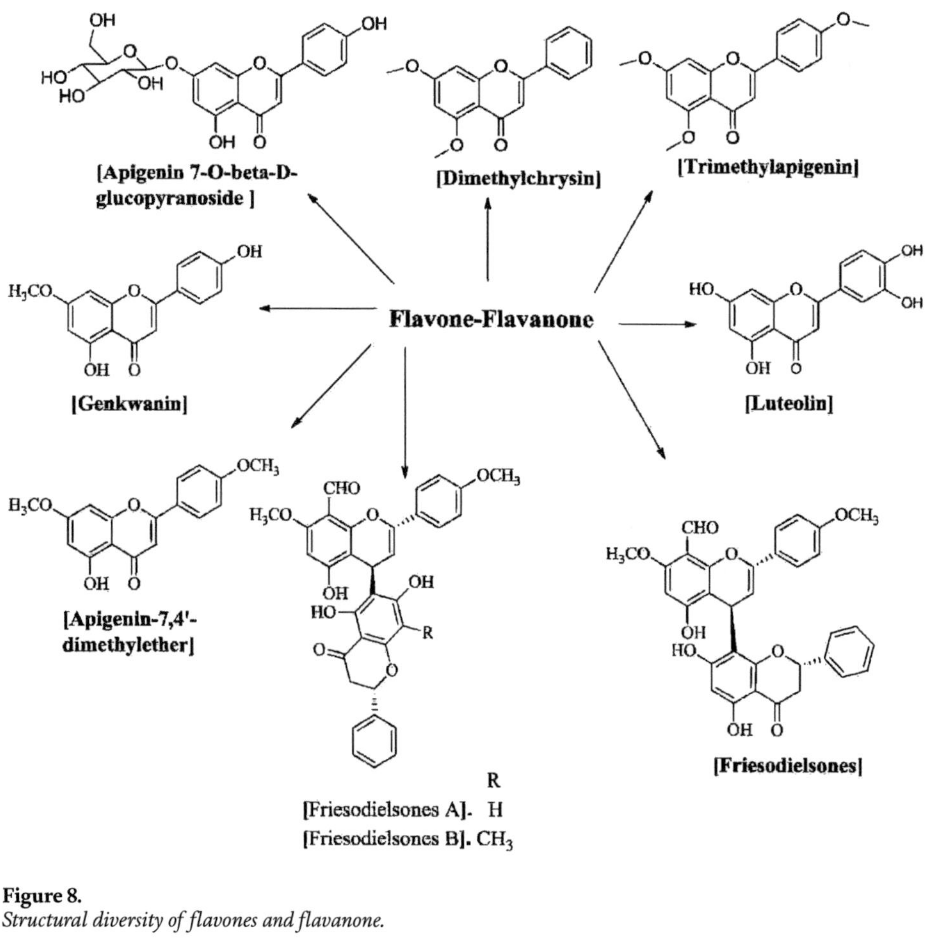

Figure 8.
Structural diversity of flavones and flavanone.

4.3 Flavone-flavanone

Many different structures of flavones and flavanone are synthesized via shikimate and flavonoids pathway. These structures of these are showed in **Figure 8**. Several type of flavones and flavanone were isolated such as apigenin 7-O-β-D-glucopyranoside, dimethylchrysin, trimethylapigenin, 5,7,3′,4′-tetrahydroxyflavone (Luteolin) from *Sterculia foetida* leaves [74], three new flavan-flavanones as friesodielsones A, friesodielsones B, friesodielsones, from *Friesodielsia desmoides* leaves [75], and flavonoids (flavones) as apigenin-7,4′-dimethylether, genkwanin from *Aquilaria sinensis* leaves [76].

4.4 Isoflavonoids

The diverse type structure of isoflavonoids was synthesized from flavanone, which have been reported several plants as corylifol A, neobavaisoflavone, and irisflorentin from *Cytisus striatus* [77], formoninetin and biochanin A from *Hylastinus obscurus* [78]. One new leptoisoflavone A (a rare 5-membered dihydrofuran ring) from *Limonium leptophyllum* [79], 2,2′-trimethoxy-6,8-dihydroxy-isoflavone from the ethanol extract of *Thespesia populnea* bark [80] and isoflavones, genistein and daidzein from *Hericium erinaceum* (**Figure 9**) [81].

4.5 Flavonol

Several type of flavonol were reported from parts of plants as myricetin 3-O-(2″,4″-di-O-acetyl)-α-L-rhamnopyranoside from *Myrsine Africana* [82], flavonoid glycoside named as 3’-*O*-methyl quercetin-3-glucose-6-gallic acid from *Cordia oblique* leaves [83], 2-(3, 4-dihydroxyphenyl)-3, 5, 7-trihydroxy-4H-chromen-4-one

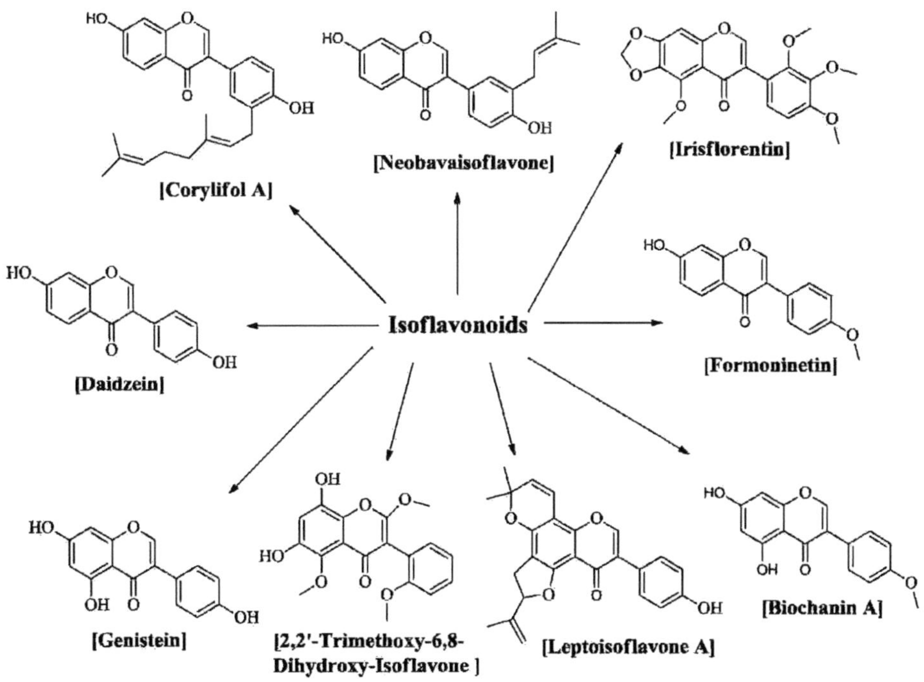

Figure 9.
Diverse structure of isoflavonoids from plants species.

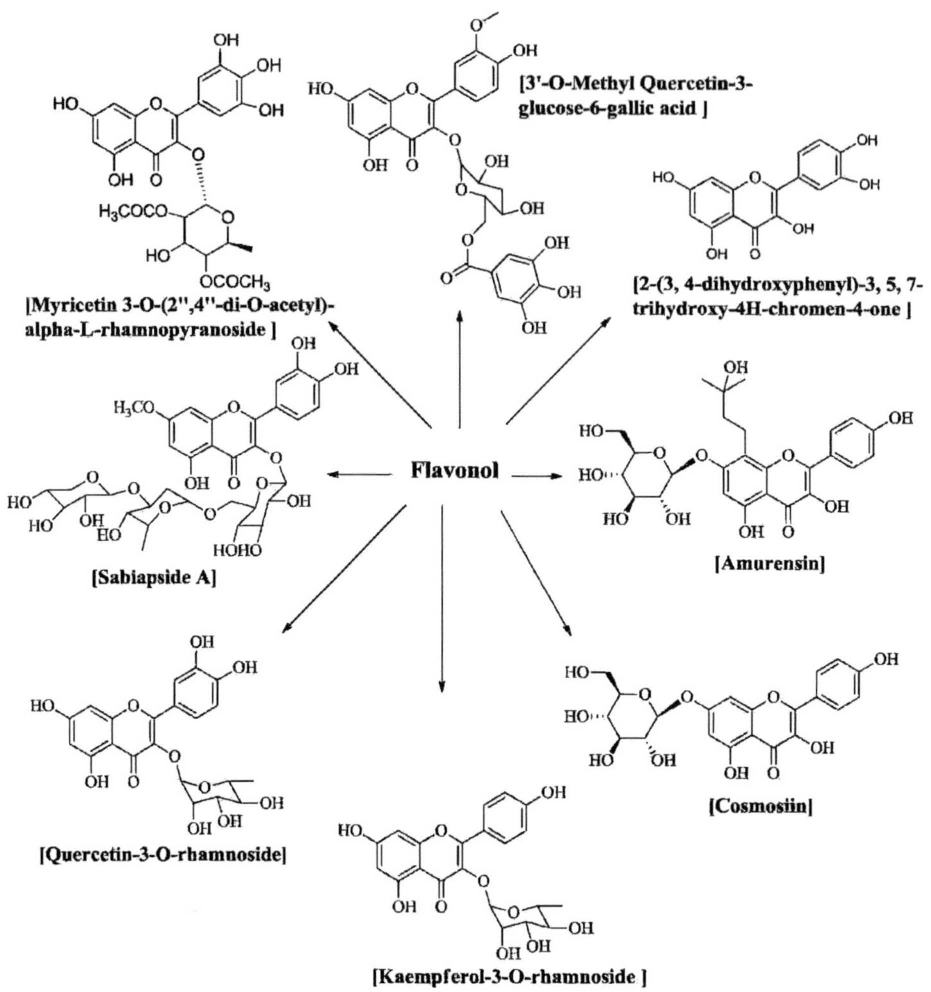

Figure 10.
Several different structures of flavonol isolated from parts of plant species.

from aerial parts of *Chenopodium album* [84], amurensin and cosmosiin from *Trigonella foenum graecum* [85], rhamnosides flavonol, kaempferol-3-*O*-rhamnoside and quercetin-3-*O*-rhamnoside from leaves of *Pometia pinnata* [86] and a new flavonol glycoside, sabiapside A from *Sabia parviflora* [87] (**Figure 10**).

5. Conclusions

Flavonoids are a large class of natural compounds, which isolated from various of plants as seed, root, bark, flower, leaves, fruit etc. and prevent from various diseases. The biosynthesis of flavonoids is highly complicated because a group of enzyme (DHAP synthase, SA kinase, EPSP synthase, PAL, 4CL, CHS, CHI, F3H, DFR) plays a key role in the pathway of flavonoids biosynthesis. These enzymes play a potential role in modification of flavonoids via isomerases, hydroxylases, reductases, and polymerises reaction. The proanthocynidins are interested natural compounds, which formed via polymerization reaction of flavonoids. The flavonoids are synthesized in various plant species via shikimate and phenylpropanoid pathway.

Acknowledgements

Author thanks to UGC, New Delhi, India for institutional fellowship. The author is also thankful to the Chairman, Department of Chemistry, AMU, Aligarh, for providing research facilities.

Conflict of interest

The author declares no conflicts of interest.

Author details

Mohd Rehan
Department of Chemistry, Aligarh Muslim University, Aligarh, India

*Address all correspondence to: rehancy3@gmail.com

References

[1] Wang T-Y, Li Q, Bi K-S. Bioactive flavonoids in medicinal plants: Structure, activity and biological fate. Asian Journal of Pharmaceutical Sciences. 2018;13:12-23. doi: https://doi.org/10.1016/j.ajps.2017.08.004

[2] Brodowska KM. Natural Flavonoids: Classification, Potential Role, and Application of Flavonoid Analogues. European Journal of Biological Research. 2017;7:108-123. doi: http://dx.doi.org/10.5281/zenodo.545778

[3] Isoda H, Motojima H, Onaga S, et al. Analysis of the erythroid differentiation effect of flavonoid apigenin on K562 human chronic leukemia cells. Chem Biol Interact. 2014;220:269-277. doi: https://doi.org/10.1016/j.cbi.2014.07.006

[4] Winkel BSJ. The Biosynthesis of Flavonoids. In: Grotewold E. (eds) The Science of Flavonoids. Springer, New York, NY; 2006. doi: https://doi.org/10.1007/978-0-387-28822-2_3

[5] Heim KE, Tagliaferro AR, Bobilya DJ. Flavonoid antioxidants: chemistry, metabolism and structure-activity relationships. J Nutr Biochem. 2002;13(10):572-584. doi: https://doi.org/10.1016/s0955-2863(02)00208-5

[6] Wang Y, Chen S, Yu O. Metabolic engineering of flavonoids in plants and microorganisms. Appl Microbiol Biotechnol. 2011;91:949-956. doi: https://doi.org/10.1007/s00253-011-3449-2

[7] Park CH, Han SE, Nam-Goong IS, Kim YI, Kim ES. Combined Effects of Baicalein and Docetaxel on Apoptosis in 8505c Anaplastic Thyroid Cancer Cells via Downregulation of the ERK and Akt/mTOR Pathways. Endocrinol Metab (Seoul). 2018;33(1):121-132. doi: https://doi.org/10.3803/EnM.2018.33.1.121

[8] Chamcheu JC, Esnault S, Adhami VM, Noll AL, Banang-Mbeumi S, Roy T, Singh SS, Huang S, Kousoulas KG, Mukhtar H. Fisetin, a 3,7,3′,4’-Tetrahydroxyflavone Inhibits the PI3K/Akt/mTOR and MAPK Pathways and Ameliorates Psoriasis Pathology in 2D and 3D Organotypic Human Inflammatory Skin Models. Cells. 2019;8:1089. doi: 10.3390/cells8091089

[9] Ahmed NZ, Ibrahim SR, Omar A. Ahmed-Farid OA. Quercetin and Apigenin of *Cymbopogon citratus* mediate inhibition of HCT-116 and PC-3 cell cycle progression and ameliorate Doxorubicin-induced testicular dysfunction in male rats. Biomedical Research and Therapy 2018;5(7):2466-2479. doi: https://doi.org/10.15419/bmrat.v5i7.457

[10] Mohammed IA, Akhtar MN, Biau FJ, Tor YS, Zareen S, Shahabudin SB, AbdHamid HB, Haq ZU, Khalil R, Khalaf RM. Isolation of Cardamonin and Pinostrobin Chalcone from the Rhizomes of Boesenbergia rotunda (L.) Mansf. and their Cytotoxic Effects on H-29 and MDA-MB-231 Cancer Cell Lines. The Natural Products Journal. 2019;9:341. doi: https://doi.org/10.2174/2210315509666190117151542

[11] Zhou Y, Wang L, Liu T, Mao Z, Ge Q, Mao J. Isolation of Homoisoflavonoids from the Fibrous Roots of Ophiopogon japonicus by Recycling High-Speed Counter-Current Chromatography and Online Antioxidant Activity Assay. Acta Chromatographica. 2019;31(4):272-279. doi: https://doi.org/10.1556/1326.2018.00509

[12] Díaz Quiroz DC, Carmona SB, Bolívar F, Escalante A. Current perspectives on applications of shikimic and aminoshikimic acids in pharmaceutical chemistry. Research and Reports in Medicinal Chemistry.

2014;4:35-46. doi: https://doi.org/10.2147/RRMC.S46560

[13] Tohge T, Watanabe M, Hoefgen R, Fernie AR. Shikimate and phenylalanine biosynthesis in the green lineage. Frontiers in plant science. 2013;4:62. doi: https://doi.org/10.3389/fpls.2013.00062

[14] Keith B, Dong XN, Ausubel FM, Fink GR. Differential induction of 3-deoxy-D-arabino-heptulosonate 7-phosphate synthase genes in *Arabidopsis thaliana* by wounding and pathogenic attack. Proceedings of the national Academy of Science, USA. 1991;88:8821-8825. doi: https://dx.doi.org/10.1073%2Fpnas.88.19.8821

[15] Gorlach J, Schmid J, Amrhein N. Differential expression of tomato (*Lycopersicon esculentum* L.) genes encoding shikimate pathway isoenzymes. II. Chorismate synthase. Plant Molecular Biology. 1993;23:707-716. doi: https://doi.org/10.1007/BF00021526

[16] Bentley R, Haslam E. The shikimate pathway-a metabolic tree with many branches. Critical Reviews in Biochemistry and Molecular Biology. 1990;25(5):307-384. doi: https://doi.org/10.3109/10409239009090615

[17] Hawkins AR, Moore JD, Adeokun AM. Characterization of the 3-dehydroquinase domain of the pentafunctional AROM protein, and the quinate dehydrogenase from *Aspergillus nidulans*, and the overproduction of the type II 3-dehydroquinase from *Neurospora crassa*. Biochemical Journal. 1993;296(2):451-457. doi: https://dx.doi.org/10.1042%2Fbj2960451

[18] Moore JD, Coggins JR, Virden R, Hawkins AR. Efficient independent activity of a monomeric, monofunctional dehydrogenate synthase derived from the N-terminus of the pentafunctional AROM protein of *Aspergillus nidulans*. Biochemical Journal. 1994;301(1):297-304. doi: https://doi.org/10.1042/bj3010297

[19] Coracini JD, Azevedo WF. Shikimate kinase, a protein target for drug design. Current Medicinal Chemistry. 2014;21:592-604. doi: https://doi.org/10.2174/09298673113206660299

[20] Blanco B, Prado V, Lence E, Otero JM, García-Doval C, van Raaij MJ, et al. *Mycobacterium tuberculosis* shikimate kinase inhibitors: Design and simulation studies of the catalytic turnover. Journal of the American Chemical Society. 2013;135:12366-12376. doi: https://doi.org/10.1021/ja405853p

[21] Funke T, Han H, Healy-Fried ML, Fischer M, Schonbrunn E (2006) Molecular basis for the herbicide resistance of Roundup Ready crops. Proc. Natl. Acad, Sci. USA, 103, 13010-13015. doi: https://doi.org/10.1073/pnas.0603638103

[22] Franz JE, Mao MK, Sikorski JA. Glyphosate A Unique Global Herbicide. Washington Americal Chemical Society. 1997.

[23] He Z, Stigers Lavoie KD, Bartlett PA, Toney MD. Conservation of mechanism in three chorismate-utilizing enzymes. J Am Chem Soc. 2004;126(8):2378-2385. doi: https://doi.org/10.1021/ja0389927

[24] Schaller A, Vanafferden M, Windhofer V, Bulow S, Abel G, Schmid J, et al. Purification and characterization of chorismate synthase from *Euglena gracilis* comparison with chorismate synthases of plant and microbial origin. Plant Physiol. 1991;97:1271-1279. doi: https://dx.doi.org/10.1104%2Fpp.97.4.1271

[25] Maeda H, Dudareva N. The shikimate pathway and aromatic Amino acid biosynthesis in plants. Annu Rev Plant Biol. 2012;63:73-105.

doi: https://doi.org/10.1146/annurev-arplant-042811-105439

[26] Andrews PR, Cain EN, Rizzardo E, Smith GD. Rearrangement of chorismate to prephenate. Use of chorismate mutase inhibitors to define the transition state structure. Biochemistry. 1977;22:4848-4852. doi: https://doi.org/10.1021/bi00641a015

[27] Gray JV, Golinelli-Pimpaneau B, Knowles JR. Monofunctional chorismate mutase from Bacillus subtilis: purification of the protein, molecular cloning of the gene, and overexpression of the gene product in Escherichia coli. Biochemistry. 1990;29:376-383. doi: https://doi.org/10.1021/bi00454a011

[28] Krappmann S, Helmstaedt K, Gerstberger T, Eckert S, Hoffmann B, Hoppert M, Schnappauf G, Braus GH. The aroC gene of Aspergillus nidulans codes for a monofunctional, allosterically regulated chorismate mutase. J Biol Chem. 1999;274:22275-22282. doi: https://doi.org/10.1074/jbc.274.32.22275

[29] Holland CK, Berkovich DA, Kohn ML, Maeda H, Jez JM. Structural basis for substrate recognition and inhibition of prephenate aminotransferase from Arabidopsis. Plant J. 2018;94(2):304-314. doi: https://doi.org/10.1111/tpj.13856

[30] Bross CD, Howes TR, Abolhassani Rad S, Kljakic O, Kohalmi SE. Subcellular localization of Arabidopsis arogenate dehydratases suggests novel and non-enzymatic roles. J Exp Bot. 2017;68(7):1425-1440. doi: https://doi.org/10.1093/jxb/erx024

[31] Cho MH, Corea OR, Yang H, Bedgar DL, Laskar DD, Anterola AM, Moog-Anterola FA, Hood RL, Kohalmi SE, Bernards MA, et al. Phenylalanine biosynthesis in *Arabidopsis thaliana*. Identification and characterization of arogenate dehydratases. J Biol Chem. 2007;282:30827-30835. doi: https://doi.org/10.1074/jbc.m702662200

[32] Jorrín J, López-Valbuena R, Tena M. Purification and properties of phenylalanine ammonia-lyase from sunflower (*Helianthus annuus* L.) hypocotyls. Biochim Biophys Acta. 1988;964:73-82. doi: https://doi.org/10.1016/0304-4165(88)90069-4

[33] Bandoni RJ, Moore K, Subba Rao PV, Towers GH. Phenylalanine and tyrosine ammonia-lyase activity in some Basidiomycetes. Phytochemistry. 1968;7:205-207. doi: https://doi.org/10.1016/S0031-9422(00)86316-5

[34] Xiang L, Moore BS. Inactivation, complementation, and heterologous expression of encP, a novel bacterial phenylalanine ammonia-lyase gene. J Biol Chem. 2002;277:32505-32509. doi: https://doi.org/10.1074/jbc.m204171200

[35] Emes AV, Vining LC. Partial purification and properties of L-phenylalanine ammonia-lyase from *Streptomyces verticillatus*. Can J Biochem. 1970;48:613-622. doi: https://doi.org/10.1139/o70-099

[36] Czichi U, Kindl H. Formation of *p*-coumaric acid and *o*-coumaric acid from L-phenylalanine by microsomal membrane fractions from potato: evidence of membrane-bound enzyme complexes. Planta. 1975;125:115-125. doi: https://doi.org/10.1007/BF00388698

[37] Young MR, Towers GH, Neish AC. Taxonomic distribution of ammonia-lyases for L-phenylalanine and L-tyrosine in relation to lignification. Can J Bot. 1966;44:341-349. doi: https://doi.org/10.1139/b66-040

[38] Liu T, Yao R, Zhao Y, Xu S, Huang C, Luo J, Kong L. Cloning, Functional Characterization and Site-Directed Mutagenesis of 4-Coumarate: Coenzyme A Ligase (4CL) Involved in

Coumarin Biosynthesis in *Peucedanum praeruptorum* Dunn. Front. Plant Sci. 2017;8:4. doi: https://doi.org/10.3389/fpls.2017.00004

[39] Brenda Winkel-Shirley. Flavonoid Biosynthesis. A Colorful Model for Genetics, Biochemistry, Cell Biology, and Biotechnology. Plant Physiology Jun 2001, 126 (2) 485-493. doi: https://doi.org/10.1104/pp.126.2.485

[40] Sanchez IJF. Polyketide synthase in *Cannabis sativa* L. PhD thesis, Leiden University, Leiden, The Netherlands; 2008.

[41] Mierziak J, Kostyn K, Kulma A. Flavonoids as important molecules of plant interactions with the environment. Molecules. 2014;19:16240-16265. doi: https://doi.org/10.3390/molecules191016240

[42] Koes RE, Quattrocchio F, Mol JNM. The flavonoid biosynthetic pathway in plants: function and evolution. BioEssays. 1994;16:123-132. doi: https://doi.org/10.1002/bies.950160209

[43] Martens S, Preuss A, Matern U. Multifunctional flavonoid dioxygenases: flavonols and anthocyanin biosynthesis in Arabidopsis thaliana L. Phytochemistry. 2010;71:1040-1049. doi: https://doi.org/10.1016/j.phytochem.2010.04.016

[44] Steele CL, Gijzen M, Qutob D, Dixon RA. Molecular characterization of the enzyme catalyzing the aryl migration reaction of isoflavonoid biosynthesis in soybean. Arch Biochem Biophys. 1999;367:146-150. doi: https://doi.org/10.1006/abbi.1999.1238

[45] Yu O, Jung W, Shi J, Croes RA, Fader GM, McGonigle B, Odell JT. Production of the isoflavones genistein and daidzein in non-legume dicot and monocot tissues. Plant Physiol. 2000;124:781-794. doi: https://doi.org/10.1104/pp.124.2.781

[46] Meng C, Zhang S, Deng Y-S, Wang G-D, Kong F-Y. Overexpression of a tomato flavanone 3-hydroxylase-like protein gene improves chilling tolerance in tobacco. Plant Physiol. Biochem. *2015;*96:388-400. doi: https://doi.org/10.1016/j.plaphy.2015.08.019

[47] Han Y, Huang K, Liu Y, Jiao T, Ma G, Qian Y, et al. Functional analysis of two flavanone-3-hydroxylase genes from *Camellia sinensis*: a critical role in flavonoid accumulation. *Genes*. 2017;8:E300. doi: 10.3390%2Fgenes8110300

[48] Holton TA, Brugliera F, Tanaka Y. Cloning and expression of flavonol synthase from *Petunia hybrida*. Plant J. 1993;4:1003-1010. doi: https://doi.org/10.1046/j.1365-313x.1993.04061003.x

[49] Owens DK, Alerding AB, Crosby KC, Bandara AB, Westwood JH, Winkel BS. Functional analysis of a predicted flavonol synthase gene family in *Arabidopsis*. Plant Physiol. 2008;147:1046-1061. doi: https://doi.org/10.1104/pp.108.117457

[50] Nielsen K, Deroles SC, Markham KR, Bradley MJ, Podivinsky E, Manson D. Antisense flavonol synthase alters copigmentation and flower color in lisianthus. Mol Breeding. 2002; 9:217-229. doi: https://doi.org/10.1023/A:1020320809654

[51] Shi M-Z, Xie D-Y. Biosynthesis and metabolic engineering of anthocyanins in Arabidopsis thaliana. Recent patents on biotechnology. 2014;8(1):47-60. doi: https://dx.doi.org/10.2174%2F1872208307666131218123538

[52] Shimada N, Sasaki R, Sato S, Kaneko T, Tabata S, Aoki T, et al. A comprehensive analysis of six dihydroflavonol 4-reductases encoded by a gene cluster of the Lotus japonicus genome. Journal of Experimental

Botany. 2005;56(419):2573-2585. doi: https://doi.org/10.1093/jxb/eri251

[53] Hua C, Linling L, Shuiyuan C, Fuliang C, Feng X, Honghui Y, et al. Molecular cloning and characterization of three genes encoding dihydroflavonol-4-reductase from *Ginkgo biloba* in anthocyanin biosynthetic pathway. PloS one. 2013;8(8):e72017–e. doi: https://doi.org/10.1371/journal.pone.0072017

[54] Ahmed NU, Park J-I, Jung H-J, Yang T-J, Hur Y, Nou I-S. Characterization of dihydroflavonol 4-reductase (DFR) genes and their association with cold and freezing stress in Brassica rapa. Gene. 2014;550(1):46-55. doi: https://doi.org/10.1016/j.gene.2014.08.013

[55] Tian J, Han Z-Y, Zhang J, Hu Y, Song T, Yao Y. The Balance of Expression of Dihydroflavonol 4-reductase and Flavonol Synthase Regulates Flavonoid Biosynthesis and Red Foliage Coloration in Crabapples. Scientific Reports. 2015;5. doi: https://doi.org/10.1038/srep12228

[56] Luo P, Ning G, Wang Z, Shen Y, Jin H, Li P, et al. Disequilibrium of Flavonol Synthase and Dihydroflavonol-4-Reductase Expression Associated Tightly to White vs. Red Color Flower Formation in Plants. Frontiers in Plant Science. 2016;6. doi: https://doi.org/10.3389/fpls.2015.01257

[57] Tanner GJ, Francki KT, Abrahams S, Watson JM, Larkin PJ, Ashton AR. Proanthocyanidin biosynthesis in plants. Purification of legume leucoanthocyanidin reductase and molecular cloning of its cDNA. J Biol Chem. 2003;278:31647-33156. doi: https://doi.org/10.1074/jbc.m302783200

[58] Xie DY, Sharma SB, Dixon RA. Anthocyanidin reductases from *Medicago truncatula* and *Arabidopsis thaliana*. Arch. Biochem Biophys. 2004;422:91-102. doi: https://doi.org/10.1016/j.abb.2003.12.011

[59] Xie DY, Sharma SB, Paiva NL, Ferreira D, Dixon RA. Role of anthocyanidin reductase, encoded by BANYULS in plant flavonoid biosynthesis. Science. 2003;299:396-399. doi: https://doi.org/10.1126/science.1078540

[60] Pang Y, Abeysinghe IS, He J, He X, Huhman D, Mewan KM, et al. Functional characterization of proanthocyanidin pathway enzymes from tea and their application in metabolic engineering. Plant Physiol. 2013;161:1103-1116. doi: https://doi.org/10.1104/pp.112.212050

[61] Gargouri M, Chaudière J, Manigand C, Maugé C, Bathany K, Schmitter JM, et al. The epimerase activity of anthocyanidin reductase from *Vitis vinifera* and its regiospecific hydride transfers. Biol Chem. 2010;391:219-227. doi: http://doi.org/10.1515/bc.2010.015

[62] Zhao J, Dixon RA. MATE transporters facilitate vacuolar uptake of epicatechin 3′-O-glucoside for proanthocyanidin biosynthesis in *Medicago truncatula* and Arabidopsis. Plant Cell. 2009;21:2323-2340. doi: https://dx.doi.org/10.1105%2Ftpc.109.067819

[63] Liu C, Jun JH, Dixon RA. MYB5 and MYB14 play pivotal roles in seed coat polymer biosynthesis in *Medicago truncatula*. Plant Physiol. 2014;165:1424-1439. doi: https://doi.org/10.1104/pp.114.241877

[64] Sanchez IJF. Polyketide synthase in *Cannabis sativa* L. PhD thesis. Leiden University, Leiden, The Netherlands; 2008.

[65] Iwashina T. The structure and distribution of the flavonoids in plants. J

Plant Res. 2000;113:287-99. doi: https://doi.org/10.1007/PL00013940

[66] Zhu F, Jiang D, Zhang M, Zhao B. 2,4-Dihydroxy-3′-methoxy-4′-ethoxychalcone suppresses cell proliferation and induces apoptosis of multiple myeloma *via* the PI3K/akt/mTOR signaling pathway. Pharm Biol. 2019;57(1):641-648. doi: https://dx.doi.org/10.1080%2F13880209.2019.1662814

[67] Escobar-Ramos A, Lobato-García CE, Zamilpa A, Gómez-Rivera A, Tortoriello J, González-Cortazar M. Homoisoflavonoids and Chalcones Isolated from Haematoxylum campechianum L., with Spasmolytic Activity. Molecules. 2017;22(9):1405. doi: https://dx.doi.org/10.3390%2Fmolecules22091405

[68] Meng C-Y, Han Y, Duan Y-X, Chen J-X, Hu Q-F, Gao X-M. Isolation of Chalcones from the Root of Codonopsis cordifolioidea and Their Antitobacco Mosaic Virus Activities. Asian Journal of Chemistry. 2013;25(17):9517-9519. doi: http://dx.doi.org/10.14233/ajchem.2013.15052A

[69] Namratha V, Rao GV. Isolation and structure determination of a new 2, 2-dimethyl chromeno dihydro chalcone from *Crotalaria ramosissima*. Int J Chem Sci. 2009;7(2):981-985.

[70] Alawaf H. Phytochemical analysis, isolation and identification of flavan-3ol from Syrian *Pinus halepensis*. Chemistry and Materials Research. 2017;9(4).

[71] Parvez MK, Al-Dosari MS, Ahmed S, Rehman MT, Al-Rehaily AJ, Alajmi MF. Oncoglabrinol C, a new flavan from *Oncocalyx glabratus* protects endothelial cells against oxidative stress and apoptosis, and modulates hepatic CYP3A4 activity. Saudi Pharmaceutical Journal. 2020;28:648-656. doi: https://doi.org/10.1016/j.jsps.2020.04.004

[72] Guyasa B, Melaku Y, Endale M. Antibacterial activity of two flavans from the stem bark of *Embelia schimperi*. Adv Pharmacol Sci. 2018;2018:5870161. doi: https://doi.org/10.1155/2018/5870161

[73] Gleńsk M, Dudek MK, Ciach M, Włodarczyk M. Isolation and structural determination of flavan-3-ol derivatives from the Polypodium vulgare L. rhizomes water extract. Natural Product Research. 2019;1-10. doi: https://doi.org/10.1080/14786419.2019.1655302

[74] DuyCuong DT, Dat HT, Duan NT, Thuong PD, Phat NT, Tri MD, Son DV, Hoa NT, KimTuyen PN, PhiPhung NK. Isolation and characterization of six flavonoids from the leaves of Sterculia foetida Linn. Vietnam J Chem. 2019;57(4):438-442. doi: https://doi.org/10.1002/vjch.201900084

[75] Meesakul P, Pudhom K, Pyne SG, Laphookhieo S. Hybrid flavan–flavanones from Friesodielsia desmoides and their inhibitory activities against nitric oxide production. RSC Adv. 2017;7:17545. doi: https://doi.org/10.1039/C7RA02528A

[76] Yang M-X, Liang Y-G, Chen H-R, Huang Y-F, Gong H-G, Zhang T-Y, Ito Y. Isolation of Flavonoids From Wild Aquilaria sinensis Leaves by an Improved Preparative High-Speed Counter-Current Chromatography Apparatus. Journal of Chromatographic Science. 2018;56(1):18-24. doi: https://doi.org/10.1093/chromsci/bmx076

[77] Abreu AC, Coqueiro A, Sultan AR, Lemmens N, Kim HK, Verpoorte R, van Wamel WJB, Simões M, Choi YH. Looking to nature for a new concept in antimicrobial treatments: isoflavonoids from Cytisus striatus as antibiotic adjuvants against MRSA. Sci Rep. 2017;7(1):3777. doi: https://doi.org/10.1038/s41598-017-03716-7

[78] Quiroz A, Mendez L, Mutis A, Hormazabal E, Ortega F, Birkett MA, Parra L. Antifeedant activity of red clover root isoflavonoids on Hylastinus obscurus. Journal of Soil Science and Plant Nutrition. 2017;17(1):231-239. doi: http://dx.doi.org/10.4067/S0718-95162017005000018

[79] Dizamatova A, Zhumanova K, Zhusupova GE, Zhussupova AI, Srivedavyasasri R, Ibrahim MA, Ross SA. A new prenylated isoflavonoid from *Limonium leptophyllum*. Natural Product Communications. 2019;1-5. doi: https://doi.org/10.1177%2F1934578X19844137

[80] Kothari M, Jain DK. Quantification of phytoconstituents and isolation of Flavonoids from procedure to extract isoflavones from soybean seeds. Journal of Cleaner Production. 2018;170:1102e1110.

[81] He J, Fan P, Feng S, Shao P, Sun P. Isolation and purification of two isoflavones from *Hericium erinaceum* mycelium by High-Speed Counter-Current Chromatography. Molecules. 2018;23(3):560. doi: https://doi.org/10.3390/molecules23030560

[82] Kishore N, Twilley D, vanStaden AB, Verma P, Singh B, Cardinali G, Kovacs D, Picardo M, Kumar V, Lall N. Isolation of Flavonoids and Flavonoid Glycosides from Myrsine africana and Their Inhibitory Activities against Mushroom Tyrosinase. J Nat Prod. 2018;81:49–56. doi: https://doi.org/10.1021/acs.jnatprod.7b00564

[83] Gupta R, Gupta GD. Isolation and characterization of flavonoid glycoside from Cordia obliqua Willd. Leaf. International Journal of Green Pharmacy. 2018;12(2). doi: http://dx.doi.org/10.22377/ijgp.v12i02.1834

[84] Arora S, Itankar P. Extraction, isolation and identification of flavonoid from Chenopodium album aerial parts. Journal of Traditional and Complementary Medicine. 2018;8(4):476-482. doi: https://doi.org/10.1016/j.jtcme.2017.10.002

[85] Ahmed SI, Hayat MQ, Zahid S, Tahir M, Mansoor Q, Ismail M, Keck K, Bates R. Isolation and identification of flavonoids from anticancer and neuroprotective extracts of *Trigonella foenum graecum*. Trop J Pharm Res. 2017;16(6):1391. doi: https://doi.org/10.4314/tjpr.v16i6.25

[86] Utari F, Itam A, Syafrizayanti S, Putri WH, Ninomiya M, Koketsu M, Tanaka K, Efdi M. Isolation of flavonol rhamnosides from *Pometia pinnata* leaves and investigation of α-glucosidase inhibitory activity of flavonol derivatives. Journal of Applied Pharmaceutical Science. 2019;9(08):053-065. doi: http://dx.doi.org/10.7324/JAPS.2019.90808

[87] Sun Q, Pan G, Xu W, Lu X, Bai C, Liu M, Chen Y. Isolation and structure elucidation of a new flavonol glycoside from Sabia Parviflora. Natural Product Research. 2019;1-6. doi: https://doi.org/10.1080/14786419.2019.1679130

Chapter 7

Biosynthesis of the Immunomodulatory Molecule Capsular Polysaccharide A from *Bacteroides fragilis*

Sunita Sharma

Abstract

Capsular Polysaccharide A (CPSA) is a polymer of a tetrasaccharide unit found on the surface of the symbiotic gut bacteria *Bacteroides fragilis*. CPSA has been suggested to be important for maintaining a natural equilibrium between Th1 and Th2 cell levels in the normal immune system of mammals. If this equilibrium is disrupted, the human body can develop different autoimmune disorders. The gene locus responsible for CPSA biosynthesis has been previously identified. The locus was proposed to encode one glycosyl-1-phosphate transferase (WcfS) and three glycosyltransferases (WcfN, -P and -Q), three sugar modifying enzymes (WcfM, WcfR and WcfO), a flippase (Wzx) and a polysaccharide polymerase (Wzy) based on homology tools. A route for the complete biosynthesis of CPSA has been elucidated. The initiating sugar transferase, WcfS has been previously identified and characterized. An *in vitro* method was used to enzymatically synthesize CPSA, which was assembled on a fluorescent analogue of the native bactoprenyl diphosphate anchor one sugar at a time. Function of the hypothesized pyruvyltransferase WcfO was also determined. This is the first study to characterize a pyruvyltransferase involved in polysaccharide biosynthesis from a prokaryote. The biosynthesis of the polysaccharide was achieved in a single pot, compared to multiple steps involved in chemical synthesis, displaying an enormous leap in the biosynthesis of complex molecules like CPSA.

Keywords: *Bacteroides fragilis*, pyruvyltransferase, glycosyltransferase, capsular polysaccharide A, biosynthesis

1. Introduction

B. fragilis is an obligate anaerobic bacterium which colonizes the intestinal tract of the human gut, and essentially all other mammals. It is an integral component of the normal gastrointestinal flora [1, 2]. It is classified as a Gram-negative, non-spore forming and anaerobic bacilli. This mammalian symbiont and opportunistic pathogen depends on its capsular layer for virulence as well as for symbiosis in the mammalian gut [3, 4]. Eight capsule polysaccharides can be expressed on its surface, depending on the environmental niche of the organism, designated as CPSA through CPSH [5–10]. Capsular polysaccharide A is one of the eight polysaccharides found on the surface of *B. fragilis,* and is the most abundant. CPSA plays a role in abscess formation when the bacterium localizes outside of its normal niche in the gastrointestinal tract or during surgical procedures [11]. However, this view has been challenged when it was found that treating the animal with the CPSA and then introducing the abscess-inducing bacteria resulted in the immune system of the animal protecting itself against the production of abscesses. Furthermore, few studies have also claimed that the abscess formation by *B. fragilis* actually prevents infection in the wound by other pathogenic bacteria [12, 13].

CPSA is a unique polymer. It has both negatively and positively charged motifs present on each repeating monomer, making it a zwitterionic molecule [7, 14] (**Figure 1**). The presence of this zwitterionic character has been attributed to the novel immunologic activity displayed by CPSA. The zwitterionic character has been shown to modulate the mammalian immune system by interacting with the adaptive immune system [15]. Elimination of either charge group in CPSA results in a lack of *in vivo* activation of the T-cells [16, 17].

CPSA modulates the immune system by its stimulation of a T-cell dependent form of immunity that provides protection against the formation of the intraabdominal abscesses. At the molecular level, CPSA interacts with the MHCII pathway similar to traditional protein antigens [18]. The first step is endocytosis of CPSA by the antigen-presenting cells like dendritic cells. Once in the endosome, CPSA is depolymerized based on the chemical reaction, deaminative cleavage [19]. This cleaving is mediated by nitric oxide, that has been generated by the upregulation of inducible nitric oxide synthase (iNOS). The 130 kDa CPSA is processed

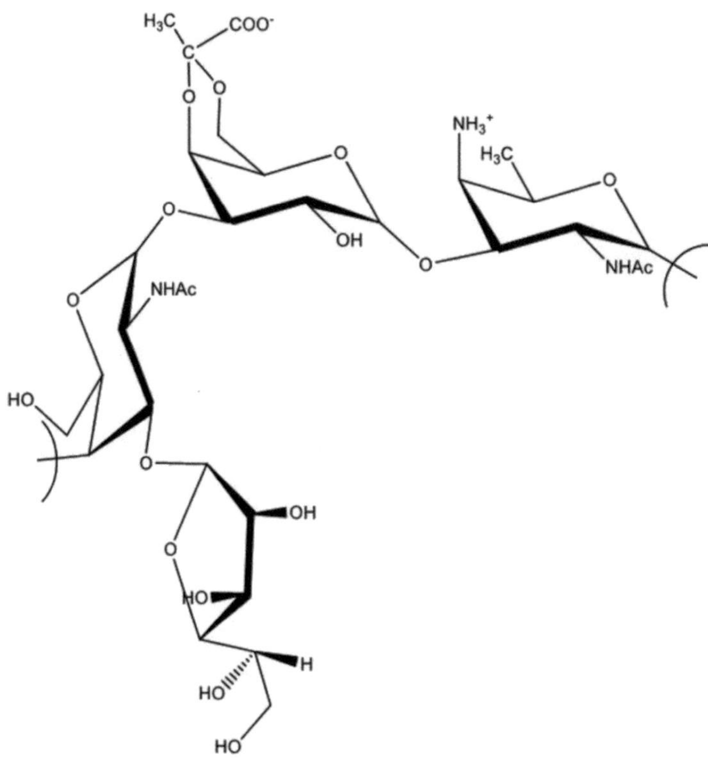

Figure 1.
Tetrameric repeat unit of the CPSA found on B. fragilis. *It consists of an acetamido-4-amino-6-deoxygalactopyranose (AADGal), 4,6-pyruvate galactose (4,6-pyr-gal), N-acetylgalactosamine (GalNAc), and a galactofuranose (Galf) sugar.*

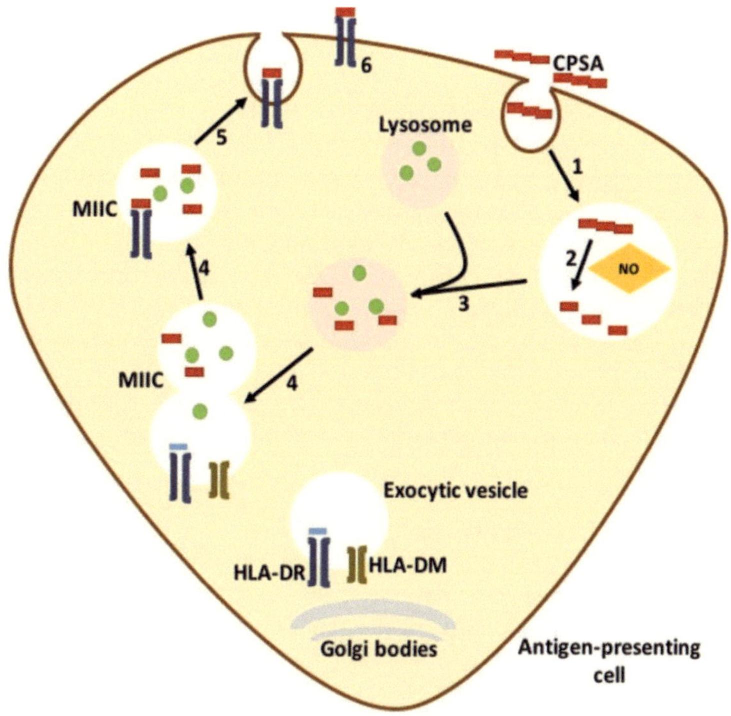

Figure 2.
Depolymerization of CPSA in antigen presenting cell. 1. Internalization of CPSA in an endosome. 2. iNOS upregulation produces NO, which cleaves 130 kDa CPSA to ~15 kDa units. 3. Endosome fuses with the lysosome. 4. Endo-lysosome fuses with exocytic vesicle to form MIIC vesicle which has HLA-DR, HLA-DM and processed polysaccharide. In here, processed polysaccharide is loaded on HLA-DR with the help of HLA-DM. 5, 6. The loaded HLA-DM is presented on the surface of the antigen presenting cell to be recognized by alpha beta TCR present on CD4+ T-cell.

to 15 kDa units. After being processed, the endosomes fuse with lysosomes and exocytic vesicles to form MIIC vesicle carrying HLA-DR and the accessory molecule HLA-DM. HLA-DM catalyzes the binding of MHCII to CPSA fragments, which is then presented to the CD4+ T cell receptor (**Figure 2**). This leads to the proliferation of the CD4+ T cell population, that produces IL-10, which is responsible for providing protection against the formation of intra-abdominal abscesses [15, 20].

CPSA can restore the immune system from a variety of autoimmune disorders, making it a promising candidate for a therapeutic drug. Colonization of nude mice with wild type *B. fragilis*, that produces the zwitterionic capsular polysaccharide A, protected animals from antibiotic induced experimental autoimmune encephalomyelitis (EAE), while animals infected with mutant *B. fragilis* deficient in the production of the polysaccharide were not protected [12, 21]. In germ free animal models of Inflammatory Bowel Disease (IBD), it was found that CPSA alone without the bacterial carrier was enough to stimulate normal immune system function and prevent intestinal inflammatory disease [22, 23]. CPSA has been given therapeutically to decrease pro-inflammatory cytokine production in an experimental model of colonic irritation [24].

2. CPSA gene locus

CPSA is a polymer of a tetramer repeated approximately 160 times. Its size is estimated to be 110 kDa [25]. The CPSA tetrameric repeat unit consists of an acetamido-4-amino-6-deoxygalactopyranose (AADGal), 4,6-pyruvate galactose (4,6-pyr-Gal), N-acetylgalactosamine (GalNAc), and a galactofuranose (Galf) sugar (**Figure 3**) [26]. The structure of CPSA has previously been well investigated using total correlated spectroscopy and NOESY NMR [27]. Three-dimensional structure of a highly related PSA2 molecule shows a right-handed helix with two repeating units per turn, and a pitch of 20 Å. The zwitterionic motif is formed with alternating anionic carboxylate lying in repeated grooves and the cationic-free amines exposed on the outer surface of the carbohydrate [12, 28].

Although the chemical composition of CPSA is known, yet the biochemical pathway involved in its production is poorly documented [29, 30]. The location of the proposed CPSA locus was knocked out, making a mutant *B. fragilis* which did not express CPSA on its surface, thereby confirming the location of the biosynthetic locus (**Figure 4**). Within the CPSA locus, there are eleven genes, of which nine express proteins similar to other proteins involved in various other polysaccharide biosynthesis (**Table 1**).

2.1 Initiating the CPSA biosynthesis

The function of the nine genes have been elucidated and a pathway has been constructed (**Figure 5**). The identity of the genes present in the CPSA gene locus

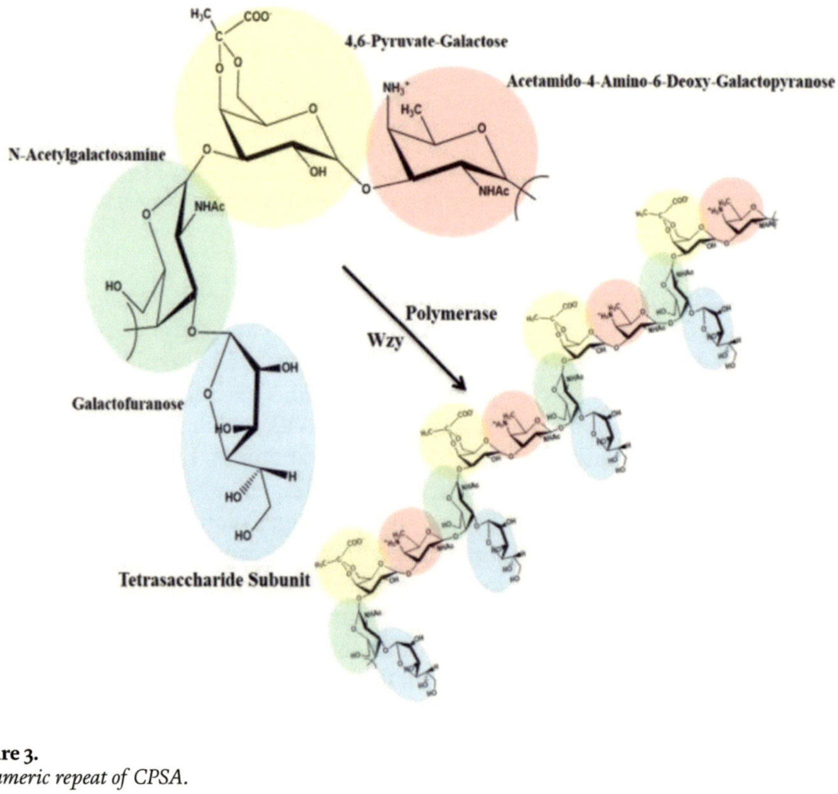

Figure 3.
Tetrameric repeat of CPSA.

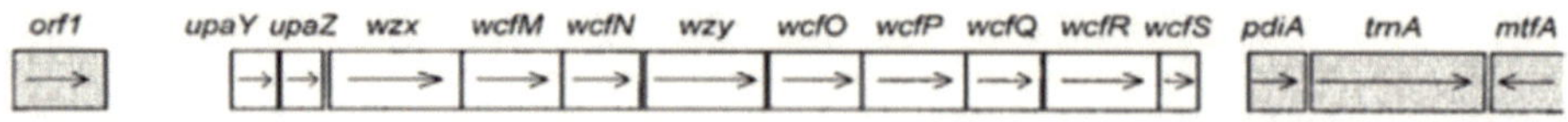

Figure 4.
CPSA locus in the B. fragilis *genome.*

ORF	Size (aa)	Size (kDa)	Role	Accession no.
wzx	482	56	flippase	AAK68914.1
wcfM	364	43	galactopyranose mutase	AAK68915.1
wcfN	291	34	glycosyltransferase	AAK68916.1
wzy	434	43	polymerase	AAK68917.1
wcfO	357	40	pyruvyltransferase	AAK68918.1
wcfP	378	44	glycosyltransferase	AAK68919.1
wcfQ	268	32	glycosyltransferase	AAK68920.1
wcfR	407	45	aminotransferase	AAK68921.1
wcfS	195	23	glycosyltransferase	AAK68922.1

Table 1.
Functions of the gene products in the CPSA biosynthesis operon.

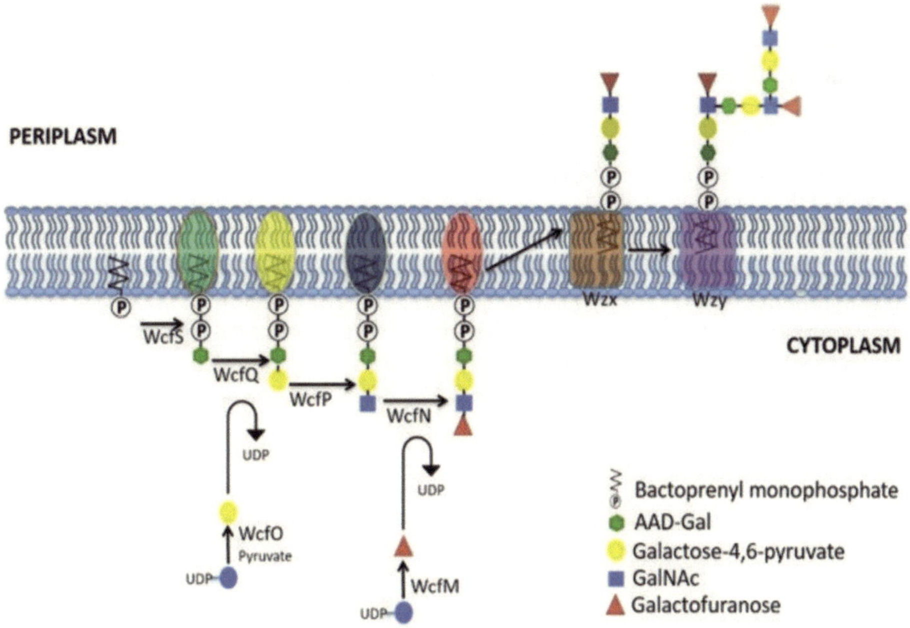

Figure 5.
Pathway of CPSA biosynthesis.

suggests that the most likely route for assembling the complex bacterial polysaccharide is a Wzy-dependent pathway in which the repeat unit oligosaccharides are assembled one sugar at a time on the cytosolic face of the bacterial inner membrane [31]. Assembly of the oligosaccharide takes place on a C55 isoprenoid bactoprenol [32]. It is a hydrophobic anchor which holds the growing polymer in the cell membrane.

The enzymes responsible for the synthesis of the first sugar, AADGal, in the tetrameric repeat, and the enzyme that catalyzes the transfer of this sugar to the bactoprenol anchor have been well characterized [33]. AADGal is synthesized by the sequential action of a dehydratase and an aminotransferase, which is then transferred to the bactoprenyl anchor by a hexose phosphate initiating transferase.

Within the CPSA biosynthesis locus, there is a predicted aminotransferase gene, *wcfR*, and a hexose phosphate initiating transferase, *wcfS*, but no predicted dehydratase was found. However, a gene encoding a potential dehydratase, *ungD2*, has been identified elsewhere in the *B. fragilis* genome. When this gene was knocked out by Coyne *et al*, they found out that, synthesis of the seven out of the eight capsular polysaccharides was stopped. Initial studies with UngD2 and WcfR did not show any promise in the synthesis of AADGal. Hence a previously well characterized dehydratase, PglF [34], from *Campylobacter jejuni* was used to provide the substrate needed for WcfR function. The coupling of these enzymes together led to the production of AADGal (**Figure 6**) [35, 36]. This also points to the notion that depending on homology alone for functional assignment of genes, is not always right, and wet lab results are needed to confirm the function of the gene product.

The synthesized UDP-AADGal was further used as a potential substrate for WcfS, identified as the initiating hexose phosphate transferase. Studies done by Mostafavi et al. demonstrated that WcfS was indeed the initiating hexose phosphate transferase, which lead to the formation of the bactoprenyl linked monosaccharide (**Figure** 7) [33].

As mentioned previously, assembly of the polysaccharides in bacterial cells is done on a C55 bactoprenyl anchor. It is produced by the condensation of farnesyl diphosphate (FPP) to eight units of isopentenyl diphosphate (IPP), done by the enzyme undecaprenyl diphosphate synthase (UPPS). A major drawback of using this compound in in vitro assays is that, it does not have easily distinguishable chromophores associated to it, hence very few rapid assays are available to detect and quantify the activity of enzymes associated with polysaccharide synthesis. To circumvent this problem, the Troutman lab developed fluorescent analogues of the native bactoprenyl, which are easily traceable [25, 37]. Assays done using these analogues take a short time to reveal valuable information about the enzymes when compared to traditional assays, which follow the more tedious route of using radioactive labeled substrates. Mostafavi et al. used a p-nitroaniline bactoprenyl phosphate analogue to find out the function of WcfS (**Figure** 7) [33].

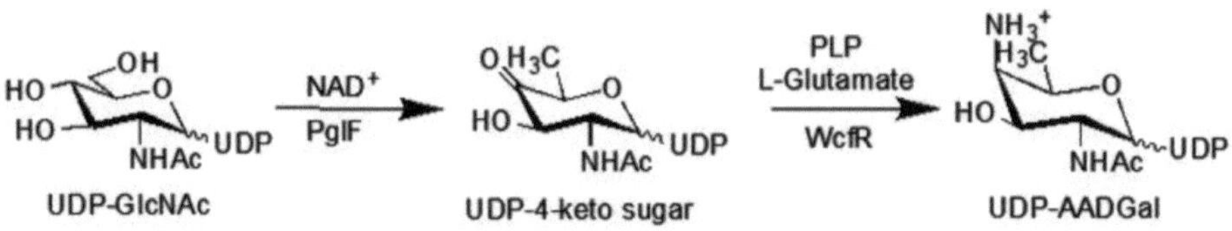

Figure 6.
Biosynthesis of AADGal.

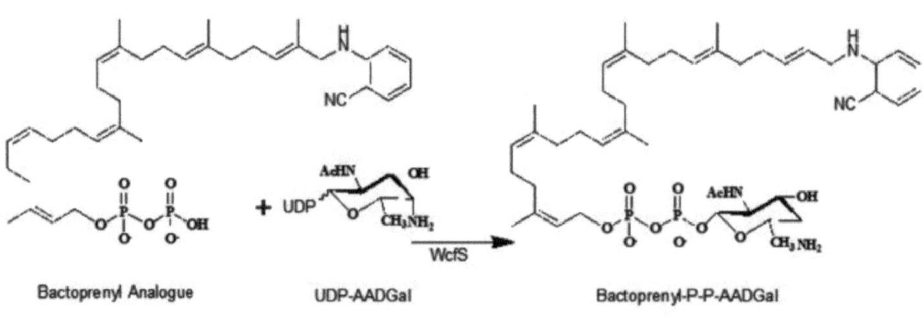

Figure 7.
Biosynthesis of bactoprenyl linked monosaccharide.

2.2 Glycosyltransferases involved in CPSA biosynthesis

Cell surface polysaccharides are nothing but complex carbohydrates. They play important roles in a number of biological processes such as cell growth, cell to cell interactions, immune response, and inflammation. The polysaccharides are synthesized by a class of enzymes known as glycosyltransferases [38]. Glycosyltransferases are an enzyme superfamily responsible for the attachment of carbohydrate moieties to a wide array of acceptors that include nucleic acids, polysaccharides, proteins, lipids, and carbohydrates. The majority of glycosyltransferases are sugar nucleotide-dependent enzymes, and utilize nucleoside diphosphate sugars (NDP-sugars) as donors for the glycosidic bond formation. In other cases, the sugar donors can also be lipid phosphates and unsubstituted phosphate [39].

The glycosyltransferases have been classified by sequence homology into 96 families in the Carbohydrate Active enZyme database (CAZy), each of which catalyze the reaction as shown in **Figure 8** [40]. Chain elongation of the oligosaccharide units in complex carbohydrates is achieved by the addition of monosaccharide units through the action of different glycosyltransferases in a specific sequence. The CAZy database provides a highly powerful predictive tool, as the structural fold and mechanism of action are invariant in most of the families [22]. Therefore, where the structure and mechanism of a glycosyltransferase member for a given family has been reported, some assumptions about other members of the family can be made. Substrate specificity, however, is more difficult to predict, and requires experimental characterization of individual glycosyltransferases.

Determining both the sugar donor and acceptor for a glycosyltransferase of unknown function can be challenging, and it is one of the reasons there are significantly fewer well characterized isoprenoid linked sugar glycosyltransferases when compared to the glycosyltransferases responsible for synthesizing disaccharides or the oligosaccharides [40]. The less reports on isoprenoid linked sugar transferases can be attributed to the fact that, a high throughput method has not yet been developed which will enable for faster characterization. Another challenge in characterizing the glycosyltransferases is the availability of rare sugars, as most of the bacterial polysaccharides contain rare sugars. Rare sugars, such as rhamnose or fucose, may provide the bacterial polysaccharides with additional biological properties compared to those composed of more common sugar monomers [23, 41]. Rare sugars are monosaccharides that are not commonly found in nature, in comparison to D-glucose, D-galactose, D-fructose, D-xylose, D-ribose, and L-arabinose which are more abundant [23]. Moreover, the traditional methods like radioisoptopic

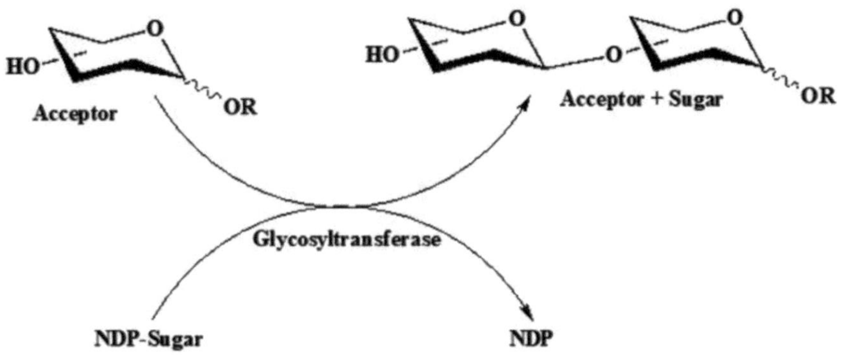

Figure 8.
General reaction scheme for a glycosyltransferase (GTs).

labelling, thin-layer chromatography (TLC) used to characterize the glycosyltransferase, often tends to be tedious and challenging in tracking the product.

Glycosyltransferases catalyze glycosidic bond formation with either overall retention or inversion of anomeric configuration when compared to the stereochemistry in the sugar donor (**Figure 9**). Inverting glycosyltransferases are generally believed to proceed via a single displacement SN2 mechanism with concomitant nucleophilic attack by the acceptor at the anomeric carbon, facilitated by proton transfer to the catalytic base, and leaving group departure [22]. Structural data have shown that several inverting glycosyltransferases, contain no obvious candidate catalytic base indicating these enzymes use an alternative mechanism [38, 39].

The reaction coordinate employed by retaining glycosyltransferases has been much debated, and it could be possible the mechanism is not conserved for all retaining enzymes. One possibility is a double displacement mechanism via a covalent mechanism, analogous to that used by glycoside hydrolases [22]. A report by Soya et al. provided mass spectrometry evidence for the formation of a covalent intermediate between the donor substrate and a cysteine, which had been substituted for the candidate catalytic nucleophile, on two retaining glycosyltransferases [42]. The more favored mechanism in the field is an SN1 or SN1-like mechanism, which involves interaction between the leaving group and attacking nucleophile on the same face. This mechanism is supported by kinetic isotope effect studies to analyze the structure of the transition state and by computational modeling [38, 39].

The CPSA gene locus has three genes, wcfQ, wcfP and wcfN, that putatively encode for glycosyltransferases [29, 30]. Each of these glycosyltransferases is expected to transfer a sugar moiety to the bactoprenyl linked monosaccharide, the disaccharide and the trisaccharide. Based on the CAZy database, and homology studies, WcfQ and WcfN are hypothesized to belong to the glycosyltransferase superfamily A, which follows the inverting mechanism in the sugar transfer. Whereas WcfP is proposed to belong to the glycosyltransferase superfamily B, which follows the retaining mechanism [40].

WcfQ, identified as the first glycosyltransferase, transfers galactose to the isoprenoid linked monosaccharide, even though it was observed by authors that, WcfQ could also transfer glucose to the bactoprenyl linked monosaccharide. This is because WcfQ required glucose in much excess when compared to galactose. It was also found out that even though WcfP had the capability of transferring galactose, WcfQ was more efficient in it, hence it was identified as the galactosyltransferase

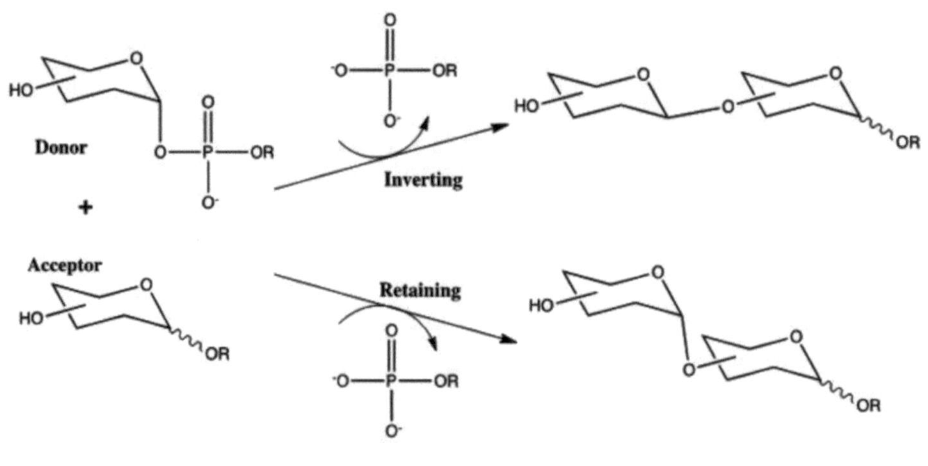

Figure 9.
Glycosyltransferases catalyze glycosyl group transfer with either inversion or retention of the anomeric stereochemistry with respect to the donor sugar.

in the CPSA biosynthetic pathway. Moreover, based on the Carbohydrate-Active enZYmes (CAZY) database the WcfQ sequence matched the GT_2 family of glycosyltransferases which invert the configuration of the anomeric carbon of the donor, while WcfP was similar to a GT_4 family glycosyltransferase, which retains the anomeric stereo-configuration of the donating sugar [43, 44]. The published structure of the CPSA tetrasaccharide unit suggests that the linkage should be in a beta configuration [27]. This supported the conclusion that WcfQ is the protein responsible for introducing galactose, and that it introduces the sugar in the appropriate beta configuration [45].

As stated before, WcfP is related to the GT_4 family of proteins suggesting that it is a retaining glycosyltransferase, it was therefore more likely that WcfP catalyzed UDP-GalNAc transfer to the galactose, but it was not known if it transferred UDP-GalNAc to the unpyruvylated disaccharide or the pyruvylated disaccharide. Both WcfN and WcfP were analyzed with the pyruvylated and the unpyruvylated disaccharides, it was demonstrated that WcfP transfers only UDP-GalNAc to the pyruvylated disaccharide.

In homology studies, WcfN was predicted to be a member of the GT_2 family, whose members have been identified to transfer furanose residues. WcfN was also hypothesized to be an inverting transferase, which inverts the stereochemistry of the anomeric carbon. Since the linkage between the third and the fourth sugar in the tetrasaccharide repeat unit is in the beta configuration, WcfN fitted the role of being the last glycosyltransferase. WcfN was found to transfer the galactofuranose to the trisaccharide, hence completing the mapping of the pathway of synthesis of the tetrasaccharide.

2.3 WcfM as the galactopyranosemutase

Polysaccharides composed of furanosyl residues are important constituents of many bacteria, protozoa, fungi, plants and archaebacteria [46, 47]. The furanosyl constituents have also been identified in glycopeptides, glycolipids as well as nucleotide sugars. D-Galactose is by far the most widespread hexose in the furanose form in naturally occurring polysaccharides, and the most impressive examples of these glycans are encountered in mycobacteria [48–50]. Galactofuranose, (Galf), which is thermodynamically less stable than galactose, is essential for the viability of several pathogenic species of bacteria and protozoa. It is absent in this form in mammalian cell structure, hence the biochemical pathways by which galactofuranose containing glycans are assembled have been attractive sites for drug action [47, 51]. This potential has led to an increased interest in the synthesis of molecules containing galactofuranose residues, and their subsequent use in studies directed towards understanding of the enzymes that process these residues and the identification of potential inhibitors of these pathways [46].

The enzyme UDP-galactopyranose mutase is central to galactofuranose metabolism. Most organisms cannot use exogenous galactofuranose, and UDP-galactofuranose appears to be the biological source of galactofuranose residues in polysaccharides [46]. The major structural component of the *Mycobacterium tuberculosis* cell wall contains a galactan chain of approximately thirty-five galactofuranose units, and the biosynthesis of the galactan is essential for viability [47]. The O-antigens of both *Escherichia coli* and *Klebsiella pneumoniae* contain galactofuranose as a component of lipopolysaccharide [47]. Several galactofuranose containing glycoconjugates have been found in *Trypanosoma cruzi*, the causative agent of Chagas disease, including glycoinositolphospholipids, lipopeptidophosphoglycans and mucin-like proteins. The galactomannan of *Aspergillus fumigatus* also contains galactofuranose, and this polysaccharide is used for clinical detection of fungal infections. Finally, it is also known that stopping galactofuranose biosynthesis in *Leishmania major* attenuates its virulence [46–48, 51]. The above-mentioned pathogenic organisms all use the same building block for synthesizing galactofuranose-containing polysaccharides: uridine diphosphogalactofuranose (UDP-galactofuranose). This sugar nucleotide is produced from UDP-Glcp by the enzymes UDP-Glucose 4-epimerase (generating UDP-Galp,) and UDP-galactopyranose mutase (UGM), which catalyzes the transformation of UDP-Galp to UDP-galactofuranose. The gene encoding UGM was first identified in *E. coli* in 1996, followed shortly by its identification in *K. pneumoniae* and *M. tuberculosis* [48, 49, 52] . More recently, UGM was identified in the eukaryotes *A. fumigatus, Cryptococcus neoformans, L. major* and *T. cruzi*.

In the past several years' major milestones have been achieved, which include an in-depth understanding of the mechanism of UDP-galactopyranose mutase (UGM), the enzyme which produces UDP-galactofuranose, and is the donor species used by galactofuranosyltransferases. A number of methods for the synthesis of galactofuranosides have also been developed [50]. UDP-galactofuranose has also been prepared by a number of approaches, and currently it appears that a chemoenzymatic approach is the most viable method for producing multi-milligram amounts of this important rare sugar [46, 50].

The biosynthetic gene operon of CPSA encodes a wcfM gene, which was found to be homologous to other galactopyranose mutases. It is homologous to two known UDP-galactopyranose mutases, one from *Streptococcus pneumonia* (Cps33fN: 66% identity and 82% similarity) and the other from *E. coli* (59% identity and 79% similarity). The gene encodes a 43 kDa protein with one potential N-terminal transmembrane domain. Like other galactopyranose mutases, the protein is hypothesized to catalyze the reaction as shown in **Figure 10**. The product of WcfM is required for the final step in the synthesis of the CPSA tetrasaccharide repeat unit. The last glycosyltransferase transfers UDP-galactofuranose to the trisaccharide.

2.4 WcfO as the pyruvyltransferase

Pyruvyltrasferases and pyruvylation have been less studied in prokaryotes, despite a burgeoning evidence of its presence in bacteria. Addition of pyruvate moiety gives a negative charge to the polymer and is utilized by the bacteria in various functions [53]. An example of this is the pyruvylation of ManNAc residue by the enzyme CsaB in the secondary cell wall polymer of *Bacillus anthracis* and *Paenibacillus Alvei* [54, 55]. This pyruvylated residue comes in use in anchoring the S-layer proteins in Gram positive bacteria by binding to the SLH domains of the S-layer proteins [56]. Knocking out the CsaB has led to a lethal phenotype, which suggests that, pyruvylation of the secondary cell wall polymer is essential to the

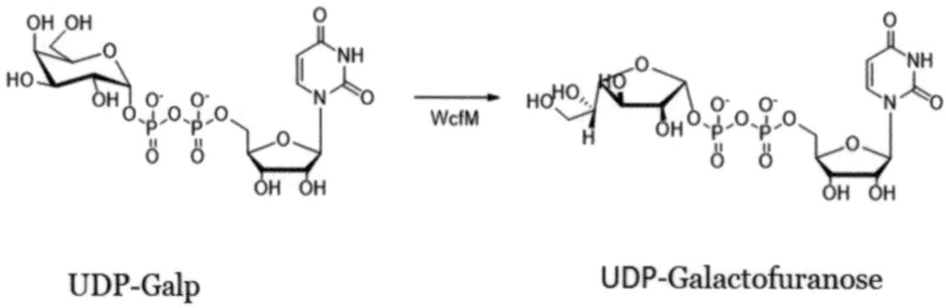

Figure 10.
Reaction catalyzed by UGM.

growth and survival of the bacteria [55]. CsaB was recently characterized by the Schaffer group [57]. Including WcfO, a total of three pyruvyltransferases have now been functionally characterized. Pvg1b is from an eukaryote, and whose crystal structure has been solved [58, 59].

Polysaccharides of various prokaryotes are covalently linked with variable combinations of sulfates and pyruvates, for example, *Rhizobium leguminosarum*: 4,6-pyrGalactose and 4,6- pyrGlucose, *Bacillus anthracis*: 4,6-pyrManNAc, and *Xanthomonas campestris*: 4,6-pyrMannose. These modifications provide a highly negative charge of these polysaccharides, which is often essential for function [60]. For example, when the pyruvyltransferase PssM, responsible for the pyruvate modification in the *R. leguminosarum* exopolysaccharide was deleted, the bacterium was found to be ineffective in infecting pea plants to initiate the formation of root nodules. This led to formation of aberrant root nodules, which were unable to fix nitrogen [61, 62]. Moreover, some studies have linked the pyruvic acetals in oligo- and polysaccharides to their immunological properties [63, 64].

Among the eleven proteins encoded in the CPSA gene operon, one of the genes transcribes a hypothesized pyruvyltransferase based on homology studies performed using pBLAST [31]. There is little sequence similarity to other known proteins with the wcfO gene product. WcfO has very minimal sequence identity to the two characterized pyruvyltransferases Pvg1p from *S. pombe* and PssM from *R. leguminosarum*. The activity of CPSA is dependent on its zwitterionic character in which the –AADGal amino group is positively charged while the pyruvate is negatively charged [16]. Due to the fact that all other sugar modifying enzymes and glycosyltransferases required for CPSA biosynthesis have been located in the CPSA biosynthesis operon, it was proposed by the authors that the wcfO gene product was likely responsible for the pyruvylation modification required for the formation of the second sugar in the CPSA tetrasaccharide repeat unit. WcfO is capable of modifying galactose or glucose when they are linked to the isoprenoid lipid carrier. This points to the direction that, there may be sub-families within the pyruvyl-transferase family that utilize different substrates. Kinetic evaluation of WcfO was performed by the authors to test if discriminated between glucose and galactose, and it apparently utilized both the substrates with equal vigor.

3. Significance of capsular polysaccharide A

Previous studies on the CPSA molecule have revealed it to be effective as a therapeutic molecule, the tetrasaccharide repeat needs to be a polymer of ten repeat units or longer. If shorter than that, it fails to activate the immune system [64, 65]. CPSA operon encodes for a flippase wzx, which takes the repeat unit and flips it from the cytoplasmic space to the periplasmic space, where the polymerase wzy, utilizes the repeat unit and polymerizes it till it reaches a length of approximately 130 repeat units [65, 66].

Recent successes in cancer vaccines and in monoclonal antibody cancer immunotherapy have given the impetus towards development of vaccines targeting cancer-associated carbohydrates. The Andreana group have been developing carbohydrate immunogens to elicit a T-cell dependent immune response. CPSA is known to stimulate a strong T-cell mediated response. They have successfully linked CPSA to the tumor-associated carbohydrate antigen (TACA), Sialyl Thomsen-nouveau (STn) and were able to obtain a robust immune response to the antigen [67–71]. They have further reported total synthesis of the CPSA unit in 19 steps with a final yield of 5% [67]. Chemoenzymatic assembly is a faster and scalable approach, that can be used as an alternative or in combination with chemical synthesis. CPSA obtained in this way, can then be linked to the antigen. The chemoenzymatic method has also been used to create capsule polysaccharide based glycoconjugates for *Neisseria meningitidis* serotypes A, C and X [72–74]. In some cases, recombinant glycosyltransferases can be used to assemble non-native carbohydrate antigens in compliant host organisms like *Escherichia coli*. This method has been successfully used by the Brendan W. Wren lab for the in vivo assembly of capsular polysaccharide from several serotypes of *Streptococcus pneumoniae*. A similar approach is also currently being applied with respect to CPSA, wherein the whole CPSA biosynthesis and assembly will be done inside *E. coli*. This will allow to have access to longer oligomers of CPSA, which can be helpful in studies towards size requirement in eliciting immune response. So far there have been no reports of CPSA unit being polymerized synthetically.

4. Conclusion

CPSA molecue has a very common modification on its surface. Pyruvylation of sugars is fairly common yet an extensive search of the literature reveals little on successful isolations of an enzyme responsible for this sugar modification. However, very recently a family of genes has been identified that appear to be involved in pyruvate transfer reactions in prokaryotes. A publication in 2013 showed successful purification of pyruvyltransferase Pvg1p from the eukaryote *Schizosaccharomyces pombe*. This group demonstrated the activity of Pvg1p on beta-nitrophenyl galactose, a substrate analogue of galactose [54]. Apart from this eukaryotic pyruvyltransferase Pvg1p and the prokaryotic pyruvyltransferase PssM from *R. leguminosarum*, no other pyruvyltransferases have been characterized [55]. More studies are needed in uncovering this family of enzymes, and also a path needs to be elucidated towards the polymerization of CPSA, to reap its full therapeutic benefits.

Acknowledgements

The author acknowledges help of Dr. Swati Singh in the preparation of the manuscript.

Conflict of interest

The authors declare no conflict of interest.

Acronyms and abbreviations

AADGal	2-N-acetylamido-4-amino-6-deoxy-galactopyranose
BPP	bactoprenyl diphosphate
BP	bactoprenyl phosphate
CAZY	Carbohydrate-Active Enzymes database
CE	capillary electrophoresis
CPS	capsular polysaccharide
CPSA	capsular polysaccharide A
E. coli.	*Escherichia coli*
EAE	experimental autoimmune encephalomyelitis
HLA-DM	human leukocyte antigen DM

HLA-DR	human leukocyte antigen DR
HR-MS	high resolution mass spectrometry
GAG	glucosaminoglycan
Galf galactofuranose	
GalNAc	N-acetylgalactosesamine
GlcNAc	N-acetylglucosamine
Galp	galactopyranose
iNOS	inducible nitric oxide synthase
IPTG	isopropylthiogalactopyranoside
LC/MS	Liquid chromatography mass spectrometry
MALDI-MS	Matrix assisted laser desorption/ionization mass spectrometry
MHCII	major histocompatibility class II
PglF	a dehydratase
PHYRE2	Protein Homology/analogy Recognition Engine v 2.0
TACA	tumor-associated carbohydrate antigen
TCR	T-cell receptor
2AA-BP	2-amideaniline bactoprenyl monophosphate
2CNA-BP	2-nitrileaniline bactoprenyl monophosphate
4,6-pyr-Gal	4,6-pyruvate-galactose
4,6-pyr-Glc	4,6-pyruvate-glucose

Author details

Sunita Sharma
Sharda University, Greater Noida, India

*Address all correspondence to: sunita.sharma@sharda.ac.in

References

[1] Round JL, Mazmanian SK. The gut microbiota shapes intestinal immune responses during health and disease. Nat Rev Immunol. 2009;9(5):313-23.

[2] Kasper DL, Hayes ME, Reinap BG, Craft FO, Onderdonk AB, Polk BF. Isolation and identification of encapsulated strains of *Bacteroides fragilis*. J Infect Dis. 1977;136(1):75-81.

[3] Tzianabos AO, Kasper DL, Onderdonk AB. Structure and function of *Bacteroides fragilis* capsular polysaccharides: relationship to induction and prevention of abscesses. Clin Infect Dis. 1995;20 Suppl 2:S132-40.

[4] Tzianabos A, Kasper D, Onderdonk AB. Structure and Function of *Bacteroides fragilis* Capsular Polysaccharides: Relationship to Induction and Prevention of Abscesses. Clinical Infectious Diseases. 1995;20.

[5] Kalka-Moll WM, Tzianabos AO, Wang Y, Carey VJ, Finberg RW, Onderdonk AB, et al. Effect of molecular size on the ability of zwitterionic polysaccharides to stimulate cellular immunity. J Immunol. 2000;164(2): 719-24.

[6] Kalka-Moll WM, Wang Y, Comstock LE, Gonzalez SE, Tzianabos AO, Kasper DL. Immunochemical and biological characterization of three capsular polysaccharides from a single *Bacteroides fragilis* strain. Infect Immun. 2001;69(4):2339-44.

[7] Kalka-Moll WM, Tzianabos AO, Bryant PW, Niemeyer M, Ploegh HL, Kasper DL. Zwitterionic polysaccharides stimulate T cells by MHC class II-dependent interactions. J Immunol. 2002;169(11):6149-53.

[8] Krinos CM, Coyne MJ, Weinacht KG, Tzianabos AO, Kasper DL, Comstock LE. Extensive surface diversity of a commensal microorganism by multiple DNA inversions. Nature. 2001;414(6863):555-8.

[9] Pantosti A, Tzianabos AO, Reinap BG, Onderdonk AB, Kasper DL. Bacteroides fragilis strains express multiple capsular polysaccharides. J Clin Microbiol. 1993;31(7):1850-5.

[10] Pantosti A, Colangeli R, Tzianabos AO, Kasper DL. Monoclonal antibodies to detect capsular diversity among *Bacteroides fragilis* isolates. J Clin Microbiol. 1995;33(10):2647-52.

[11] Tzianabos AO, Onderdonk AB, Zaleznik DF, Smith RS, Kasper DL. Structural characteristics of polysaccharides that induce protection against intra-abdominal abscess formation. Infect Immun. 1994;62(11):4881-6.

[12] Troy EB, Kasper DL. Beneficial effects of *Bacteroides fragilis* polysaccharides on the immune system. Front Biosci. 2010;15:25-34.

[13] Kasper DL, Onderdonk AB, Crabb J, Bartlett JG. Protective efficacy of immunization with capsular antigen against experimental infection with *Bacteroides fragilis*. J Infect Dis. 1979;140(5):724-31.

[14] Choi YH, Roehrl MH, Kasper DL, Wang JY. A unique structural pattern shared by T-cell-activating and abscess-regulating zwitterionic polysaccharides. Biochemistry-Us. 2002;41(51):15144-51.

[15] Avci FY, Kasper DL. How bacterial carbohydrates influence the adaptive immune system. Annu Rev Immunol. 2010;28:107-30.

[16] Cobb BA, Kasper DL. Zwitterionic capsular polysaccharides: the new

MHCII-dependent antigens. Cell Microbiol. 2005;7(10):1398-403.

[17] Cobb BA, Kasper DL. Coming of age: carbohydrates and immunity. Eur J Immunol. 2005;35(2):352-6.

[18] Cobb BA, Wang Q, Tzianabos AO, Kasper DL. Polysaccharide processing and presentation by the MHCII pathway. Cell. 2004;117(5):677-87.

[19] Duan J, Avci FY, Kasper DL. Microbial carbohydrate depolymerization by antigen-presenting cells: deamination prior to presentation by the MHCII pathway. Proc Natl Acad Sci U S A. 2008;105(13):5183-8.

[20] Duan J, Kasper DL. Regulation of T cells by gut commensal microbiota. Curr Opin Rheumatol. 2011;23(4):372-6.

[21] Tzianabos AO, Gibson FC, 3rd, Cisneros RL, Kasper DL. Protection against experimental intraabdominal sepsis by two polysaccharide immunomodulators. J Infect Dis. 1998;178(1):200-6.

[22] Gloster TM. Advances in understanding glycosyltransferases from a structural perspective. Curr Opin Struct Biol. 2014;28:131-41.

[23] Roca C, Alves VD, Freitas F, Reis MA. Exopolysaccharides enriched in rare sugars: bacterial sources, production, and applications. Front Microbiol. 2015;6:288.

[24] Mazmanian SK. Capsular polysaccharides of symbiotic bacteria modulate immune responses during experimental colitis. J Pediatr Gastroenterol Nutr. 2008;46 Suppl 1:E11-2.

[25] Mostafavi AZ, Lujan DK, Erickson KM, Martinez CD, Troutman JM. Fluorescent probes for investigation of isoprenoid configuration and size discrimination by bactoprenol-utilizing enzymes. Bioorg Med Chem. 2013;21(17): 5428-35.

[26] Tzianabos AO, Pantosti A, Baumann H, Brisson JR, Jennings HJ, Kasper DL. The capsular polysaccharide of *Bacteroides fragilis* comprises two ionically linked polysaccharides. J Biol Chem. 1992;267(25):18230-5.

[27] Tzianabos AO, Pantosti A, Baumann H, Michon F, Brisson JR, Jennings HJ, et al. Structural characterization of two surface polysaccharides of *Bacteroides fragilis*. Trans Assoc Am Physicians. 1991;104:285-95.

[28] Baumann H, Tzianabos AO, Brisson JR, Kasper DL, Jennings HJ. Structural elucidation of two capsular polysaccharides from one strain of *Bacteroides fragilis* using high-resolution NMR spectroscopy. Biochemistry-Us. 1992;31(16):4081-9.

[29] Coyne MJ, Kalka-Moll W, Tzianabos AO, Kasper DL, Comstock LE. Bacteroides fragilis NCTC9343 produces at least three distinct capsular polysaccharides: cloning, characterization, and reassignment of polysaccharide B and C biosynthesis loci. Infect Immun. 2000;68(11):6176-81.

[30] Coyne MJ, Tzianabos AO, Mallory BC, Carey VJ, Kasper DL, Comstock LE. Polysaccharide biosynthesis locus required for virulence of *Bacteroides fragilis*. Infect Immun. 2001;69(7):4342-50.

[31] Altschul SF, Gish W, Miller W, Myers EW, Lipman DJ. Basic Local Alignment Search Tool Journal of Molecular Biology. 1990;215:7.

[32] Troy F. The chemistry and biosynthesis of selected bacterial capsular polymers. Annual Review of Microbiology. 1979;33.

[33] Mostafavi AZ, Troutman JM. Biosynthetic assembly of the *Bacteroides fragilis* capsular polysaccharide A precursor bactoprenyl diphosphate-linked acetamido-4-amino-6-deoxygalactopyranose. Biochemistry-Us. 2013;52(11):1939-49.

[34] Coyne MJ, Chatzidaki-Livanis M, Paoletti LC, Comstock LE. Role of glycan synthesis in colonization of the mammalian gut by the bacterial symbiont *Bacteroides fragilis*. Proc Natl Acad Sci U S A. 2008;105(35):13099-104.

[35] Olivier NB, Chen MM, Behr JR, Imperiali B. In Vitro Biosynthesis of UDP-N,N' Diacetylbacillosamine by Enzymes of the Campylobacter jejuni General Protein Glycosylation System. Biochemistry-Us. 2006;45:13659-69.

[36] Olivier NB, Chen MM, Behr JR, Imperiali B. In Vitro Biosynthesis of UDP-N,N' Diacetylbacillosamine by Enzymes of the Campylobacter jejuni General Protein Glycosylation System. Biochemistry-Us. 2006;45:13659-69.

[37] Lujan DK, Stanziale JA, Mostafavi AZ, Sharma S, Troutman JM. Chemoenzymatic synthesis of an isoprenoid phosphate tool for the analysis of complex bacterial oligosaccharide biosynthesis. Carbohydr Res. 2012;359:44-53.

[38] Wagner GK, Pesnot T. Glycosyltransferases and their assays. Chembiochem. 2010;11(14):1939-49.

[39] Lairson LL, Henrissat B, Davies GJ, Withers SG. Glycosyltransferases: structures, functions, and mechanisms. Annual review of biochemistry. 2008;77:521-55.

[40] Lombard V, Golaconda Ramulu H, Drula E, Coutinho PM, Henrissat B. The carbohydrate-active enzymes database (CAZy) in 2013. Nucleic Acids Res. 2014;42(Database issue):D490-5.

[41] Emmadi M, Kulkarni SS. Recent advances in synthesis of bacterial rare sugar building blocks and their applications. Nat Prod Rep. 2014;31(7):870-9.

[42] Soya N, Fang Y, Palcic MM, Klassen JS. Trapping and characterization of covalent intermediates of mutant retaining glycosyltransferases. Glycobiology. 2011;21(5):547-52.

[43] Kapitonov D, Yu RK. Conserved domains of glycosyltransferase. Glycobiology. 1999;9(10).

[44] Coutinho PM, Deleury E, Davies GJ, Henrissat B. An Evolving Hierarchical Family Classification for Glycosyltransferases. Journal of Molecular Biology. 2003;328(2):307-17.

[45] Troutman JM, Sharma S, Erickson KM, Martinez CD. Functional identification of a galactosyltransferase critical to *Bacteroides fragilis* Capsular Polysaccharide A biosynthesis. Carbohydr Res. 2014;395:19-28.

[46] Richards MR, Lowary TL. Chemistry and biology of galactofuranose-containing polysaccharides. Chembiochem. 2009;10(12):1920-38.

[47] Peltier P, Euzen R, Daniellou R, Nugier-Chauvin C, Ferrieres V. Recent knowledge and innovations related to hexofuranosides: structure, synthesis and applications. Carbohydr Res. 2008;343(12):1897-923.

[48] Guan S, Clarke AJ, Whitfield C. Functional analysis of the galactosyltransferases required for biosynthesis of D-galactan I, a component of the lipopolysaccharide O1 antigen of *Klebsiella pneumoniae*. J Bacteriol. 2001;183(11):3318-27.

[49] Nassau PM, Martin SL, Brown RE, Weston A, Monsey D, McNeil MR, et al.

Galactofuranose Biosynthesis in *Escherichia coli* K-12: Identification and Cloning of UDP-Galactopyranose Mutase. Journal of Bacteriology. 1996;178(4):6.

[50] Poulin MB, Nothaft H, Hug I, Feldman MF, Szymanski CM, Lowary TL. Characterization of a bifunctional pyranose-furanose mutase from Campylobacter jejuni 11168. J Biol Chem. 2010;285(1):493-501.

[51] Koplin R, Brisson JR, Whitfield C. UDP-galactofuranose Precursor Required for Formation of the Lipopolysaccharide O Antigen of *Klebsiella pneumoniae* Serotype O1 Is Synthesized by the Product of the rfbDKPO1 Gene. Journal of Biological Chemistry. 1997;272(7):8.

[52] Lee R, Monsey D, Weston A, Duncan K, Rithner C, McNeil M. Enzymatic Synthesis of UDP-Galactofuranose and an Assay for UDP-Galactopyranose Mutase Based on High-Performance Liquid Chromatography. Analytical Biochemistry. 1996;242:7.

[53] Cava F, de Pedro MA, Schwarz H, Henne A, Berenguer J. Binding to pyruvylated compounds as an ancestral mechanism to anchor the outer envelope in primitive bacteria. Mol Microbiol. 2004;52(3):677-90.

[54] Ahlgren JA. Purification and Characterization of a Pyruvated-Mannose-Specific Xanthan Lyase from Heat-Stable, Salt-Tolerant Bacteria. Appl Environ Microbiol. 1991;57(9):2523-8.

[55] Forsberg LS, Abshire TG, Friedlander A, Quinn CP, Kannenberg EL, Carlson RW. Localization and structural analysis of a conserved pyruvylated epitope in *Bacillus anthracis* secondary cell wall polysaccharides and characterization of the galactose-deficient wall polysaccharide from avirulent *B. anthracis* CDC 684. Glycobiology. 2012;22(8):1103-17.

[56] Mesnage S, Fontaine T, Mignot T, Delepierre M, Mock M, Fouet A. Bacterial SLH domain proteins are non covalently anchored to the cell surface via a conserved mechanism involving wall polysaccharide pyruvylation. The EMBO Journal. 2000;19(17): 4473-84.

[57] Hager FF, Lopez-Guzman A, Krauter S, Blaukopf M, Polter M, Brockhausen I, et al. Functional Characterization of Enzymatic Steps Involved in Pyruvylation of Bacterial Secondary Cell Wall Polymer Fragments. Front Microbiol. 2018;9:1356.

[58] Yoritsune K, Matsuzawa T, Ohashi T, Takegawa K. The fission yeast Pvg1p has galactose-specific pyruvyltransferase activity. FEBS Lett. 2013;587(7):917-21.

[59] Higuchi Y, Yoshinaga S, Yoritsune K, Tateno H, Hirabayashi J, Nakakita S, et al. A rationally engineered yeast pyruvyltransferase Pvg1p introduces sialylation-like properties in neo-human-type complex oligosaccharide. Sci Rep. 2016;6:26349.

[60] Yang FL, Yang YL, Liao PC, Chou JC, Tsai KC, Yang AS, et al. Structure and immunological characterization of the capsular polysaccharide of a pyrogenic liver abscess caused by *Klebsiella pneumoniae*: activation of macrophages through Toll-like receptor 4. J Biol Chem. 2011;286(24):21041-51.

[61] Ivashina TV, Fedorova EE, Ashina NP, Kalinchuk NA, Druzhinina TN, Shashkov AS, et al. Mutation in the pssM gene encoding ketal pyruvate transferase leads to disruption of Rhizobium leguminosarum bv. viciae-Pisum sativum symbiosis. J Appl Microbiol. 2010;109(2):731-42.

[62] Ivashina TV, Ksenzenko VN. Exopolysaccharide Biosynthesis in Rhizobium leguminosarum: From Genes to Functions. 2012.

[63] Ziegler T. Pyruvated Saccharides - Novel Strategies for Oligosaccharide Synthesis.

[64] Mazmanian SK, Kasper DL. The love-hate relationship between bacterial polysaccharides and the host immune system. Nat Rev Immunol. 2006;6(11):849-58.

[65] Mazmanian SK, Liu CH, Tzianabos AO, Kasper DL. An immunomodulatory molecule of symbiotic bacteria directs maturation of the host immune system. Cell. 2005;122(1):107-18.

[66] Stingele F, Corthesy B, Kusy N, Porcelli SA, Kasper DL, Tzianabos AO. Zwitterionic polysaccharides stimulate T cells with no preferential V beta usage and promote anergy, resulting in protection against experimental abscess formation. J Immunol. 2004;172(3): 1483-90.

[67] Eradi P, Ghosh S, Andreana PR. Total Synthesis of Zwitterionic Tetrasaccharide Repeating Unit from *Bacteroides fragilis* ATCC 25285/NCTC 9343 Capsular Polysaccharide PS A1 with Alternating Charges on Adjacent Monosaccharides. Org Lett. 2018;20(15):4526-30.

[68] De Silva RA, Wang Q, Chidley T, Appulage DK, Andreana PR. Immunological Response from an Entirely Carbohydrate Antigen: Design of Synthetic Vaccines Based on Tn-PS A1 Conjugates. Journal of American Chemical Society. 2009;131:9622-3.

[69] Shi M, Kleski KA, Trabbic KR, Bourgault JP, Andreana PR. Sialyl-Tn Polysaccharide A1 as an Entirely Carbohydrate Immunogen: Synthesis and Immunological Evaluation. J Am Chem Soc. 2016;138(43):14264-72.

[70] Nishat S, Andreana PR. Entirely Carbohydrate-Based Vaccines: An Emerging Field for Specific and Selective Immune Responses. Vaccines (Basel). 2016;4(2).

[71] Ghosh S, Nishat S, Andreana PR. Synthesis of an Aminooxy Derivative of the Tetrasaccharide Repeating Unit of *Streptococcus dysgalactiae* 2023 Polysaccharide for a PS A1 Conjugate Vaccine. J Org Chem. 2016;81(11):4475-84.

[72] Fiebig T, Freiberger F, Pinto V, Romano MR, Black A, Litschko C, et al. Molecular cloning and functional characterization of components of the capsule biosynthesis complex of Neisseria meningitidis serogroup A: toward in vitro vaccine production. J Biol Chem. 2014;289(28):19395-407.

[73] Fiebig T, Romano MR, Oldrini D, Adamo R, Tontini M, Brogioni B, et al. An efficient cell free enzyme-based total synthesis of a meningococcal vaccine candidate. NPJ Vaccines. 2016;1:16017.

[74] McCarthy PC, Saksena R, Peterson DC, Lee CH, An Y, Cipollo JF, et al. Chemoenzymatic synthesis of immunogenic meningococcal group C polysialic acid-tetanus Hc fragment glycoconjugates. Glycoconj J. 2013;30(9):857-70.

Chapter 8

Flavonoids: Understanding Their Biosynthetic Pathways in Plants and Health Benefits

InnocentMary IfedibaluChukwu Ejiofor and Maria-Goretti Chikodili Igbokwe

Abstract

Flavonoids are polyphenolic compounds and are one of the most abundant secondary metabolites present in plants. They are found in almost all vegetables and fruits. Flavonoids are of essence to plants and to man as well, due to their Medicinal and Pharmaceutical importance. Explicit understanding of the biosynthetic pathway of flavonoids is very essential. This will provide a stepwise explanation of the processes and mechanisms through which different forms of flavonoids are synthesized in plants, including the enzyme(s) responsible for each step. The importance in plants, medicine and pharmacy, of all the product(s) of each step will be emphasized.

Keywords: flavonoid, biosynthesis, plant, phenylpropanoid

1. Introduction

Flavonoids represent an important class of natural products; mainly, they are of the family of secondary plant metabolites having a multi-phenolic structure, found commonly in fruits, vegetables and certain beverages. They have various favorable biochemical, and antioxidant effects associated with various diseases such as cancer, Alzheimer's disease (AD), atherosclerosis and other reported pharmacological effects [1–3]. Flavonoids are associated with a wide spectrum of health-promoting effects and are crucial component in various nutraceutical, pharmaceutical, medicinal and cosmetic applications. This broad spectrum of health-promoting effect is due to their antioxidative, anti-inflammatory, anti-mutagenic and anti-carcinogenic properties coupled with their ability for cellular enzyme functions modulation [4].

In discussing the understanding of biosynthesis of flavonoids and their health benefits, we will be looking at it based on the sub headings of the enzymes involved in the biosynthetic pathway.

2. Phenylalanine ammonia lyase

Lyases are group of enzymes that catalyzes the removal of a functional group or a moiety from a compound by cleaving a carbon–carbon, carbon–oxygen, phosphorous-oxygen, and carbon-nitrogen bonds by mechanism of reaction other than hydrolysis, or oxidation. This removal of a functional group or a moiety often leads to the formation of a new double bond or ring structure. The bonds cleaving usually occur by means of elimination reaction [5]. In the synthesis of flavonoids, an ammonia-lyase, Phenylalanine ammonia-lyase (PAL) is the primary enzyme in the pathway for the synthesis of phenol [6]. At the initial synthesis of flavonoids, the conversion of Phenylalanine to trans-cinnamic acid is catalyzed by Phenylalanine ammonia lyase through a mechanism of non-oxidative deamination [7] as shown below. This Phenylalanine that is deaminated is a product of another pathway, shikimic acid pathway. Shikimic acid pathway produces three amino acid; L-Tyrosine, L-Phenylalanine and L-Tryptophan. The phenylpropanoid pathway takes its root from the L-Phenylalanine generated from the shikimic acid pathway.

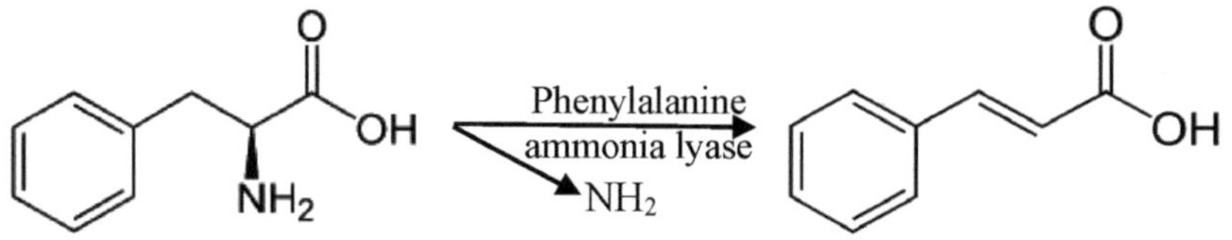

L-Phenylalanine *Trans*-cinnamic acid

Cinnamic acids belong to a group of aromatic carboxylic acids (C6–C3), which occurs naturally in plant kingdom. In the biochemical process that leads to the formation of lignin, which is the naturally occurring polymeric material that is responsible for providing mechanical support to plant cell wall, cinnamic acids are produced [8]. In all green plants cinnamic acids occur [9]. They are covalently bound to cell walls in minute quantities [10]. They are also found in the reproductive organs of flowering plants [11].

Coffee beans, tea, cocoa, apples, pears, berries, citrus, brassicas vegetables, spinach, beetroot, artichoke, potato, tomato, celery, faba beans, grape and cereals also contain cinnamic acids [12]. Cinnamic acids, with quinic acid, usually appears as conjugates known as the chlorogenic acids. With other acids, sugars or lipids they can also form esters. They can also with aromatic and aliphatic amines, form amides. Some cinnamic-related molecules have been shown in literature those possess the following pharmacological properties; anticancer [13], antituberculosis [14], antimalarial [15], antifungal [16], antimicrobial [17], antiatherogenic [18] and antioxidant [17] activities.

Also, various surveys directed towards the synthetic procedure for cinnamic acid preparation and related atoms have been shown in the literature [19–21]. Medicinal chemists have done the alteration of potency, permeability, solubility or other parameters of a preferred drug or pharmacophore with the aid of cinnamic acids [22].

Cinnamic acid exists in two isomeric forms; trans and cis. Most often available in nature and commercially is the trans form. Cinnamic acid can be obtained from cinnamon bark and balsam resins such as storax.

In the flavors, dyes, and pharmaceuticals production, trans-Cinnamic acid is utilized. The principal use of trans-cinnamic acid is in manufacturing of its methyl, ethyl, and benzyl esters, which are an essential component of perfumes. Also, in the production of the sweetener aspartame, the acid serves as a precursor [23].

3. Cinnamate-4-hydroxylase (C4H)

Hydroxylases are enzymes which add hydroxyl group to organic compounds. C4H found in plants is a cytochrome P450 that catalyzes trans-cinnamic acid conversion to

DOI: http://dx.doi.org/10.5772/intechopen.96715

p-coumaric acid and is the first hydroxylation step of lignin, flavonoids and hydroxycinnamic acid ester biosynthetic pathway correlating with lignifications [24–27]. Generally, in the synthesis of flavonoids, Cinnamate-4-hydroxylase catalyzes the addition of hydroxyl group to the trans-cinnamic acid generated from the deamination of the L-Phenylalanine, leading to the production of p-coumaric acid as shown below.

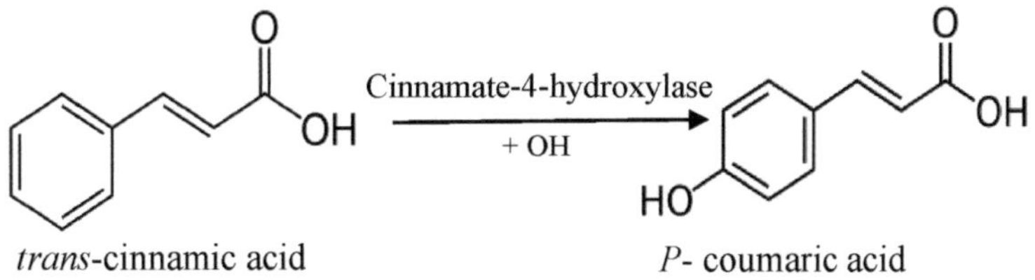

p-coumaric acid also known as 4-Hydroxycinnamic acid or p-Hydroxycinnamic acid, is a hydroxycinnamic acid, and also organic compound which is a hydroxy derivative of cinnamic acid. p-coumaric acid exists naturally in three isomers, namely; ortho-, meta- and para- coumaric acid [28].

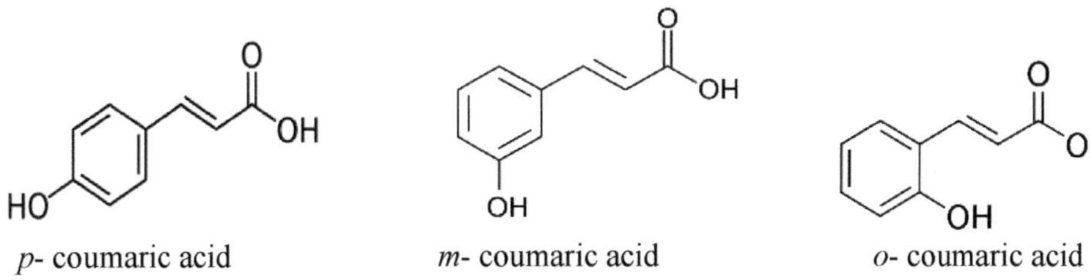

As can be observed from above, these isomers differ from each other differ by the position of the hydroxy group substitution on the phenyl group. p-coumaric acid (4-hydroxy-cinnamic acid) occurs widely in the cell walls of graminaceous plants and is the most abundant of the three isomers [29]. It decreases low-density lipoprotein (LDL) peroxidation [30, 31], antimicrobial activities [32, 33] and plays a vital role in human health. Coumaric acids have been shown to possess radical-scavenging effect [34–40] which reduces stomach cancer risk by suppressing carcinogenic nitrosamines formation [41–43].

p-coumaric acid effectively suppressed endothelial cell migration, tube formation, and rat aorta ring sprouting [44]. It reduces intracellular and mitochondrial reactive oxygen species production [44]. In vivo p-coumaric acid significantly suppressed tumor growth in vivo by blocking angiogenesis. p-coumaric acid is found in various edible plants, such as carrots, tomatoes and cereals [44].

4. 4-coumarate CoA ligase

Ligases are a class of enzymes capable of catalyzing the merging of two large compounds by forming a new chemical bond, mostly accompanied by a small chemical group's hydrolysis on one of the large compounds or commonly causing the linkage of two mixtures together. Ligases are classified under EC 6 primary class of enzymes. They are also further group into six subclasses which are known as ligases that create carbon–oxygen bonds, carbon-sulfur bonds, carbon-nitrogen bonds, carbon–carbon bonds, phosphoric–ester bonds, and nitrogen–metal bonds [45, 46].

Coenzyme A (CoA) is a type of coenzyme that contains pantothenic acid, adenosine 3-phosphate 5-pyrophosphate, and cysteamine; which take part in the transfer of acyl groups, notably in transacetylations [47]. Coenzymes can be defined as organic molecules or compounds that many enzymes required to elicit a catalytic effect [48].

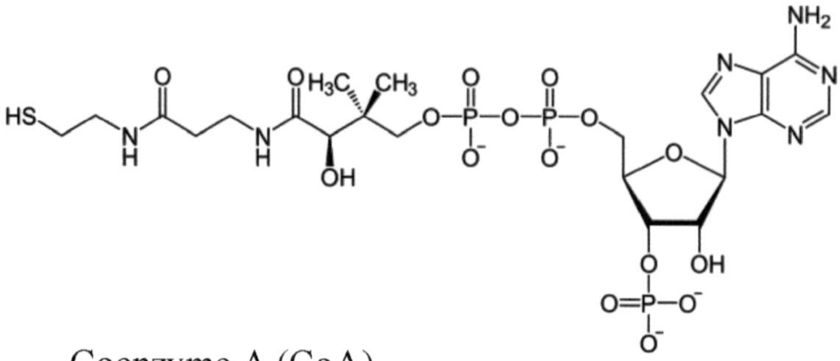

Coenzyme A (CoA)

4-coumarate-CoA ligase (4CL) is essential to the general phenylpropanoid pathway and takes part in monolignol biosynthesis through the production of p-coumaroyl-CoA, a precursor for the biosynthesis of p-coumaryl alcohol and coniferyl alcohol in conifers. Essentially, p-coumaroyl-CoA is also involved in the production of other metabolites of plant as a precursor, including stilbenes and flavonoids [49].

4-coumarate-CoA ligase causes the joining of Coenzyme A to *p*-coumaric acid, leading to the formation of *p*- coumaroyl-CoA.

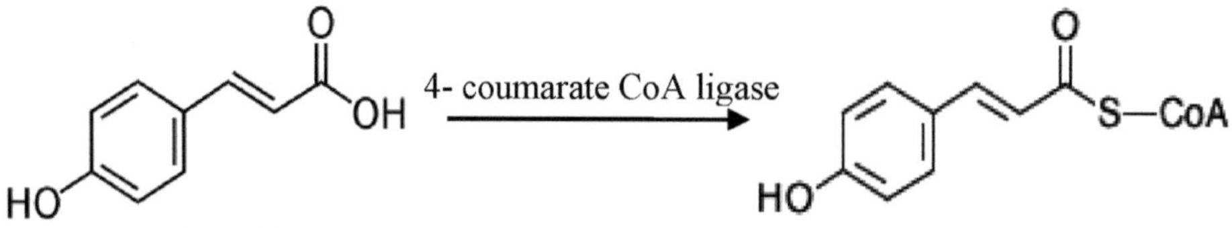

Coumaroyl-coenzyme A is a molecule or compound present in plants. It is the THIOESTER of coenzyme-A and coumaric acid. Coumaroyl-coenzyme A is a basic or fundamental intermediate in the biosynthesis of various natural products found in plants [50].

5. Chalcone synthase

Synthases are enzymes that catalyze the formation of a particular compound. Chalcone synthase catalyzes the production of chalcone, in phenylpropanoid metabolic pathway.

Chalcone synthase (CHS) or naringenin-chalcone synthase is a member of the plant polyketide synthase superfamily, which also includes stilbene synthase (STS), acridone synthase, pyrone synthase, bibenzyl synthase, and p-coumaroyl triacetic acid synthase [51]. Polyketides are a ubiquitous group of secondary metabolites which contain either alternating carbonyl and methylene groups (-CO-CH2-)

or are derived from precursors which have alternating carbonyl and methylene groups [52].

Chalcone synthases, the most well-known representatives of this family, make available the materials needed to initiate various sets of metabolites (flavonoids). These metabolites have important diverse role to play in flowering plants, like the provision of floral pigments, antibiotics, UV protectants and insect repellents [53].

In the production of chalcone, one molecule of p-coumaroyl-CoA and three malonyl-CoA molecules condensation is required. This is catalyzed by the enzyme, chalcone synthase. This process starts with the transfer of a coumaroyl moiety from a p-coumaroyl-CoA which is the starter molecule to an active site cysteine (Cys164) [54]. The next that follows is the series of condensation reactions involving three acetate units obtained from three malonyl-CoA molecules, each proceeding through an acetyl-CoA carbanion derived from malonyl-CoA decarboxylation, extends the polyketide intermediate. Following generation of the thioester-linked tetraketide, a regiospecific intramolecular Claisen condensation forms a new ring system to yield chalcone [55, 56].

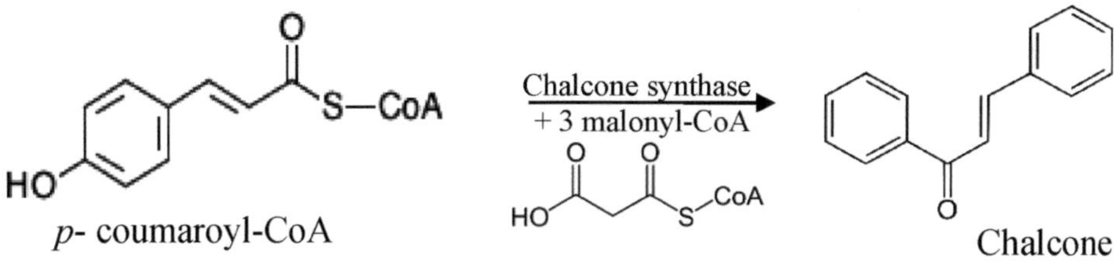

Malonyl-CoA is the starting molecule for the synthesis of fatty acid and its elongation. Malonyl-CoA is one of the foundations for the biosynthesis of some phytoalexins, flavonoids, and many malonylated compounds [57]. In plants and also in animals, malonyl-CoA is almost entirely obtained from acetyl-CoA by acetyl-CoA carboxylase [58].

Chalcone is an essential and resourceful molecule. It is a biogenetic precursor for flavonoids and isoflavonoids, which are bountiful in consumable plants. Chalcone contains two aromatic rings, linked together by a three-carbon-α, β unsaturated carbonyl system, i.e., 1,3-diphenyl-2-propen-1-one derivative. Structurally, chalcones are one of the most divergent forms of flavonoids. Chalcone derivatives exhibit a wide range of therapeutic activities which include anticancer [58–62] antioxidants [63–67], anti-inflammatory [68–73], antihypertensive [74], antimalarial [75], antiulcer [76, 77], antiviral [78–81], antiprotozoal [82], cardiovascular activity [83] and mutagenic properties [84], and many other pharmacological properties.

6. Chalcone isomerase

Isomerases are class of EC 5 primary enzymes that catalyze the reactions that involve the rearrangement of a molecule structure [46].

The first detected flavonoid pathway enzyme was Chalcone isomerase. It catalyzes the stereospecific cyclization of chalcones to (2S)-flavanones, which were found to be the exclusive substrates for the reactions to the formation of other classes of flavonoids [85].

Although this type of isomerization reaction can go on spontaneously, the rate of turnover can be increased by 10^7 fold in the presence of CHI [86]. The CHIs present in plants can be divided into four types (type I to type IV). This division depends on its phylogenetic relationships [87]. Type I and type II are considered bona fide catalysts with representative CHI enzymatic activity. Type I CHIs are mostly found in vascular plants; they are responsible for forming general flavonoids [88, 89].

In comparison with Type I CHIs, type II CHIs have wider substrate acceptability, besides making use of naringenin chalcone as substrate, they also undertake the conversion of isoliquiritigenin to isoflavonoid which appear to be the specific metabolites in legume [90, 91]. Both type III and type IV CHIs do not participate in chalcone cyclization activity, unlike type I and type II CHI proteins. Due to this, they are termed CHI-like proteins (CHIL). Type III CHIs, which is extensively dispersed in land plants and green algae, have been shown to be fatty acid-binding proteins that function to influence the synthesis and storage of fatty acid in plants [92]. Nonetheless, the action of type IV CHIs which completely lose the bona fide CHI activity is not well known, yet new studies have revealed type CHI-fold proteins might serve as the enhancer of colouration of flowers and production of flavonoid in diverse plant species [93]. All CHIs have a similar backbone arrangement and type III CHIs are thought to be the common forebear of bona fide CHIs [92, 94]. CHI, also regarded as chalcone flavonone isomerase.

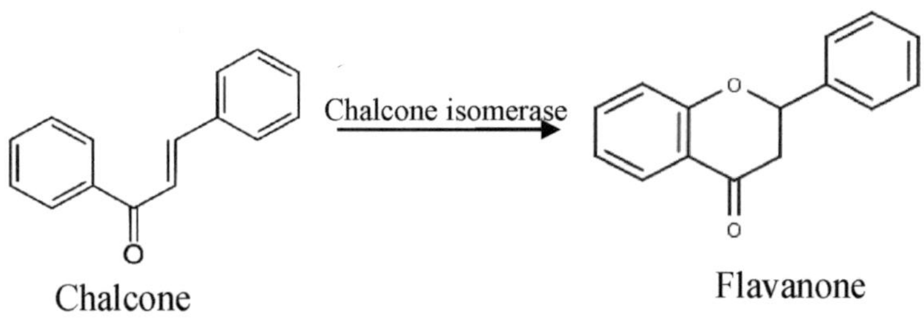

Flavanones are primarily found in about 42 larger plant families, specifically in Compositae, Leguminosae, and Rutaceae. Depending on the type of plants, flavanones can be found in all of the parts above and below ground, from vegetative parts to generative organs: branches, bark, stem, leaves, roots, flowers, fruits, seeds, rhizomes, peels, and others [95].

Flavanones show strong antioxidant and radical scavenging activity [96–103] and appear to be associated with a reduced risk of certain chronic diseases [98, 99] the prevention of some cardiovascular disorders [104–107] and certain kinds of cancer [108–112]. Flavanones also exhibit antiviral [113], antimicrobial, [114] and anti-inflammatory activities, [115] beneficial effects on capillary fragility, [116] and an ability to inhibit human platelet aggregation, [117] anti-ulcer [118, 119] and anti-allergenic [120] properties.

From flavanones other classes of flavonoids are biosynthesized with the aid of specific enzymes.

a. **Flavones**

Flavones are biosynthesized from flavanone with the aid of flavone synthase as catalyst. This enzyme catalyzes a double bond formation between C2 and C3 of flavanones.

Two FNS (I and II) enzyme systems have been described in dicots for flavone biosynthesis. FNSI is a soluble 2-oxoglutarate-dependent dioxygenase (2-ODD), while FNSII is a cytochrome P-450-dependent monooxygenase enzyme system. FNSI and most FNSII enzymes convert flavanones to flavones directly [121].

Flavones are one of the largest subgroups of flavonoids [121]. Flavones are involved in various interactions with microbes, insects, and other plants [122–124]. In addition to their extensive functions in plants' biochemistry and physiology, flavones are also essential for human nutrition and health [121, 125]. Their pharmacological effects, such as antioxidant, antiviral, anti-inflammatory activities and potential, have made these compounds increasingly popular as dietary constituents or supplements [121].

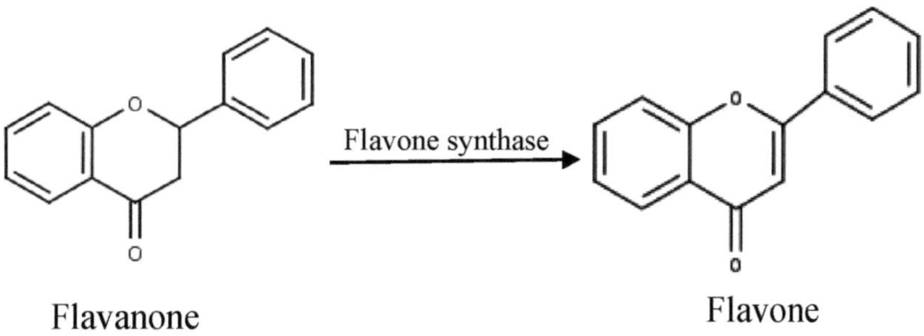

b. **Flavonol**

Flavonols are a major class of the family of flavonoids, molecules that have interesting biological activity such as antioxidant, antimicrobial, hepatoprotective, anti-inflammatory, and vasodilatation effects, and they have been considered as potential anticancer agents [126, 127]. Examples of flavonol include fisetin, quercetin, kaempferol, myricetin etc.

The biosynthetic pathway for the synthesis of flavanols is shown below.

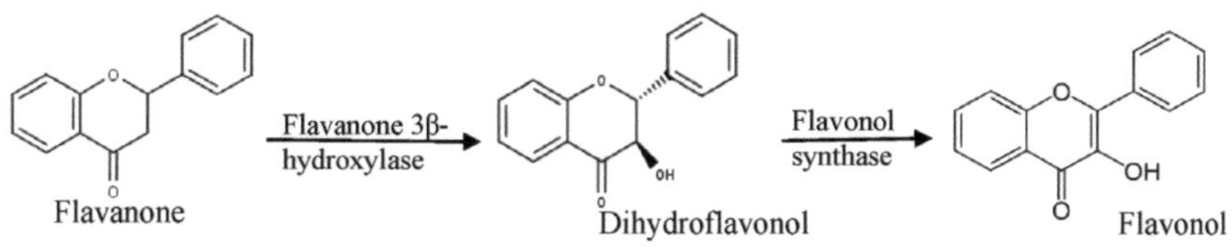

c. **Isoflavones**

Isoflavones are a polyphenol class usually found in legumes, including soybeans, chickpeas, fava beans, pistachios, peanuts, and other fruits and nuts [128]. Soybeans are the richest source of isoflavones, and soy foods and ingredients contain varying concentrations of isoflavones [129]. Isoflavones can be biosynthesized from flavanone with the aid of isoflavone synthase as a catalyst, as shown below.

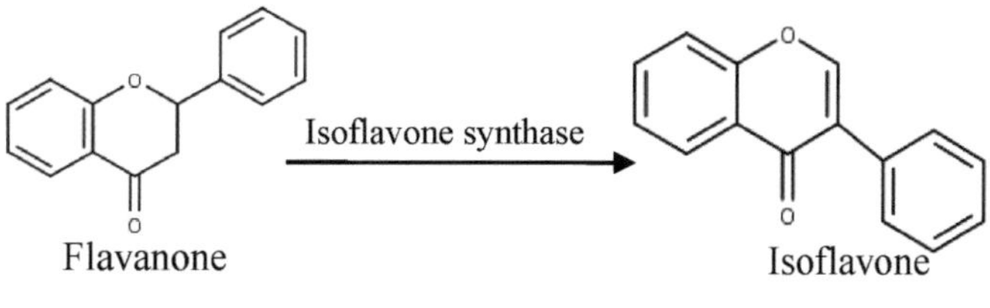

Common isoflavones include daidzin, genistin, biochanin A, and formononetin [130]. Isoflavones exhibit antioxidant, anticancer, antimicrobial, anti-inflammatory, antiosteoporotic, and estrogenic properties [131–136]. Several studies have also shown that isoflavonoids may contribute to other multiple additional health benefits by reducing cardiovascular risk, osteoporosis, and decreasing the intensity of bone resorption [137].

d. **Anthocyanidins**

Anthocyanidins are a group of phytochemicals, which are natural pigments responsible for blue, red, purple and orange colors present in many fruits and vegetables and many fruits- and vegetable-based food products. Over and above 500 different anthocyanidins are known and described in the literature [138, 139]. This flavonoid group dominates teas, honey, fruits, vegetables, nuts, olive oil, cocoa and cereals. They can also be found in berries (e.g. black currant, blueberries, strawberries, elderberries), their juices, as well as red wine [140]. Cyanidin, pelargonidin, delphinidin, malvidin, petunidin and peonidin are the most common anthocyanidins present in fruits and vegetables.

The number and position of the hydroxyl and methoxyl moiety is the determinant of different types of anthocyanidins [141]. Anthocyanidins have been reported to have some essential pharmacological role in cardiovascular disease, cholesterol decomposition, visual acuity, as well as antioxidant efficacy, and cytotoxicity [142]. Anthocyanidins can be synthesized as shown below.

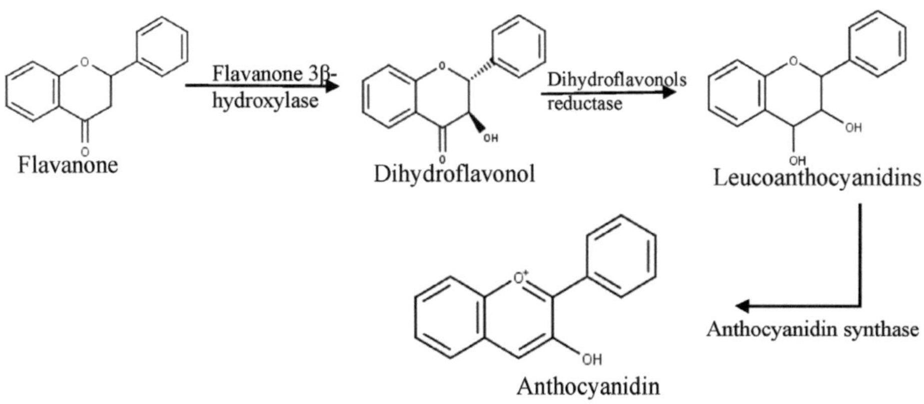

e. **Flavan-3-ol**

The most common flavonoids in the diet, flavan-3-ols are considered functional ingredients of beverages, fruits and vegetables, food grains, herbal remedies, dietary supplements, and dairy products. Flavan-3-ols have been reported to exhibit several pharmacological effects by acting as an antioxidant, anticarcinogen, cardio-preventive, antimicrobial, anti-viral, and neuroprotective agents [143]. Flavan-3-ol can be synthesized as shown below, from dihydroflavanol.

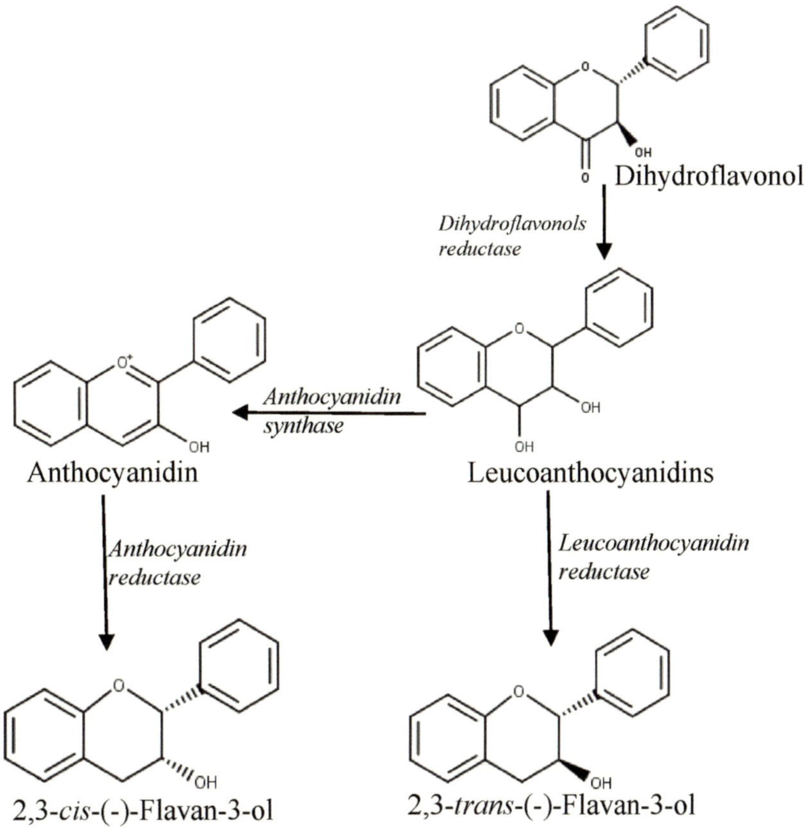

Flavonoids are an essential group of secondary metabolites that play so much role for the benefit of plants in which they exist or in their surroundings and for the health benefit of humanity. Understanding the biosynthetic processes of flavonoids and their pharmacological effects will aid proper utilization of flavonoids for health benefits and also draw more attention to the development of a synthetic laboratory process for the synthesis of flavonoid classes.

Author details

InnocentMary IfedibaluChukwu Ejiofor* and Maria-Goretti Chikodili Igbokwe
Department of Pharmacognosy and Traditional Medicine,
Faculty of Pharmaceutical Sciences, Nnamdi Azikiwe University,
Awka, Anambra State, Nigeria

*Address all correspondence to: ii.ejiofor@unizik.edu.ng

References

[1] Burak M & Imen Y. Flavonoids and their antioxidant properties. Turkiye Klin Tip Bil Derg, 1999, 19, 296-304.

[2] Ovando C, Hernandez D, Hernandez E, et al. Chemical studies of anthocyanins: a review. Food Chem, 2009, 113, 859-871.

[3] Lee Y, Yuk D, Lee J, et al. Epigallocatechin-3-gallate prevents lipopolysaccharide-induced elevation of β-amyloid generation and memory deficiency. Brain Res, 2009, 1250, 164-174.

[4] Panche A. N., Diwan A. D., and Chandra S. R. Flavonoids: an overview. Journal of Nutritional Science, 2016, 5.

[5] Creative Enzymes. Lyase Introduction. (https://www.creative enzymes.com/resource/lyase-introduction 22.html, accessed on 01/02/2021

[6] G.W. Cheng, P.J. Breen, Activity of phenylalanine ammonia-lyase (PAL) and concentrations of anthocyanins and phenolics in developing strawberry fruit, J. Amer. Soc. Hort. Sci. 116 (1991) 865-869.

[7] Yuri Dyakov, Vitaly Dzhavakhiya, Timo Korpela, 2007, Comprehensive and Molecular Phytopathology, Elsevier Science.

[8] Xu Z., Zhang D, Hu J, Zhou X, Ye X, Reichel K, Stewart N, Syrenne R, Yang X, Gao P, et al. Comparative genome analysis of lignin biosynthesis gene families across the plant kingdom. BMC Bioinform, 2009, 10, S3.

[9] Lichtenthaler, H.K.; Schweiger, J. Cell wall bound ferulic acid, the major substance of the blue-green fluorescence emission of plants. J. Plant Physiol. 1998, 152, 272-282.

[10] Kroon P.A., Williamson G., Hydroxycinnamates in plants and food: current and future perspectives. J. Sci. Food Agric. 1999, 79, 355-361.

[11] Martin-Tanguy J, Cabanne F, Perdrizet E. Martin, C. The distribution of hydroxycinnamic acid amides in flowering plants. Phytochemistry 1978, 17, 1927-1928.

[12] Clifford M.N. Chlorogenic acids and other cinnamates—nature, occurrence and dietary burden. J. Sci. Food Agric. 1999, 79, 362-372.

[13] De P., Baltas M., Bedos-Belval F. Cinnamic acid derivatives as anticancer agents-a review. Curr. Med. Chem. 2011, 18, 1672-1703.

[14] De P., Veau D., Bedos-Belval F., Chassaing S., Baltas M. Cinnamic derivatives in tuberculosis. In Understanding Tuberculosis—New Approaches to Fighting against Drug Resistance; Cardona, P.-J., Ed.; InTech Publishing: Rijeka, Croatia, 2012; Chapter 15.

[15] Wiesner J., Mitsch A., Wißner P., Jomaa H., Schlitzer M. Structure–activity relationships of novel anti-malarial agents. Part 2: Cinnamic acid derivatives. Bioorg. Med. Chem. Lett. 2001, 11, 423-424.

[16] Tawata S.; Taira S., Kobamoto N., Zhu J., Ishihara M.; Toyama S. Synthesis and antifungal activity of cinnamic acid esters. Biosci. Biotechnol. Biochem. 1996, 60, 909-910.

[17] Sova M. Antioxidant and anti-microbial activities of cinnamic acid derivatives. Mini Rev. Med. Chem. 2012, 12, 749-767.

[18] Lapeyre C., Delomenède M., Bedos-Belval F., Duran H., Nègre-Salvayre A., Baltas M. Design, synthesis, and evaluation of pharmacological properties of cinnamic

derivatives as antiatherogenic agents. J. Med. Chem. 2005, 48, 8115-8124.

[19] Simonyan A. Activity of cinnamic acid derivatives and new methods for their synthesis (review). Pharm. Chem. J. 1993, 27, 92-100.

[20] Wall V.M., Eisenstadt A., Ager D. J., Laneman S.A. The Heck reaction and cinnamic acid synthesis by heterogeneous catalysis. Platin. Metals Rev. 1999, 43, 138-144.

[21] Sharma P. Cinnamic acid derivatives: A new chapter of various pharmacological activities. J. Chem. Pharm. Res. 2011, 3, 403-423.

[22] Guzman J. D. Natural Cinnamic Acids, Synthetic Derivatives and Hybrids with Antimicrobial Activity. Molecules 2014, 19, 19292-19349.

[23] American Chemical Society. Molecule of the Week Archive, *trans*-Cinnamic acid, 2014. https://www.acs.org/content/acs/en/molecule-of-the-week/archive/c/trans-cinnamicacid.html, accessed on 10/01/2021.

[24] Koopmann E., Logemannm, E. & Hahlbrock K. Regulation and functional expression of cinnamate 4-hydroxylase from parsley. Plant Physiol., 1999, 119, 49-56.

[25] Blount J. W. et al. Altering Expression of Cinnamic Acid 4-Hydroxylase in Transgenic Plants Provides Evidence for a Feedback Loop at the Entry Point into the Phenylpropanoid Pathway. Plant Physiol., 2000, 122, 107-116.

[26] Blee K. A. et al. Antisense and sense expression of cDNA coding for CYP73A15, a class II cinnamate 4-hydroxylase, leads to a delayed and reduced production of lignin in tobacco. Phytochemistry, 2001, 57, 1159-1166.

[27] Ro D. K., Mah N., Ellis B. E. & Douglas C. J. Functional characterization and subcellular localization of poplar (Populus trichocarpa£Populus deltoids) cinnamate 4-hydroxylase. Plant Physiol., 2001, 126, 317-329.

[28] p-Coumaric acid' 2021. *Wikipedia*. Available at https://en.wikipedia.org/wiki/P-Coumaric_acid, Accessed : 02 February 2021.

[29] Huseyin Boz. p-Coumaric acid in cereals: presence, antioxidant and antimicrobial effects, International Journal of Food Science and Technology 2015, 50, 2323-2328.

[30] Garrait G., Jarrige J. F., Blanquet S., Beyssac E., Cardot J.M. & Alric M. Gastrointestinal absorption and urinary excretion of trans-cinnamic and p-coumaric acids in rats. Journal of Agriculture and Food Chemistry, 2006, 54, 2944-2950.

[31] Zilic S., Sukalovic V. H-T., Dodig D., Maksimovic V., Maksimovic M. & Basic Z. Antioxidant activity of small grain cereals caused by phenolics and lipid soluble antioxidants. Journal of Cereal Science, 2011, 54, 417-424.

[32] Lou Z., Wang H., Rao S., Sun J., Ma C. & Li J. p-Coumaric acid kills bacteria through dual damage mechanisms. Food Control, 2012, 25, 550-554.

[33] Sanchez-Maldonado A. F., Schieber A. & G€anzle M.G. Structure-function relationships of the antibacterial activity of phenolic acids and their metabolism by lactic acid bacteria. Journal of Applied Microbiology, 2011, 111, 1176-1184.

[34] Halliwell B. & Ahluwalia S. Hydroxylation of p-coumaric acid by horseradish peroxidase: the role of superoxide and hydroxylradicals. Biochemical Journal, 1976, 153, 513-518.

[35] Castelluccio C., Paganga G., Melikian N. et al. Antioxidant potential

of intermediates in phenylpropanoid metabolism in higher plants. FEBS Letters, 1995, 368, 188-192.

[36] Zang L.Y., Cosma G., Gardner H., Shi X., Castranova V. & Vallyathan V. Effect of antioxidant protection by p-coumaric acid on low-density lipoprotein cholesterol oxidation. American Journal of Physiology – Cell Physiology, 2000, 279, 954-960.

[37] Rao R. S. P. and Muralikrishna G. Non-starch polysaccharidephenolic acid complexes from native and germinated cereals and millet. Food Chemistry, 2004, 84, 527-531.

[38] Kovacova M. & Malinova E. Ferulic and coumaric acids, total phenolic compounds and their correlation in selected oat genotypes. Czech Journal of Food Science, 2007, 25, 325-332.

[39] Gani A., Wani S.M., Masoodi F.A. & Hameed G. Wholegrain cereal bioactive compounds and their health benefits: a review. Journal of Food Process and Technology, 2012, 3, 146.

[40] Yoon S.-A., Kang S.-I., Shin H.-S. et al. p-Coumaric acid modulates glucose and lipid metabolism via AMP-activated protein kinase in L6 skeletal muscle cells. Biochemical and Biophysical Research Communications, 2013, 432, 553-557.

[41] Biswick T., Park D., Shul Y.G. & Choy J.H. p- Coumaric acid-zinc basic salt nanohybrid for controlled release and sustained antioxidant activity. Journal of Physics and Chemistry of Solids, 2010, 71, 647-664.

[42] Shabir G., Anwar F., Sultana B. et al. Antioxidant and antimicrobial attributes and phenolics of different solvent extracts from leaves, flowers and bark of gold mohar [*Delonix regia* (Bojerex Hook.) Raf.]. Molecules, 2011, 16, 7302-7319.

[43] Navaneethan D. & Rasool M. p-Coumaric acid, a common dietary polyphenol, protects cadmium chloride-induced nephrotoxicity in rats. Renal Failure, 2014, 36, 244-251.

[44] Kong C. S., Jeong C. H., Choi J. S., Kim K. J. and Jeong J. W. Antiangiogenic Effects of P-Coumaric Acid in Human Endothelial Cells. Phytother. Res. 2012.

[45] Lehninger AL, Nelson DL, Cox MM, editors. Lehninger's Principles of Biochemistry, 4 ed., New York, NY, USA, 2004.

[46] Manubolu M., Goodla L., Pathakoti K. and Malmlo¨f K. (2018). Enzymes as direct decontaminating agents—mycotoxins. In C. S. Nunes and V. Kumar (Eds). *Enzymes in Human and Animal Nutrition* (pp. 312-330). USA: Elsevier Academic press.

[47] National Cancer Institute Thesaurus. Coenzyme A. https://ncit.nci.nih.gov/ncitbrowser/ConceptReport.jsp?dictionary=NCI_Thesaurus&ns=NCI_Thesaurus&code=C384, Accessed : 02 February 2021.

[48] Palmer T. and Bonner P. L. (2007). Enzymes Biochemistry, Biotechnology, Clinical Chemistry (pp. 204-220). United Kingdom: Woodhead Publishing Limited.

[49] Boudet A. Evolution and current status of research in phenolic compounds. Phytochemistry, 2007, 68, 2722-2735.

[50] Vogt, T. Phenylpropanoid Biosynthesis. Molecular Plant, 2010, 3(1), 2-20.

[51] Sanchez IJF. (2008) Polyketide synthase in *Cannabis sativa* L. PhD thesis, Leiden University, Leiden, The Netherlands.

[52] IUPAC, Compendium of Chemical Terminology, 2nd ed. (the "Gold Book") (1997). Online corrected version: (2006) "Polyketides.

[53] Hahlbrock K., Scheel D. (1989) Physiology and molecular biology of phenylpropanoid metabolism. Annu Rev Plant Physiol Plant Mol Biol 40:347-369.

[54] Lanz T., Tropf S., Marner FJ, Schröder J, Schröder G. The role of cysteines in polyketide synthases: site-directed mutagenesis of resveratrol and chalcone synthases, two key enzymes in different plant-specific pathways. J Biol Chem, 1991, 266, 9971-9976.

[55] Hahlbrock K., Zilg H., Grisebach H. Stereochemistry of the enzymatic cyclisation of 4,2′, 4′-trihydroxychalcone to 7,4′-dihydroxyflavanone by isomerases from Mung Bean seedlings. Eur J Biochem, 1970. 15, 13-18.

[56] Jez JM, Austin MB, Ferrer JL, Bowman ME, Schröder J, Noel JP. Structural control of polyketide formation in plant-specific polyketide synthases. Chem Biol, 2000, 40, 1-12.

[57] Chen H., Kim H. U., Weng H., and Browse J. Malonyl-CoA Synthetase, Encoded by ACYL ACTIVATING ENZYME13, is Essential for Growth and Development of Arabidopsis. The Plant Cell, 2011, 23, 2247-2262.

[58] Wu W., Ye H., Wan L., Han X., Wang G., Hu J., Tang M., Duan X., Fan Y., He S., Huang L. Millepachine, a novel chalcone, induces G 2/M arrest by inhibiting CDK1 activity and causing apoptosis via ROS-mitochondrial apoptotic pathway in human hepatocarcinoma cells in vitro and in vivo. Carcinogenesis, 2013, 34, 1636-1643.

[59] Shibata S., Inoue H., Iwata S., Ma R., Yu L., Ueyama H., Takayasu J., Hasegawa T., Tokuda H., Nishino A., Nishino H. (1991) Inhibitory efects of licochalcone A isolated from Glycyrrhiza infata root on infammatory ear edema and tumor promotion in mice. Planta Med, 1991, 57, 221-224.

[60] Edwards ML, Stemerick DM, Sunkara PS. Chalcones: a new class of antimitotic agents. J Med Chem, 1990, 33, 1948-1954.

[61] Wang Y., Chan FL, Chen S., Leung LK (2005) The plant polyphenol butein inhibits testosterone-induced proliferation in breast cancer cells expressing aromatase. Life Sci, 2005 77, 39-51.

[62] Yit CC, Das NP. Cytotoxic efect of butein on human colon adenocarcinoma cell proliferation. Cancer Lett, 1994, 82, 65-72.

[63] Anto RJ, Sukumaran K., Kuttan G., Rao MNA, Subbaraju V., Kuttan R. Anticancer and antioxidant activity of synthetic chalcones and related compounds. Cancer Lett, 1995, 97, 33-37.

[64] Hatano T., Takagi M., Ito H., Yoshida T. (1997) Phenolic constituents of liquorice. VII. A new chalcone with a potent radical scavenging activity and accompanying phenolics from liquorice. Chem Pharm Bull, 1997, 45, 1485-1492.

[65] Koorbanally NA, RandrianarivelojosiaM., MulhollandDA, van Uford LQ, van den Berg AJ. Chalcones from the seed of *Cedrelopsis grevei* (Ptaeroxylaceae). Phytochemistry, 2003, 62:1225-1229.

[66] Cioffi G., Escobar LM, Braca A., De Tommasi N. (2003) Antioxidant chalcone glycosides and favanones from *Maclura* (Chlorophora) *tinctoria*. J Nat Prod, 2003, 66, 1061-1064.

[67] Doan TN, Tran DT. Synthesis, antioxidant and antimicrobial activities of a novel series of chalcones, pyrazolic chalcones, and allylic chalcones. Pharmacol Pharm, 2011, 2, 282.

[68] Ballesteros JF, Sanz MJ, Ubeda A., Miranda MA, Iborra S., Paya M., Alcaraz MJ. Synthesis and pharmaco-logical evaluation of 2′-hydroxychalcones

and favones as inhibitors of inflammatory mediators generation. J Med Chem, 1995, 38, 2794-2797.

[69] Hirai S., Kim YI, Goto T., Kang MS, Yoshimura M., Obata A., Yu R., Kawada T. Inhibitory effect of naringenin chalcone on inflammatory changes in the interaction between adipocytes and macrophages. Life Sci, 2007, 81, 1272-1279.

[70] Funakoshi TM, Okamoto K., Izumi R., Tago K., Yanagisawa K., Narukawa Y., Kiuchi F., Kasahara T., Tamura H. Anti-infammatory activity of favonoids in Nepalese propolis is attributed to inhibition of the IL-33 signaling pathway. Int Immunopharmacol, 2015, 25, 189-198.

[71] Yang EB, Zhang K., Cheng LY, Mack P. Butein, a specifc protein tyrosine kinase inhibitor. Biochem Biophys Res Commun, 1998, 245, 435-438.

[72] Zhao F., Nozawa H., Daikonnya A., Kondo K., Kitanaka S. Inhibitors of nitric oxide production from hops (*Humulus lupulus* L.). Biol Pharm Bull, 2003, 26, 61-65.

[73] Piramal Life Sciences Ltd. Preparation of heterocyclic benzene compounds as cytokine inhibitors for treating inflammatory disorders, 2011, WO2011/121505.

[74] Won SJ, Liu CT, Tsao LT, Weng JR, Ko HH, Wang JP, Lin CN. Synthetic chalcones as potential anti-infammatory and cancer chemopreventive agents. Eur J Med Chem, 2005, 40:103-112.

[75] Dominguez JN, Leon C., Rodrigues J., de Dominguez NG, Gut J., Rosenthal PJ. Synthesis and antimalarial activity of sulfonamide chalcone derivatives. Il Farmaco, 2005, 60, 307-311.

[76] Kyogoku K., Hatayama K., Yokomori S., Saziki R., Nakane S., Sasajima M., Sawada J., Ohzeki M., Tanaka I. Anti-ulcer efect of isoprenyl favonoids. II. Synthesis and anti-ulcer activity of new chalcones related to sophoradin. Chem Pharm Bull, 1979, 27, 2943-2953.

[77] Ming DS, Lopez A., Hillhouse BJ, French CJ, Hudson JB, Towers GN. Bioactive Constituents from Iryanthera megistophylla. J Nat Prod, 2002, 65, 1412-1416.

[78] Sasajima M., Nakane S., Saziki R., Saotome H., Hatayama K., Kyogoku K., Tanaka I. Studies on the anti-ulcer efects of isoprenyl favonoids (1). The anti-ulcer efects of isoprenyl chalcone extracted from Sophora subprostrata. Folia Pharmacol Japonica, 1978, 74, 897-905.

[79] Ishitsuka H., Ninomiya YT, Ohsawa C., Fujiu M., Suhara Y. Direct and specifc inactivation of rhinovirus by chalcone Ro09-0410. Antimicrob Agents Chemother, 1982, 22, 617-621.

[80] Kaul TN, Middleton E Jr, Ogra PL. Antiviral efect of flavonoids on human viruses. J Med Vir, 1985, 15, 71-79.

[81] Mustafa KA, Kjaergaard HG, Perry NB, Weavers RT. Hydrogenbonded rotamers of 2′,4′,6′-trihydroxy-3′-formyldihydrochalcone, an intermediate in the synthesis of a dihydrochalcone from *Leptospermum recurvum*. Tetrahedron, 2003, 59, 6113-6120.

[82] Chen M, Christensen SB, Blom J, Lemmich E, Nadelmann L, Fich K, Theander TG, Kharazmi A (1993) Licochalcone A, a novel antiparasitic agent with potent activity against human pathogenic protozoan species of Leishmania. Antimicrob Agents Chemother, 1993, 37, 2550-2556.

[83] Cheng ZJ, Lin CN, Hwang TL, Teng CM. Broussochalcone A, a potent antioxidant and efective suppressor of inducible nitric oxide synthase in lipopolysaccharide-activated

macrophages. Biochem Pharmacol, 2001, 61, 939-946.

[84] Rashid KA, Mullin CA, Mumma RO. Structure-mutagenicity relationships of chalcones and their oxides in the Salmonella assay. Mutat Res, 1986, 169, 71-79.

[85] Forkmann G. and Heller W. (1999). Polyketides and Other Secondary Metabolites Including Fatty Acids and Their Derivatives. In D. Barton, K. Nakanishi and O. Meth-Cohn (Eds) Comprehensive Natural Products Chemistry, Vol. 1 (pp. 714-743). Elsevier Science Ltd.

[86] Cheng A., Zhang X., Han X., Zhang Y., Gao S., Liu C., et al. Identification of chalcone isomerase in the basal land plants reveals an ancient evolution of enzymatic cyclization activity for synthesis of flavonoids. New Phytol, 2018, 217, 909-924.

[87] Ralston L., Subramanian, S., Matsuno, M., and Yu, O. Partial reconstruction of flavonoid and isoflavonoid biosynthesis in yeast using soybean type I and type II chalcone isomerases. Plant Physiol, 2005, 137, 1375-1388.

[88] Jez J. M., Bowman M. E., Dixon R. A., and Noel J. P. Structure and mechanism of the evolutionarily unique plant enzyme chalcone isomerase. Nat. Struct. Biol, 2000, 7, 786-791.

[89] Jez J. M., Bowman M. E., and Noel J. P. Role of hydrogen bonds in the reaction mechanism of chalcone isomerase. J. Biol. Chem, 2002, 41, 5168-5176.

[90] Dixon R. A., Blyden E. R., Robbins M. P., Tunen A. V., and Mol J. N. Comparative biochemistry of chalcone isomerases. Phytochemistry, 1988, 27, 2801-2808.

[91] Ngaki M. N., Louie G. V., Philippe R. N., Manning G., Pojer F., Bowman M. E., et al. Evolution of the chalcone-isomerase fold from fatty-acid binding to stereospecific catalysis. Nature, 2012, 485, 530-533.

[92] Morita Y., Takagi K., Fukuchi-Mizutani M., Ishiguro K., Tanaka Y., Nitasaka E., et al. (2014). A chalcone isomerase like protein enhances flavonoid production and flower pigmentation. Plant J, 2014, 78, 294-304.

[93] Das A. B., Goud V. V., Das C. (2019) Phenolic Compounds as Functional Ingredients in Beverages. In A. Grumezescu, A. M. Holban (Eds). The Science of Beverages; Value-added Ingredients and Enrichments of Beverages. Elsevier Science & Technology.

[94] Wang H, Cao G, Prior RL. Total antioxidant capacity of fruits. J Agric Food Chem 1996, 44(3), 701-5.

[95] Cao G, Sofic E, Prior RL. Antioxidant and prooxidant behavior of flavonoids: structure-activity relationships. Free Radic Biol Med, 1997, 22(5), 749-60.

[96] Pietta PG. Flavonoids as antioxidants. J Nat Prod, 2000, 63 (7):1035-42.

[97] Rice-Evans C. Flavonoid antioxidants. Curr Med Chem, 2001, 8 (7), 797-807.

[98] Kaur C, Kapoor HC. Antioxidants in fruits and vegetables-the millennium's health. Int J Food Sci Tech, 2001, 36(7), 703-25.

[99] Burda S, Oleszek W. Antioxidant and antiradical activities of flavonoids. J Agric Food Chem 2001;49(6):2774-9.

[100] Sa'nchez-Moreno C, Plaza L, De Ancos B, Cano MP. Quantitative bioactive compounds assessment and their relative contribution to the antioxidant capacity of commercial orange juices. J Sci Food Agric, 2003, 83(5), 430-9.

[101] Zhang HY, Yang DP, Tang GY. Multipotent antioxidants: from screening to design. Drug Discov Today, 2006, 11(15-16):749-54.

[102] Drewnowski A, Gomez-Carneros C. Bitter taste, phytonutrients and the consumer: a review. Am J Clin Nutr, 2000, 72 (6), 1424-35.

[103] Kris-Etherton PM, Lefevre M, Beecher GR, Gross MD, Keen CL, Etherton TD. Bioactive compounds in nutrition and healthresearch methodologies for establishing biological function: the antioxidant and anti-inflammatory effects of flavonoids on atherosclerosis. Ann Rev Nutr, 2004, 24, 511-38.

[104] Hertog MG. Flavonoid intake and long-term risk of coronary heart-disease and cancer in the seven countries study. Arch Intern Med, 1995, 155(4), 381-6.

[105] Knekt P, Jarvinen R, Reunanen A, Maatela J. Flavonoid intake and coronary mortality in Finland: a cohort study. Brit Med J, 1996, 312(7029), 478-81.

[106] Yochum L., Kushi LH, Meyer K., Folsom AR. Dietary flavonoid intake and risk of cardiovascular disease in postmenopausal women. Am J Epidemiol, 1999, 149(10), 943-9.

[107] Gross M. Flavonoids and cardio-vascular disease. Pharm Biol, 2004, 42(Suppl.), 21-35.

[108] Steinmetz KA, Potter JD. Vegetables, fruit, and cancer prevention: a review. J Am Diet Assoc, 1996, 96(10), 1027-39.

[109] Le Marchand L, Murphy SP, Hankin JH, Wilkens LR, Kolonel LN. Intake of flavonoids and lung cancer. J Natl Cancer Inst, 2000, 92(2), 154-60.

[110] Kris-Etherton PM, Hecker KD, Bonanome A, Coval SM, Binkoski AE, Hilpert KF, et al. Bioactive compounds in foods: their role in the prevention of cardiovascular disease and cancer. Am J Med, 2002, 113(Suppl. 9B), 71-88.

[111] Nichenametla SN, Taruscio TG, Barney DL, Exon JH. A review of the effects and mechanism of polyphenolics in cancer. Crit Rev Food Sci, 2006, 46(2), 161-83.

[112] Moon YJ, Wang X, Morris ME. Dietary flavonoids: effects on xenobiotic and carcinogen metabolism. Toxicol In Vitro, 2006, 20 (2), 187-210.

[113] Asres K, Seyoum A, Veeresham C, Bucar F, Gibbons S. Naturally derived anti-HIV agents. Phytother Res, 2005, 19(7), 557-81.

[114] Cushnie TPT, Lamb AJ. Anti-microbial activity of flavonoids. Int J Antimicrob Agent, 2005, 26(5), 343-56.

[115] Kim HP, Son KH, Chang HW, Kang SS. Anti-inflammatory plant flavonoids and cellular action mechanisms. J Pharmacol Sci, 2004, 96(3), 229-45.

[116] Benavente-Garcı'a O, Castillo J, Marı'n FR, Ortun˜o A, Del Rı'o JA.mUses and properties of Citrus flavonoids. J Agric Food Chem, 1997, 45(12), 4505-15.

[117] Tijburg LBM, Mattern T., Folts JD, Weisgerber UM, Katan MB. Tea flavonoids and cardiovascular diseases: a review. Crit Rev Food Sci, 1997, 37(8), 771-85.

[118] Borrelli F., Izzo AA. The plant kingdom as a source of anti-ulcer remedies. Phytother Res, 2000, 14(8), 581-91.

[119] Wightman JD. Red berries and their health benefits. In: Nutraceutical Beverages: Chemistry, Nutrition, and Health Effects (ACS Symp. Ser. 871), 2004, pp. 123-132.

[120] Middleton E, Kandaswami C. Effects of flavonoids on immune and

inflammatory cell functions. Biochem Pharmacol, 1992, 43(6), 1167-79.

[121] Martens S., & Mithöfer A. Flavones and flavone synthases. Phytochemistry, 2005, 66(20), 2399-2407.

[122] Fisher R. F. & Long S. R. Rhizobium-plant signal exchange. Nature, 1992, 357, 655-660.

[123] Simmonds M. S. Flavonoid-insect interactions: recent advances in our knowledge. Phytochemistry, 2003, 64, 21-30.

[124] Soriano I., Asenstorfer R., Schmidt O. and Riley I. Inducible flavone in oats (*Avena sativa*) is a novel defense against plant-parasitic nematodes. Phytopathology, 2004, 94, 1207-1214.

[125] Pandey A. et al. AtMYB12 expression in tomato leads to large scale differential modulation in transcriptome and flavonoid content in leaf and fruit tissues. Sci. Rep. 2015, -Uk 5.

[126] Cushnie T. P. T., & Lamb A. J. Antimicrobial activity of flavonoids. International Journal of Antimicrobial Agents, 2005, 26(5), 343-356.

[127] Erlund I. (2004). Review of the flavonoids quercetin, hesperetin, and naringenin. Dietary sources, bioactivities, bioavailability, and epidemiology. Nutrition Research, 24(10), 851-874.

[128] United States Department of Agriculture. (2008). A Database for the Isoflavone Content of Selected Foods, Release 2.0. September

[129] Thrane M., Paulsen P. V., Orcutt M. W. and Krieger T. M. (2017). Soy Protein: Impacts, Production, and Applications. In S. Nadathur, J. Wanasundara, L. Scanlin (Eds). *Sustainable Protein Sources* (pp. 23-45). Elsevier Inc.

[130] COT (committee on toxicity), Committee on toxicity of chemicals in food, consumer products and the environment. Phytoestrogens and health; 2003, London, UK: Crown Copyright. Available at http://www.food.gov.uk/science/ouradvisors/toxicity/COTwg/wg_phyto

[131] Jang C.H.; Lim J.K.; Kim J.H.; Park C.S.; Kwon D.Y.; Kim Y.S.; Shin D.H.; Kim J.S. Change of Isoflavone Content during Manufacturing of Cheonggukjang, a Traditional Korean Fermented Soyfood. Food Sci. Biotechnol. 2006, 15, 643-646.

[132] Iwasaki M., Inoue M., Otani T., Sasazuki S., Kurashi N., Miura T., Yamoto S., Tsugane, S. Plasma isoflavone level and subsequent risk of breast cancer among Japanese women: A nested case-control study from Japan Public Health Center-base prospective study group. J. Clin. Oncol. 2008, 26, 1677-1683.

[133] Bezek S., Ujhazy E., Mach M., Navarova J., Dubovicky M. Developmental origin of chronic diseases: Toxicological implication. Interdisc. Toxicol. 2008, 1, 29-31.

[134] Ørgaard A., Jensen L. The effects of soy isofl avones on obesity. Exp. Biol. Med. 2008, 233, 1066-1080.

[135] Sabudak T., Guler N., Trifolium L. A review on its phytochemical and pharmacological profile. Phytother. Res. 2009, 23, 439-446.

[136] Choi E.J., Kim G.H. Antiproliferative activity of daidzein and genistein may be related to ERα/c-erbB-2 expression in human breast cancer cells. Mol. Med. Rep. 2013, 7, 781-784.

[137] Rochfort S., Panozzo J. Phytochemicals for health, the role of pulses. J. Agric. Food Chem. 2007, 55, 7381-7994.

[138] de Pascual-Teresa S., Moreno DA, Garcia-Viguera C. Flavanols and anthocyanins in cardiovascular health:

a review of current evidence. Int J Mol Sci, 2010, 11, 1679-1703.

[139] de Pascual-Teresa S., Sanchez-Ballesta MT. Anthocyanins: from plant to health. Phytochem Rev, 2008, 7, 281-299.

[140] Roy HJ, Lundy S., Eriksen C., Kalicki B. Anthocyanins. Pennigton Nutrition Series, 2009.

[141] de Pascual-Teresa S., Santos-Buelga C., RivasGonzalo JC. Quantitative analysis of flavan-3ols in Spanish foodstuffs and beverages. J Agric Food Chem, 2000, 48, 5331-5337.

[142] Bagchi D., Sen CK, Bagchi M., Atalay M. Antiangiogenic, antioxidant, and anti-carcinogenic properties of a novel anthocyanin-rich berry extract formula. Biochem, 2004, 69(1), 75-80.

[143] Aron P. M., & Kennedy J. A. Flavan-3-ols: Nature, occurrence and biological activity. Molecular Nutrition & Food Research, 2008, 52(1), 79-104.

Chapter 9

Sterols Biosynthesis in Algae

Mariane Bittencourt Fagundes and Roger Wagner

Abstract

Sterols are secondary metabolites, they are considered bioactive, due to their recognized activity as antioxidants, anticarcinogenic, cardiovascular protectors, and antiviral capacity. These triterpenoids can be found in a wide range of concentrations in different algae strains, being the variations related to external factors. In the world, there are millions of algae, some strains have the ability to produce high-value phytosterols, like stigmasterol, and sitosterol, however, others could lead to cholesterol production. For this reason, understand the principal factors involved in sterols biosynthesis, allows us to appoint the algae strain for industrial application and escalating these specific compounds production. Some algae are capable to produce sterols from mevalonic acid pathway, other strains present the methylerythritol 4-phosphate (MEP), or 1-deoxy-D-xylulose-5-phosphate (DOXP) as the main pathway, each one is responsible for the production of plans of intermediary compounds. In this sense, this chapter summarizes current knowledge of the biosynthetic pathways responsible for different sterols formation, as well as, describe main sterols that could be isolated from algae metabolism.

Keywords: macroalgae, microalgae, cyanobacteria, phytosterols

1. Introduction

Marine biota has a diversified metabolism, possessing worldwide most complex and unexplored organisms, and maybe the richest source of important compounds, bioactive molecules, that could lead in benefits for distinct areas in human life [1]. In this way, exploring these microorganisms in the context of their biochemistry is an important step, not only for drug discovery, or nutraceuticals, but also to understand their evolution. This affirmative comes from a question never totally elucidated about the molecular origin, and its association with algae sterol metabolism, named as "sterolomic". This approach could present important information's about the cell membrane, without them does not exist cellular protection and organization [2].

Cellular membrane composition is major composed by phospholipids, and between sterols cholesterol, in terms of animal cell organization, however, plants possess phytosterols replacing cholesterol, and the most interesting information in the microalgae metabolic system is associated with the capability of some strains producing both classes of sterols. In this chapter, we are going to synthesize aspects about algae principal sterols metabolic pathways, and the ways that they can be manipulated to produce specific compounds.

2. Algae metabolism: sterols discovery

The literature brings information's about diverse algae sterolomic profile, so in this chapter let us begin with the most curious and strong algae, considered the earliest life forms in the world, the prokaryotes microalgae (cyanobacteria). These strains are also known as blue-green algae, they are widely distributed in the world, due to their robustness. Cyanobacteria are considered by biologists a variation from bacteria and eucaryotic strains, which could lead in a production of sterols related with vegetal, and also animal kingdom [3].

Cyanobacteria for this reason, can occur in marine environments with a huge salt variation, in cold waters as Antarctic system, and hot waters, could also proliferate in desert sand and rocks, providing a major response from their metabolic systems modifications according to the natural evolution. These cyanobacteria can produce different metabolites according to the habitat that they are living, for this reason, merging the information's we can understand that they can present many metabolic pathways leading to different end-sterols products. Their resistance comes from their plasmatic membrane associated mostly with structures named hopanoids, that are very similar to sterols, and are responsible for the flexibility of cyanobacteria cellular membrane [4].

The major discussion on the literature is the unknown ability of these organisms producing sterols. Many years ago, some researches described the possibility to exist only hopanoids in their structure, in fact, with the advance in tandem mass spectrometry, nuclear magnetic resonance analysis associated with new extraction techniques it was discovered the presence of sterols in their membrane. Thus, metabolism involved in sterols biosynthesis by cyanobacteria are not totally elucidated.

In the history context, the first works showing sterols production in cyanobacteria were in a filamentous cyanobacteria named as *Phormidium luridum* in 1968 [5], in this study it was isolated unsaturated sterols, like as 24 ethyl sterols, following this research's other studies investigated a way to produce this metabolite in large scale, considering the fact that this cyanobacterium has resistance in front of other microorganisms, inferring a remarkable capability for industrial application.

In the ninety's the researchers Sallal, Nimer, and Radwan [6] studied other cyanobacteria strains, and verified that after dark incubation, sterols concentration increased. In in agreement to this study, Fagundes et al. [7] showed higher concentrations of sterols (β-sitosterol, stigmasterol, and cholesterol), for *Phormidium autumnale* cultured in heterotrophic system, being the inoculum without the presence of these compounds. In general, cyanobacteria are manly photosynthetic, but some strains can growth in heterotrophic conditions, in this context, it can be concluded that more studies on this particular area are necessary for further acquire more comprehension for biotechnological application.

Eukaryotic microalgae are reported in the literature as the most prominent strains for sterols production, and they are important to make feasible membrane cell permeability, and maintain structural protection [8, 9]. In this sense, the study of sterols biosynthesis started in eukaryotic cells, standing out in numerous hypothesis, and one of them is related to life adaptation on earth, showing that these molecules were produced in this cell as a protective response to reactive species of oxygen [10]. The first study in eukaryotic microalgae was in 1960 with *Scenedesmus*, showing as the major compound chondrillasterol [11], years later the same researchers Iwata and Sakurai [12] reported ergosterol as the most abundant sterol for *Chlorella*. In terms of macroalgae, the (brown) species *Ulva lactuca*, and *Cytoceira adriatica* from Adriatic Sea, were analyzed by the authors Kapetanovic et al. [13], showing that these species were the main sterols cholesterol and fucosterol for both algae.

In summary algae strain choice directly reflects in their potential for commercial application, for this reason, the knowledge of algal productivity, and the biotechnological treatment applied for each alga is important. So, understand the metabolic pathways for the full comprehension of sterols, and their intermediary metabolites formed provides important information for future culture modifications enhancing specific compounds [14]. For this, depending on the triterpenoid produced they can be applied for medical proposes, which is a great alternative since in the last decade we have the challenge for the isolation of new compounds, in front of many problems associated nowadays with diseases' outbreaks. Algae possess a diverse metabolic system; their sterol composition is interesting due to the fact that they show in their composition unconventional structural variations [15]. The main structure consist of a tetracyclic, with a fused-ring skeleton, with the presence of a hydroxyl group at the carbon 3 (head group- 3β), and biochemical modifications at the carbon C24 (in sterol side chain), besides modifications found in the tetracyclic nuclei, and also their side chain with different alkylation's patterns [15].

Nowadays, there are studies focusing on unconventional sterols bioactivity like the sterols isolated from *Isochrysis galbana*, being cholest-5-24-1,3-(acetyloxy)-3β-ol, ergost-5-en-3-β-ol, and 24-oxocholesterol acetate. Other study identified unconventional sterols in *Sargassum fusiforme:* saringosterol, 24-hydroperoxy-24-vinyl-cholesterol, 29-hydroperoxy-stigmasta-5,24 (28)-dien-3β-ol, 24-methylene-cholesterol, 24-keto-cholesterol, and 5α, 8α-epidioxyergosta-6,22-dien-3β-ol all associated with anti-atherosclerotic function [16].

Industrial initiative for algae biomass application started in 20 centuries with the investment in many programs for algae research. The principal countries producing algae biomass and their products are shown in the **Figure 1**. Their major focus are on biofuels, or commercializing the biomass powder, and in terms of fine-chemicals the market is based on pigments, being only two sterols commercially produced from algae, fucosterol and desmosterol [17]. With this in mind, is important highlight that sterols are important bioactive metabolites that are normally isolated from non-renewable source, comprehend the metabolic sterols pathways and the ways to modify their production, presenting algae as a new source of sterols to the world, could lead to a sustainable sterols production.

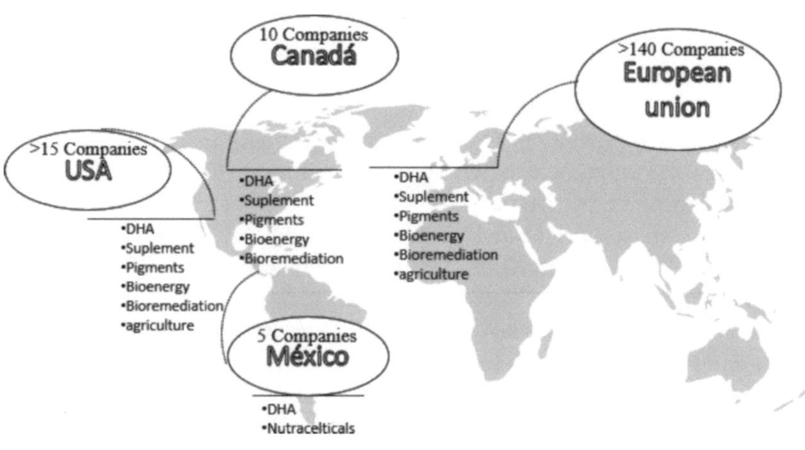

Figure 1.
Principal countries with important algae biotechnology companies' and their products. DHA - docosahexaenoic acid.

3. Algae sterols metabolic pathways

Sterols biosynthesis started by two main pathways the mevalonic acid (MVA), and by the 1-deoxy-D-xylulose-5-phosphate/2-C-methyl-D-erythritol-4-phosphate (MEP), recently discovered [18], also known as non-mevalonate pathway. The objective of these two pathways is produce an isoprenoid structure, a molecule of 5 carbons isopentenyl diphosphate (IPP), and dimethylallyl pyrophosphate (DMAPP), that are considered the sterol building block. MVA pathway occurs in cytosol, when MEP in the plastids, however the pathways activation are different according to the algae classification, being that some algae with the presence of both pathways' biochemical machinery MEP and MVA and others with only one of them active [19].

Understand the pathways involved for sterols production in algae is difficult, due to a huge phylogenetic heterogeneity found in strains. Since today still have research's showing for the first time the active pathway in some algae, like the observed by Scodelaro Bilbao et al. [20] studying *Haematococcus pluvialis*. A deeper discussion about numerous algae and the two possible active pathways can be found at the review from the authors Lohr et al. [21].

The prokaryotic cell, are known for possess MEP as the active isoprenoid producer, and for the ancestor reason, probably they were responsible for introducing this metabolism in eukaryotic strains. The MEP pathway is described as the major used for sterols production in algae, being green algae (*Chlorophyta*), with only MEP active for sterol production due to the loss of MVA pathway in the algae cellular evolution [21], as in many algae system both pathways occur, for this reason the pathways are depicted in the **Figure 2**.

The pathways are divided in two segments, the first one can be observed at the **Figure 2A**, which represents the transformation of DMAPP and IPP to squalene, this step consists in the MVA, and MEP. MVA pathway occurs in the cell cytosol until a condensation of two molecules of acetyl-CoA with the catalysis of acetoacetyl-CoA thiolase, after occurs other condensation forming 3-(*s*)-hydroxy-3-methylglutaril-CoA (HMG-CoA) by the action of 3-(s)-hydroxy-3-methylglutaril coenzyme A synthase. After that, the conversion to 3-(*R*)-mevalonate trough a reduction occurred by a nicotinamide adenine dinucleotide phosphate (NADPH)-dependent reductive diacylation by the enzyme HMG-CoA reductase [22]. The following steps consist in the conversion of MVA to mevalonate-5-diphosphate, catalyzed by mevalonate kinase (MK), and mevalonate-5-diphosphate kinase (MVADP), with the insertion of two ATP molecules, being the last step the conversion by isopentenyl diphosphate isomerase to the formation of DMAPP.

In terms of MEP pathway, the first step is a thiamin diphosphate-dependent condensation between D-glyceraldehyde 3-phosphate and pyruvate forming 1-Deoxy-D-xylulose-5-phosphate by the enzyme 1-deoxy-d-xylulose-5-phosphate synthase (DXS), following an isomerization to 2-*C*-methyl-o-erythritol-4-phosphate (MEP) by the enzyme 1-Deoxy-D-xylulose-5-phosphate (DXR) reducto-isomerase [18]. After, MEP and cytidine 5′-triphosphate are coupled, being catalyzed by 4-diphosphocytidyl-2-C-methylerythritol (MCT) synthetase, forming methylerythritol cytidyl diphosphate. The other enzymes involved in MEP pathway are in the sequence: 4-(cytidine 5′-diphospho)-2-*C*-methyl-D-erythritol kinase (CMK) for 4-(cytidine 5′-diphospho)-2-*C*-methyl-D-erythritol 2-phosphate formation. After, 2-*C*-methyl-D-erythritol-2,4-cyclodiphosphate synthase (MDS) which forms the 2-*C*-methyl-D-erythritol-2,4-cyclic diphosphate, 4-hydroxy-3-methylbut-2-enyl diphosphate synthase (HMBPP- synthase), leading to (2E)-4-hydroxy-3-methylbut-2-enyl diphosphate, and 4-hydroxy-3-methylbut-2-enyl diphosphate

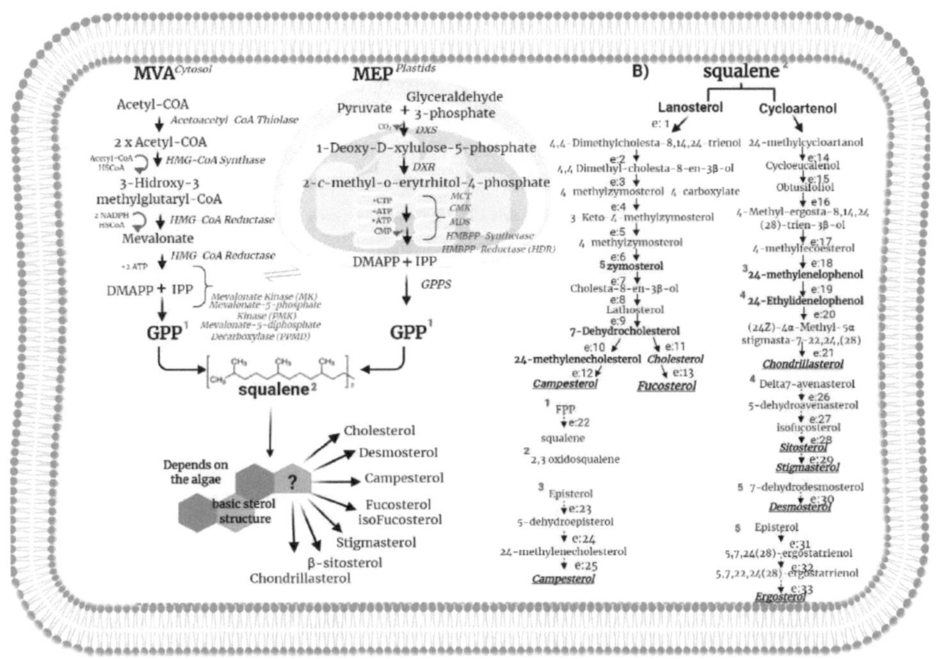

Figure 2.
Algae sterols different pathways, a: Mevalonic acid pathway, and B: Non-mevalonic acid pathway(methylerythritol 4-phosphate): HMG-CoA: Beta-Hydroxy-beta-methylglutaryl-coenzyme a, ATP: Adenosine triphosphate, GPP: Geranyl pyrophosphate, FPP: Farnesyl pyrophosphate, e:1: Lanosterol synthase, e:2: Sterol 14alpha-demethylase, e:3: Methylsterol monooxygenase, e:4: Sterol-4alpha-carboxylate 3-dehydrogenase, e:5: 3-keto steroid reductase, e:6: Methylsterol monooxygenase, e:7: Cholestenol Delta-isomerase, e:8: Cholestenol delta-isomerase, e:9: Delta7-sterol 5-desaturase, e:10: Delta7-sterol C5 desaturase, e:11: 7-dehydrocholesterol reductase, e:12: Delta-24-sterol reductase, e:13: 24-sterol reductase, e:14: 3-beta-hydroxysteroid 3-dehydrogenase, e:15: Cycloeucalenol cycloisomerase, e:16: Sterol 14alpha-demethylase, e:17: Delta14-sterol reductase, e:18: Cholestenol Delta-isomerase, e:19: 24-methylenesterol C-methyltransferase, e:20: 4-alpha-monomethylsterol monooxygenase, e:21: 7-dehydrocholesterol reductase, e:22: Farnesyl-diphosphate farnesyltransferase, e:23: Delta24(24(1))-sterol reductase, e:24: 7-dehydrocholesterol reductase, e:25: Delta24-sterol reductase, e:26: 3-beta-hydroxysteroid 3-dehydrogenase, e:27: 7-dehydrocholesterol reductase, e:28: Delta24-sterol reductase, e:29: Sterol 22-desaturase, e:30: 7-dehydrocholesterol reductase,e:31: Delta7-sterol 5-desaturase, e:32: Sterol 22-desaturase, e:33: Delta-24(24(1))-sterol reductase.

reductase (HMBPP-Reductase) being formed (2E)-4-hydroxy-3-methylbut-2-enyl diphosphate. The last step consists in the building blocks IPP and DMAPP and their coupling through isopentenyl-diphosphate isomerase [18, 23].

In the literature, there are numerous data, in which sometimes contrast about the biosynthesis of the isoprene units. MEP pathway was detected for the first time in bacteria, however further evidence has shown that in eukaryotes which performs photosynthesis found compounds from this metabolic pathway [24]. Normally a cyanobacteria which possess a metabolic system similar to bacteria produce phytosterols by MEP pathway, and also other authors describe that photosynthetic eukaryotic strain produce phytosterols only from MEP pathway [25]. On the other hand, MVA pathway normally is used for the production of cholesterol in animals, and also the green macroalgae sterols, in last case it occurs due to their metabolic similarity with higher plants, differently occurred with green microalgae from Chlorophyceae as described by Volkman [8, 9].

Geranyl pyrophosphate (GPP) is formed by the isoprenoids DMAPP and IPP, and through the diverse condensations leading to a presqualene compound, followed by the formation of squalene trough farnesyl-diphosphate farnesyltransferase, and trough squalene monooxygenase, or an alternative squalene epoxidase newly discovered [26]. These two pathways transform squalene into squalene

2,3-epoxide which is the lanosterol or cycloartenol intermediary, formed when squalene is oxidized by the enzyme squalene monooxygenase.

The following stages for different sterols isolated in algae are presented at the **Figure 2B**, being considered the anaerobic postsqualene pathway step. The biosynthesis occurs through cycloartenol pathway, however some strains produce cholesterol by lanosterol pathway. In the case of ergosterol the same pathway is activated for other microorganisms, but it is different for algae, starting their pathway by cycloartenol as observed in a study performed with *Chlamydomonas reinhardtii* [27]. Fucosterol is produced manly by lanosterol pathway as observed by Gallo et al. [28] in diatoms, and sitosterol followed by a C22 desaturation leading to stigmasterol both produced until cycloartenol pathway, the same occurs with desmosterol and chondrillasterol. Cholesterol is represented in the pathway figure produced by lanosterol, however there is research proving that this compound production also occurs by cycloartenol-dependent pathway [29].

4. Ways to manipulate sterol biosynthesis

Algae sterols can be easily manipulated to enhance their concentration, however, only few studies show the culture manipulation for this objective. In the algae metabolism commonly, the major changes occur when algae are cultured by nutrient limitation/modification. Photosynthetic system modifications consists in changing light intensity, and carbon dioxide amount, in terms of heterotrophic culture the exogenous carbon source can be considered the most important influence in sterols biosynthesis activation, salinity can be other factor important to sterol enhancer in algae [14].

For this reason, algae culture nutrient changes for phytosterols production have been mostly reported as phosphorous and nitrogen concentration. In relation to nitrogen, Zhang, Sachs, & Marchetti [30] analyzed freshwater and marine algae and they showed a reduction of 20% in sterols production when observed a nitrogen limitation for *Eudorina unicocca* and *Volvox aureus*, the reduction was similarly was observed in *Botryococcus braunii* [31], and for *Schizochytrium* sp. [32]. On the other hand, phosphorous modifications in the culture lead to a different result, the authors Piepho et al. [33] studied concentrations of 50 mM as the highest phosphorous concentration, and 10 mM as the lowest phosphorous amount. However, the phosphorus concentration was different according to the strain, being the low phosphorous concentration 1 mM for *Scenedesmus*, 5 mM for *Cryptomonas* and *Chlamydomonas* and 10 mM P for *Cyclotella,* due to each specie requirements, being the major sterol concentration found in a high-phosphorous culture system [33].

In the same line, the authors Chen et al. [34], verify for the strains *Thalassiosira oceanica*, *Rhodomonas salina*, *Isochrysis galbana*, and *Acartia tonsa*, the effect of different iron concentration added to the culture system, in fact in this experiment it was observed that the highest levels of Fe were capable to increase the total sterols, with the exception of *Isochrysis galbana.*

The effect of salt stress showed that the concentration of total free sterols increased with higher levels of NaCl in *Nitzschia laevis* [35], being the same observed in *Dunaliella salina* [36, 37]. The same comportment was observed in *Pavlova lutheri*, the changes were not observed in their total sterol composition, but in the individual sterols concentration, the enhance of salt modify the algae membrane, avoiding an excessive flux of $Na +$ and $Cl -$ ions into cells by increasing the membrane rigidity, helping the microorganism increasing high salt concentrations [38]. The nutrient composition from the culture as already mention has a huge influence on sterols, in another study the authors Fagundes et al. [7, 39], showed that *Phormidium autumnale* cultured with different carbon sources, glucose, sucrose, and different industrial wastes can accumulate more sterols, compared to the inoculum, and that each culture system shows a diverse composition.

Other factor of influence in sterols composition is the UV–C radiation doses, Ahmed and Schenk [40] proved that for *Pavlova lutheri* algae the sterols increase occurred by treating the algae with UV–C radiation, however the insertion of hydrogen peroxide does not show any effect. With regards to the photosynthetic system, there is few studies showing that after high light intensities the cell sterols content increase in three microalgae [33, 41].

The authors Pereira et al. [42], also showed that light intensities of 30, 60, 140, 230, and 490 mmol photons m^{-2} s^{-1} were tested for two Chlorophyceae *Scenedesmus quadricauda*, *Chlamydomonas globose*, *Cryptophyceae Cryptomonas ovata*, and the Mediophyceae (Bacillariophyta) *Cyclotella meneghiniana,* showing the best production in the highest sterol intensity. The authors explained this increase by some theories, being correlated with the algae species, as described in the biosynthesis topic some algae produce sterols from MVA pathway, and others from MEP, according to the study green algae that uses only MEP for sterols synthesis, being MEP linked to the chloroplast. For this reason, hypothetically related to the photosynthesis, being the explanation for the higher intensities of sterols found in *S. quadricauda* and the diatom *C. meneghiniana,* for this more studies needs to be performed with different strains to understand sterols metabolism.

Genetically modify strains to produce sterols are gaining attention, but also is a new strategy to turn these metabolic rich systems a source of sterols. According to D'Adamo et al. [43], they introduced in *Phaeodactylum tricornutum* three enzymes from a plant *Lotus japonicus*, the modifications were responsible for mRNA expression levels, increasing the expression of the native mevalonate and, consequently sterol biosynthesis pathway was estimuled, being responsible for the expression of important triterpenoids.

5. Final considerations

Algae sterols are a new segment for being studied, they are different according to the strain, and their environment, due to the fact that external factors affect the cellular membrane, as so, the sterol concentration. In this chapter, the most important sterols end-pathway products described are: Fucosterol, β-sitosterol, stigmasterol, ergosterol, cholesterol, chondrillasterol, and desmosterol. Still today there are research's discovering pathways for algae, due to the fact that algae are spread through the world, and can be isolated in simple access places or complex ones, being responsible for the metabolic variations. The studies involving algae sterols are ascending for industrial application, so, understand their origin is an important factor for future prospective.

Author details

Mariane Bittencourt Fagundes[1*] and Roger Wagner[2]

1 Department of Food Science and Technology, Santa Maria, Brazil

2 Department of Food Science and Technology, Federal University of Santa Maria, Santa Maria, Brazil

*Address all correspondence to: mari.bfagundes@gmail.com

References

[1] Volkman, John K. (2016). The Physiology of Microalgae. The Physiology of Microalgae. https://doi.org/10.1007/978-3-319-24945-2

[2] Deamer, D. (2017). The Role of Lipid Membranes in Life's Origin. Life (Basel, Switzerland), *7*(1), 5. https://doi.org/10.3390/life7010005

[3] Zahra, Z., Choo, D. H., Lee, H., & Parveen, A. (2020). Cyanobacteria: Review of current potentials and applications. Environments - MDPI, *7*(2). https://doi.org/10.3390/environments7020013

[4] Belin, B. J., Busset, N., Giraud, E., Molinaro, A., Silipo, A., & Newman, Di. K. (2018). Hopanoid lipids: From membranes to plant-bacteria interactions. Nature Reviews Microbiology, *16*(5), 304-315. https://doi.org/10.1038/nrmicro.2017.173

[5] de Souza, N. J., & Nes, W. R. (1968). Sterols: Isolation from a Blue-Green Alga. *Science*, *162*(3851), 363 LP – 363. https://doi.org/10.1126/science.162.3851.363

[6] Sallal, A. K., Nimer, N. A., & Radwan, S. S. (1990). Lipid and fatty acid composition of freshwater cyanobacteria. Microbiology, *136*(10), 2043-2048. https://doi.org/https://doi.org/10.1099/00221287-136-10-2043

[7] Fagundes, M. B., Falk, R. B., Facchi, M. M. X., Vendruscolo, R. G., Maroneze, M. M., Zepka, L. Q., ... Wagner, R. (2019). Insights in cyanobacteria lipidomics: A sterols characterization from Phormidium autumnale biomass in heterotrophic cultivation. Food Research International, *119*, 777-784. https://doi.org/10.1016/J.FOODRES.2018.10.060

[8] Volkman, J. (2003a). Sterols in microorganisms. Applied Microbiology and Biotechnology, *60*(5), 495-506. https://doi.org/10.1007/s00253-002-1172-8

[9] Volkman, J. K. (2003b). Sterols in microorganisms. Applied Microbiology and Biotechnology, *60*(5), 495-506. https://doi.org/10.1007/s00253-002-1172-8

[10] Galea, A. M., & Brown, A. J. (2009). Special relationship between sterols and oxygen: Were sterols an adaptation to aerobic life? Free Radical Biology and Medicine, *47*(6), 880-889. https://doi.org/https://doi.org/10.1016/j.freeradbiomed.2009.06.027

[11] Iwata, I., Nakata, H., Mizushima, M., & Sakurai, Y. (1961). Lipids of Algae. Agricultural and Biological Chemistry, *25*(4), 319-325. https://doi.org/10.1080/00021369.1961.10857810

[12] Iwata, I., & Sakurai, Y. (1963). Lipids of Algae. Agricultural and Biological Chemistry, *27*(4), 253-264. https://doi.org/10.1080/00021369.1963.10858097

[13] KAPETANOVIC, R., SLADIC, D., POPO, S., ZLATOVIC, M., KLJAJIC, Z., & GASIC, M. J. (2005). *Sterol composition of the Adriatic Sea algae Ulva lactuca, Codium dichotomum, Cystoseira adriatica and Fucus virsoides*. Serbian Chemical Society.

[14] Kumari, P., Kumar, M., Reddy, C. R. K., & Jha, B. (2013). 3 - Algal lipids, fatty acids and sterols. In H. B. T.-F. I. from A. for F. and N. Domínguez (Ed.), *Woodhead Publishing Series in Food Science, Technology and* Nutrition (pp. 87-134). https://doi.org/https://doi.org/10.1533/9780857098689.1.87

[15] Minale, L., & Sodano, G. (1977). *Non-Conventional Sterols of Marine Origin BT - Marine Natural Products Chemistry* (D. J. Faulkner & W. H. Fenical, Eds.). https://doi.org/10.1007/978-1-4684-0802-7_8

[16] Chen, Z., Liu, J., Fu, Z., Ye, C., Zhang, R., Song, Y., ... Liu, H. (2014). 24(S)-saringosterol from edible marine seaweed sargassum fusiforme is a novel selective LXRβ agonist. Journal of Agricultural and Food Chemistry, *62*(26), 6130-6137. https://doi.org/10.1021/jf500083r

[17] Bajpai, R., Prokop, A., & Zappi, M. (2014). Algal biorefineries: Volume 1: Cultivation of cells and products. Algal Biorefineries: Volume 1: Cultivation of Cells and Products, 1-324. https://doi.org/10.1007/978-94-007-7494-0

[18] Zhao, L., Chang, W., Xiao, Y., Liu, H., & Liu, P. (2013). Methylerythritol phosphate pathway of isoprenoid biosynthesis. Annual Review of Biochemistry, *82*, 497-530. https://doi.org/10.1146/annurev-biochem-052010-100934

[19] Felföldi-Gáva, A., Szarka, S., Simándi, B., Blazics, B., Simon, B., & Kéry, Á. (2012). Supercritical fluid extraction of Alnus glutinosa (L.) Gaertn. Journal of Supercritical Fluids, *61*, 55-61. https://doi.org/10.1016/j.supflu.2011.10.003

[20] Scodelaro Bilbao, P. G., Garelli, A., Díaz, M., Salvador, G. A., & Leonardi, P. I. (2020). Crosstalk between sterol and neutral lipid metabolism in the alga Haematococcus pluvialis exposed to light stress. Biochimica et Biophysica Acta - Molecular and Cell Biology of Lipids, *1865*(10), 158767. https://doi.org/10.1016/j.bbalip.2020.158767

[21] Lohr, M., Schwender, J., & Polle, J. E. W. (2012). Isoprenoid biosynthesis in eukaryotic phototrophs: A spotlight on algae. Plant Science, *185-186*, 9-22. https://doi.org/10.1016/j.plantsci.2011.07.018

[22] Hemmerlin, A., Harwood, J. L., & Bach, T. J. (2012). A raison d'être for two distinct pathways in the early steps of plant isoprenoid biosynthesis? Progress in Lipid Research, *51*(2), 95-148. https://doi.org/10.1016/j.plipres.2011.12.001

[23] Banerjee, A., & Sharkey, T. D. (2014). Methylerythritol 4-phosphate (MEP) pathway metabolic regulation. Nat. Prod. Rep., *31*(8), 1043-1055. https://doi.org/10.1039/C3NP70124G

[24] Bode, H. B., Zeggel, B., Silakowski, B., Wenzel, S. C., Reichenbach, H., & Müller, R. (2003). Steroid biosynthesis in prokaryotes: Identification of myxobacterial steroids and cloning of the first bacterial 2,3(S)-oxidosqualene cyclase from the myxobacterium Stigmatella aurantiaca. Molecular Microbiology, *47*(2), 471-481. https://doi.org/10.1046/j.1365-2958.2003.03309.x

[25] Hannan, M. A., Sohag, A. A. M., Dash, R., Haque, M. N., Mohibbullah, M., Oktaviani, D. F., ... Moon, I. S. (2020). Phytosterols of marine algae: Insights into the potential health benefits and molecular pharmacology. Phytomedicine, *69*, 153201. https://doi.org/10.1016/j.phymed.2020.153201

[26] Pollier, J., Vancaester, E., Kuzhiumparambil, U., Vickers, C. E., Vandepoele, K., Goossens, A., & Fabris, M. (2019). A widespread alternative squalene epoxidase participates in eukaryote steroid biosynthesis. Nature Microbiology, *4*(2), 226-233. https://doi.org/10.1038/s41564-018-0305-5

[27] Brumfield, K. M., Laborde, S. M., & Moroney, J. V. (2017). A model for the ergosterol biosynthetic pathway in Chlamydomonas reinhardtii. European Journal of Phycology, *52*(1), 64-74. https://doi.org/10.1080/09670262.2016.1225318

[28] Gallo, C., Landi, S., d'Ippolito, G., Nuzzo, G., Manzo, E., Sardo, A., & Fontana, A. (2020). Diatoms synthesize sterols by inclusion of animal and fungal genes in the plant pathway. Scientific Reports, *10*(1), 4204. https://doi.org/10.1038/s41598-020-60993-5

[29] Randhir, A., Laird, D. W., Maker, G., Trengove, R., & Moheimani, N. R. (2020). Microalgae: A potential sustainable commercial source of sterols. Algal Research, *46*(July 2019), 101772. https://doi.org/10.1016/j.algal.2019.101772

[30] Zhang, Z., Sachs, J. P., & Marchetti, A. (2009). Hydrogen isotope fractionation in freshwater and marine algae: II. Temperature and nitrogen limited growth rate effects. Organic Geochemistry, *40*(3), 428-439. https://doi.org/10.1016/j.orggeochem.2008.11.002

[31] Zhila, N. O., Kalacheva, G. S., & Volova, T. G. (2005). Effect of Nitrogen Limitation on the Growth and Lipid Composition of the Green Alga Botryococcus braunii Kutz IPPAS H-252. Russian Journal of Plant Physiology, *52*(3), 311-319. https://doi.org/10.1007/s11183-005-0047-0

[32] Sun, L., Ren, L., Zhuang, X., Ji, X., Yan, J., & Huang, H. (2014). Differential effects of nutrient limitations on biochemical constituents and docosahexaenoic acid production of Schizochytrium sp. Bioresource Technology, *159*, 199-206. https://doi.org/https://doi.org/10.1016/j.biortech.2014.02.106

[33] Piepho, M., Martin-Creuzburg, D., & Wacker, A. (2011). Simultaneous Effects of Light Intensity and Phosphorus Supply on the Sterol Content of Phytoplankton. PLOS ONE, *5*(12), e15828. Retrieved from https://doi.org/10.1371/journal.pone.0015828

[34] Chen, X., Wakeham, S. G., & Fishera, N. S. (2011). Influence of iron on fatty acid and sterol composition of marine phytoplankton and copepod consumers. Limnology and Oceanography, *56*(2), 716-724. https://doi.org/10.4319/lo.2011.56.2.0716

[35] Chen, G.-Q., Jiang, Y., & Chen, F. (2008). Salt-induced alterations in lipid composition of diatom nitzschia laevis (bacillariophyceae) under heterotrophic culture condition1. Journal of Phycology, *44*(5), 1309-1314. https://doi.org/https://doi.org/10.1111/j.1529-8817.2008.00565.x

[36] Francavilla, M., Trotta, P., & Luque, R. (2010). Phytosterols from Dunaliella tertiolecta and Dunaliella salina: A potentially novel industrial application. Bioresource Technology, *101*(11), 4144-4150. https://doi.org/https://doi.org/10.1016/j.biortech.2009.12.139

[37] Peeler, T. C., Stephenson, M. B., Einspahr, K. J., & Thompson, G. A. (1989). Lipid Characterization of an Enriched Plasma Membrane Fraction of Dunaliella salina Grown in Media of Varying Salinity. Plant Physiology, *89*(3), 970-976. https://doi.org/10.1104/pp.89.3.970

[38] Ahmed, F., Zhou, W., & Schenk, P. M. (2015). Pavlova lutheri is a high-level producer of phytosterols. Algal Research, *10*, 210-217. https://doi.org/https://doi.org/10.1016/j.algal.2015.05.013

[39] Fagundes, M. B., Alvarez-Rivera, G., Vendruscolo, R. G., Voss, M., da Silva, P. A., Barin, J. S., ... Wagner, R. (2020). Green microsaponification-based method for gas chromatography determination of sterol and squalene in cyanobacterial biomass. *Talanta*, 121793. https://doi.org/10.1016/j.talanta.2020.121793

[40] Ahmed, F., & Schenk, P. M. (2017). UV–C radiation increases sterol production in the microalga Pavlova lutheri. Phytochemistry, *139*, 25-32. https://doi.org/https://doi.org/10.1016/j.phytochem.2017.04.002

[41] Orcutt, D. M., & Patterson, G. W. (1974). Effect of light intensity upon lipid composition ofNitzschia closterium (Cylindrotheca fusiformis). Lipids, *9*(12), 1000-1003. https://doi.org/10.1007/BF02533825

[42] Pereira, C. M. P., Nunes, C. F. P., Zambotti-Villela, L., Streit, N. M.,

Dias, D., Pinto, E., ... Colepicolo, P. (2017). Extraction of sterols in brown macroalgae from Antarctica and their identification by liquid chromatography coupled with tandem mass spectrometry. Journal of Applied Phycology, *29*(2), 751-757. https://doi.org/10.1007/s10811-016-0905-5

[43] D'Adamo, S., Schiano di Visconte, G., Lowe, G., Szaub-Newton, J., Beacham, T., Landels, A., ... Matthijs, M. (2019). Engineering the unicellular alga Phaeodactylum tricornutum for high-value plant triterpenoid production. Plant Biotechnology Journal, *17*(1), 75-87. https://doi.org/https://doi.org/10.1111/pbi.12948

Chapter 10

Arginine Metabolism: An Enlightening Therapeutic Attribute for Cancer Treatment

Kapil Singh Narayan and Reenu Kashyap

Abstract

Arginine is well known semi-essential amino acid used in protein biosynthesis through several metabolic pathways. It is majorly obtained from nutrients sources and synthesized by the urea cycle in the body using citrulline. Arginine found to be involved in several mechanisms including; hormone synthesis, cell division, activation of the immune system, ammonia disposal and wound healing and also in the production of nitric oxide (NO) and polyamines. During cancer persistence, the biosynthesis of arginine is not sufficient to compensate for their higher nutritional requirements but extracellular availability of arginine is also required. Therefore, the consequences of arginine deprivation may represent a novel targeting therapy to cure cancer. The impact of different arginine deprivation agents and their mechanism of action always found to be correlated with NO and polyamine levels. Arginine deprivation strategy to hamper the proliferation of cancerous cells and their migration is represented as a new approach to cure cancer by inhibiting the argininosuccinate synthetase1 (ASS1) expression and NO and polyamines production. ASS1 is the first key enzyme that converts citrulline to arginine and numerous tumors such as hepatocellular carcinoma, melanoma, mesotheliomas and renal cancer do not express ASS1 and main focused enzyme for cancer treatment. Degradation of arginine by the enzyme arginine deiminase (ADI) specifically triggers the arginine elimination and inhibition of cancer migration. Though, ADI is a microbial enzyme but has a high affinity to arginine and converts arginine into citrulline and NH_3. This produced citrulline can be recycled back to arginine in normal cells where ASS1 expression is very high in comparison to ASS1-negative tumor cells. A modified form of ADI with pegylate (ADI-PEG20) has been formulated which showed both in-vivo and in-vitro activity against hepatocellular carcinoma and melanoma by inducing apoptosis. In this chapter, we have majorly discussed arginine production with different pathways and how its degradation into other metabolic active compounds involved in cancer treatment. Moreover, how arginine deprivation is directly taking part in the inhibition of cancer cell proliferation and its migration.

Keywords: arginine, metabolic pathways, arginine deprivation, cancer therapy

1. Introduction

Arginine is essential for microbes and semi-essential for eukaryotes which play numerous crucial roles in cellular metabolism. The impact of arginine always considered as a nonessential amino acid because a cell can synthesize arginine its own as per the requirement. But, during its limitation in the cells, it is necessary to acquire arginine from outside environment and this condition denoted arginine as a conditionally essential amino acid. Majorly, arginine is produced by two ways; from food sources and biosynthesized through urea cycle in the kidney [1]. The biosynthesis of arginine represented the conversion of citrulline to arginine by the enzymes arginosuccinate synthetase1 (ASS1) and arginosuccinate lyase (ASL). The role of enzyme ASS1 is the conversion of citrulline and aspartic acid to arginosuccinate, which then directly converted to arginine and fumaric acid by the enzyme ASL [2]. In case of bacteria, ornithine also indicated as a substrate to synthesize arginine by the enzyme ornithine transcarbamylase (OCT) [3]. Arginine is a precursor molecule for the formation of amino acids such as proline, glutamate and arginine itself and several other components like succinate, nitrate, nitrite, nitric oxide (NO), ammonia and CO_2. It acts as an intermediate in urea cycle and precursor molecules for polyamine, creatine and proteins biosynthesis [4]. Arginine becomes necessary for growth and promotes wound healing by stimulating the release of growth hormones such as insulin-like growth factor-1, insulin and prolactin and also has several immunomodulatory effects such as stimulation of T cells, natural killer cell and enhances pro-inflammatory cytokine levels [5]. Thus, arginine deprived cancer cells can be rescued by activating immunity and increasing the flux of arginine through urea cycle [6]. When a cell is under stress or need to proliferate like tumor cell, then the requirement of cellular components such as citrulline, nitric oxide and polyamine levels get increase. Therefore arginine synthesis and degradation tremendously increase in cancer cells [7]. Arginine depletion is one of the most accepted way to cure tumor cells which are auxotrophic (dependent on uptake of extracellular arginine) to arginine. Some tumor cells adapted with downregulated arginine metabolizing enzymes for inhibiting the production of arginine from the substrates and become arginine auxotrophic [8]. Therefore, during cancer some nonessential amino acids turned in to the essential and cancer cell becomes auxotrophic for these [9]. As we all know that cancerous cells are associated with very high survival rates, therefore, some significant improvements are required for early detection and treatment of cancer. The idea for cancer treatment open the door for some most advanced approaches including; hormone therapy, stem cell therapy, immunotherapy and amino acid deprivation therapy [6, 10, 11]. One of the most capable amino acid deprivation therapy is arginine deprivation where arginine-depleting agents are the main focused and depletion of arginine harms the ability of cancer cell metastasize. The mechanisms of arginine impairment are still not clear hence, in this chapter we will try to give a brief discussion about the different biosynthesis and catabolic pathways of arginine. How arginine deprivation can be focused for cancer therapy for both arginine auxotrophic and non-auxotrophic cancerous cells with different mechanism of actions. Moreover, we will discuss the impact of arginine deprivation in cell migration through different intermediates production such as polyamines and NO.

2. Arginine biosynthesis pathways

Arginine is synthesized from citrulline by the key enzymes ASS1and ASL of the urea cycle which also called ornithine cycle and then released into the bloodstream (**Figure 1a**). In large animals, citrulline is produced majorly from NH_3, CO_2 and ornithine by the enzymes OTC and carbamylphosphate synthetase I (CPS1) in the small intestine. Citrulline is also recycled to arginine when both argininosuccinate ASS1and ASL are present in the same cell and take part in to the citrulline-nitric oxide cycle [12]. In contrast arginase and nitric oxide synthetase use arginine as a common substrate and always compete for this substrate [13]. Arginine biosynthesis exhibits diverse pattern of gene organization in bacteria, mammals and plants and uses different set of enzymes which catalyze reactions for the formation of a key intermediate "ornithine". Additionally, glutamate is also utilized as the precursor for ornithine synthesis using some intermediates of the urea cycle [13, 14]. Extracellular arginine is also a source for ornithine synthesis in cells by enzyme arginase 1 [12]. In bacteria and plants, ornithine is synthesized from glutamate in five enzymatic steps initiated by the acetylation of glutamate by N-acetylglutamate synthase and called N-acetylglutamate synthase pathway (**Figure 1b**) [15]. Here, first ornithine is converted to citrulline by ornithine carbamoyltransferase.

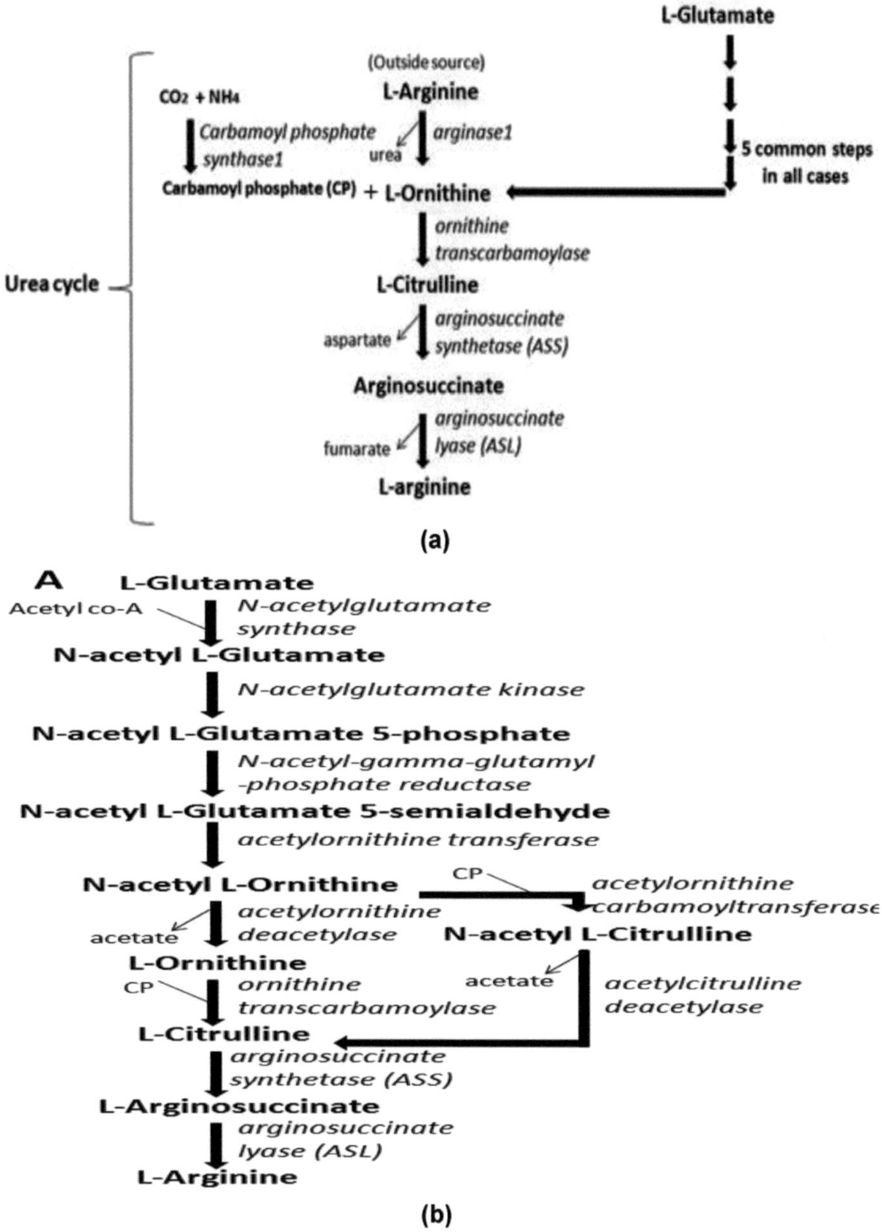

Figure 1.
(a) Arginine biosynthesis from glutamine and arginine itself in the urea cycle. (b) Key steps of the arginine biosynthesis form glutamine.

Enzyme ASS1 catalyzes the conversion of citrulline to aspartate and argininosuccinate which is further converted into arginine and fumarate by ASL [16, 17]. Ornithine can also be converted back to citrulline by arginine deiminase (ADI) pathway in bacteria [18] and by arginase1 pathway in mammals [19]. In both the cases citrulline is recycled back to arginine by ASS enzyme [15]. The ability to generate arginine from citrulline depends on the activity of ASS and ASL [20]. These two enzymes are tightly coupled for sensitivity of cells to arginine deprivation and their activity depends on their ability to regenerate arginine from the alternative sources [21].

3. Arginine catabolic pathways

There are enumerating pathways and enzymes to degrade arginine into other biomolecules and intermediates. Five main pathways including; arginine succinyltransferase (AST) (**Figure 2A**), arginine decarboxylase (ADC) (**Figure 2B**), Arginase1 (**Figure 2C**), citrulline- NO ((**Figure 2D**) and arginine deiminase (ADI) (**Figure 3**) were found to degrade arginine. These pathways are mainly focused by the researchers to study arginine degradation and find out its role in different cellular activities and ADI pathway has higher affinity for arginine among all of these pathways [22]. The essential site for arginine degradation in ureotelic organisms is the liver and second main site is the kidney where arginine is major converted into the polyamines, urea, creatine phosphate and NO and transported through bloodstream into the cells by cationic amino acid transporters (Melis et al, 2008). In bacteria arginine is degraded via three key pathways; (i) ADC pathway, here, arginine degradation is initiated by decarboxylation of arginine and form agmatine which further converted into putrescine by enzyme agmatine ureohydrolase. Putrescine is converted into γ- aminobutyric acid by putrescine transaminase and pyrroline dehydrogenase and ultimately converted into glutamate and succinate [23].

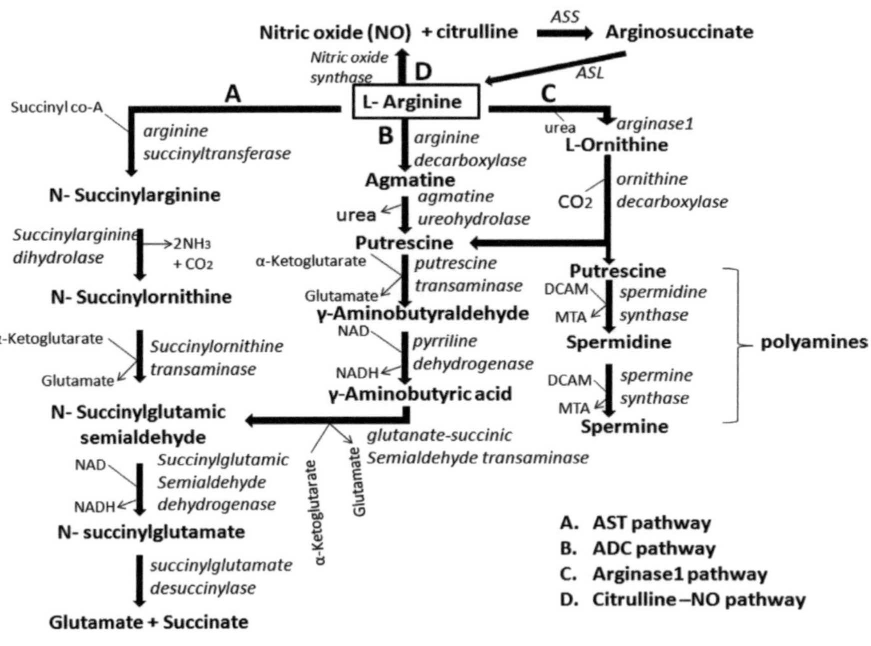

Figure 2.
Arginine biosynthesis by different metabolic pathways such as AST pathway (A), ADC pathway (B), arginase1 pathway (C) and citrulline-NO pathway (D).

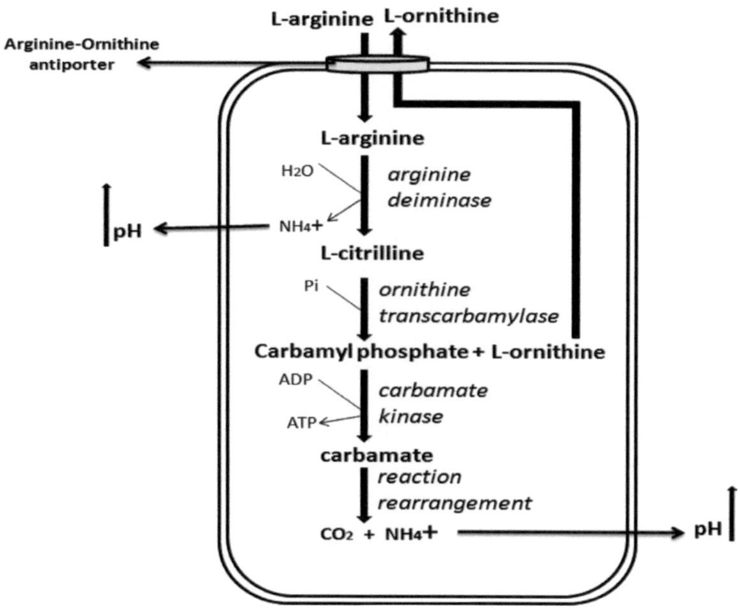

Figure 3.
degradation of arginine through ADI pathway in the bacterial system.

The enzyme arginine decarboxylase (ADC) is also considered as an important enzyme in bacteria [24], plants [25] and in mammalian systems [26]. The infusion of agmatine in the cerebral ventricles increases blood pressure and regulates the angiogenic activities [27]. (ii) AST pathway; here, arginine is degraded into glutamate, succinate and other intermediates. AST pathway is mainly activated for arginine degradation when nitrogen is limited for growth and contributes into the production of amino acids [28]. (iii) Arginase1 pathway; this pathway is activated when arginine concentration is excess in the media and urea and ornithine are produced by enzyme arginase1 during first enzymatic reaction [29]. Here, urea does not metabolize further and rapidly excreted into the medium. In arginase1 pathway, arginine used as the nitrogen and carbon sole sources and less than 3% of consumed arginine results in the formation of urea and 36% consumed by the route of putrescine and polyamine synthesis [30]. Polyamines produced by this pathway are polycations and interact with negatively charged molecules, such as DNA, RNA and also with proteins and involved in cellular growth, survival and proliferation [31]. Polyamines such as putrescine, spermidine and spermine are very tightly regulated by polyamine metabolic pathway [32]. These metabolites used by *H. pylori* to retard the expression of pro-inflammatory cytokines and prevent the immune response in stimulated macrophages [33] and also maintain the micro-environment around their cell in acidic condition for their survival using arginine [34]. Cancer and proliferative cells show high levels of polyamines and with this feature cancer cells maintain their proliferative properties [32] and high levels of polyamines were observed in cancerous cells [35]. It is proposed that both Gram negative and positive bacterial cells which contain unusually high AST and ADI level grow anaerobically in a complex acidic medium and both the enzymes help to raise the pH for the cell survival in the acidic environment [36]. Last but not least, arginine deiminase (ADI) pathway degrades arginine to ornithine, ammonia, and carbon dioxide and generates one mol of ATP by utilization of per mol of arginine [37]. A variety of bacterial cells; both gram positive and gram negative can catabolized arginine through ADI pathway [18]. Enzyme activity of ADI has been detected in several lactic acid bacteria (LAB), *bacilli, clostridia, pseudomonads, aeromonads, mycoplasmas, halobacteria, and cyanobacteria* [36]. ADI pathway is completed by three key enzymes: arginine deiminase (ADI), ornithine transcarbamoylase (OTC), and carbamate kinase (CK) as shown in **Figure 3**. Moreover, in *Pseudomonas aeruginosa*, a fourth gene that encodes a transport protein to exchange arginine and ornithine for this pathway has been identified [37]. ADI pathway is most important for the bacterial cell survival in the acidic environmental condition because arginine degradation by ADI pathway produced ammonia that raises the cytoplasmic and extracellular pH and produced ATP use as the energy source for cell survival. In the absence of carbohydrate bacteria preferred arginine and utilize it by ADI pathway as an alternate energy source to engender energy for cellular growth [20, 38]. ADI pathway is regulated at transcriptional level and regulated by transcriptional regulator ArgR [3, 37]. Moreover, carbon catabolite repression (CCR) has also been confirmed for the expression of ADI pathway in various bacteria. CCR regulates the expression of arc operon with glucose and catabolite control protein A (CcpA) [39]. CcpA is a transcriptional regulators belonging to the Crp/Fnr family and regulates the expression by the binding with regulatory proteins to the cis-acting catabolite response elements (cre) located in the promoter regions [20].

4. Impact of arginine degradation in cancer therapy

Cancer cells need excess quantities of specific amino acids for their diverse metabolism rate for higher proliferation and become resist for some cell death signals. Identification of the metabolic dissimilarly between cancer cells and normal cells, cellular metabolism of cancer cells is a therapeutic target and focusing field of cancer research [40]. The deprivation of arginine inside cancer cells, which makes the cells auxotrophic, has been centered one of the novel approach for cancer treatment [22]. There are several targets have been reported which directly take part in cancer mitigation as discussed below;

5. Citrulline-NO cycle

Citrulline is well known as a byproduct of NO synthetase enzyme and can be recycled to arginine by the key enzymes ASS1 and ASL. Both these enzymes are strongly expressed in liver and kidney then the other cells and tissues. The citrulline-nitric oxide cycle stimulates the activation of cytokines such as interferon (IFN) [41] and enhances the expression level of ASS1 enzyme as noticed in mouse microglial cells [42] and human tumor cell lines [43]. Impaired NO production from citrulline has been reported as a vital factor for the abnormal proliferation of keratinocytes in psoriasis epidermis. Higher arginase I with induced NO synthetase inhibits the keratinocyte proliferation by eliminating the arginine availability [44]. Enzymes for arginine metabolism are the potent therapeutic targets to control NO and cancerous cell proliferation as shown in **Figure 4**. In citrulline-NO pathway, NO is synthesized from arginine by the three nitric oxide synthase (NOS) isoforms; endothelial NOS (eNOS), neuronal NOS (nNOS) and inducible NOS (iNOS) and maintain the citrulline -NO cycle in the functional cells [45]. NOS and arginase1 use arginine as same substrate but arginase1 down-regulate because NO production by competing with NOS for arginine [46]. Remarkably, iNOS and arginase1 activities are reciprocally regulated in the cancerous cells by the involvement of cytokines and this can be guaranteed for the optimum production of NO but not in immunostimulated macrophages [47].

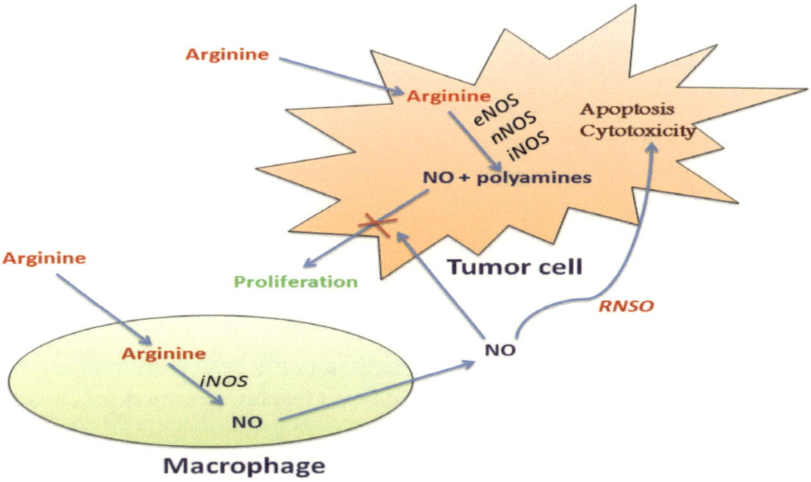

Figure 4.
Role of nitric oxide for tumor alimination.

6. Inhibition of ASS1 activity

The mechanisms which exhibited the loss of ASS1 activity are specifically depend on the type of cancerous cell and availability of arginine. ASS1 is a rate-limiting enzyme involved in arginine biosynthesis and has been investigated in numerous cancerous conditions such as melanoma [48], hepatocellular carcinoma [49] and pancreatic cancers [50], and the. ASS1-negative cancer cells are auxotrophic for arginine and exhibit sensitivity to arginine deprivation [51]. Cancerous cells have lack expression of ASS1 enzyme required for arginine biosynthesis which is an exogenous source for proteins synthesis and cellular growth [52]. Less ASS1 expression was recorded as a biomarker in cancer cell and for overall cellular functioning. The ASS1-deficient cancers with arginine auxotrophy have been initiated as the development of therapeutics by depriving arginine through degradation and trigger the apoptosis in arginine auxotrophic cancerous cells [53] as shown in **Figure 5**. The low levels of acetylated polyamine metabolites were found in arginosuccinate

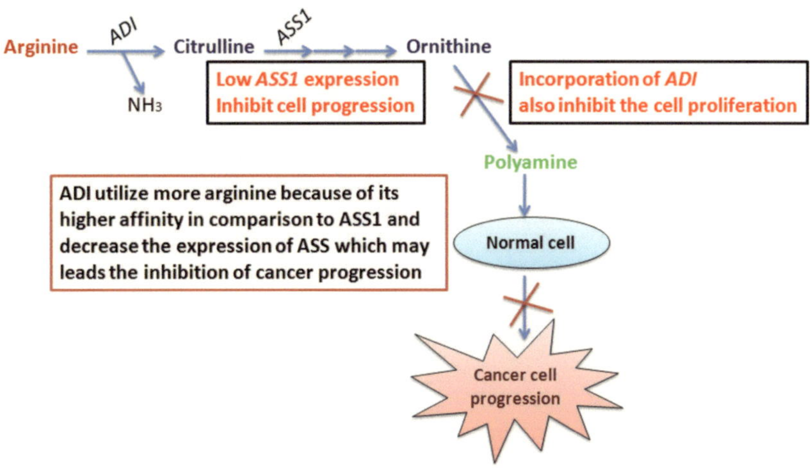

Figure 5.
Impact of gene ASS1 expression for cancer progression and proliferation and its down-regulation by ADI enzyme.

synthetase-deficient cells, pinpointing the reduction in catabolism and increase the expression of polyamine biosynthetic enzymes. This metabolic reprogramming elucidates a synthetic lethal interaction between arginosuccinate synthetase loss and polyamine metabolism, which could potentially be exploited for the treatment of arginosuccinate synthetase-negative cancers [54]. The reason for down-regulation of arginosuccinate synthetase in cancer cells is not cleared properly but always remains the center of interest among the cancer researchers [55].

7. ADI obstructed the angiogenetic activity

The enzyme ADI inhibits the tumor growth not only by depletion of arginine but also by suppression the angiogenic activity via less NO production [20] as shown **Figure 6**. ADI has an also strong capability to deplete arginine from plasma and inhibit NO production which resulting an effective inhibitory role of ADI in NO-mediated angiogenesis [56]. During in vitro study, the anti-angiogenic activity of ADI to inhibit micro vessel tube formation and migration in endothelial cell cultures was reported by Beloussow et al. [57]. Arginine depletion with the treatment of ADI enzyme also alters the level of proline, polyamines, glutamate and succinate. Polyamines are essential for tumor proliferation and their less production directly affects angiogenesis. Mycoplasma-derived ADI-PEG20 is majorly focused and most commonly used as a potential therapeutic agent for clinical investigation with different anti-neoplastic activity [58]. Mechanistically, ADI is capable of inhibiting the metabolic activity of cancerous cells and take parts in autophagy and apoptosis of auxotrophic cells. [59].

Induction of apoptosis

Arginine limitation has also been recorded to induce apoptosis which leads cell death in ASS1-negative tumor [60]. Even though, the signaling pathway for apoptosis is not clear yet, but it has been reported that apoptosis induced by arginine deprivation can be activated via caspase-dependent/independent pathways [1]. The limitation of arginine in ASS1-negative mesothelioma cells induced apoptosis via

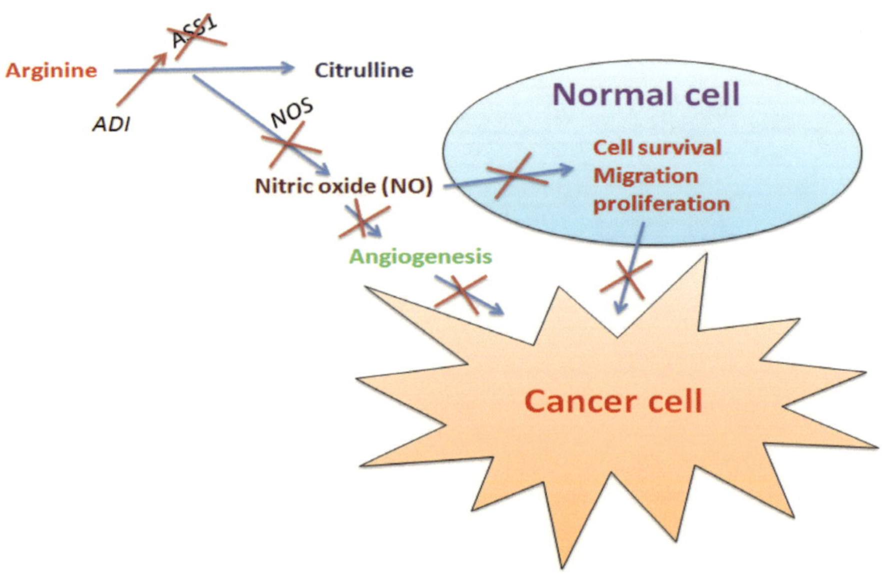

Figure 6.
Role of nitric oxide for angiogenesis and cancer proliferation and migration.

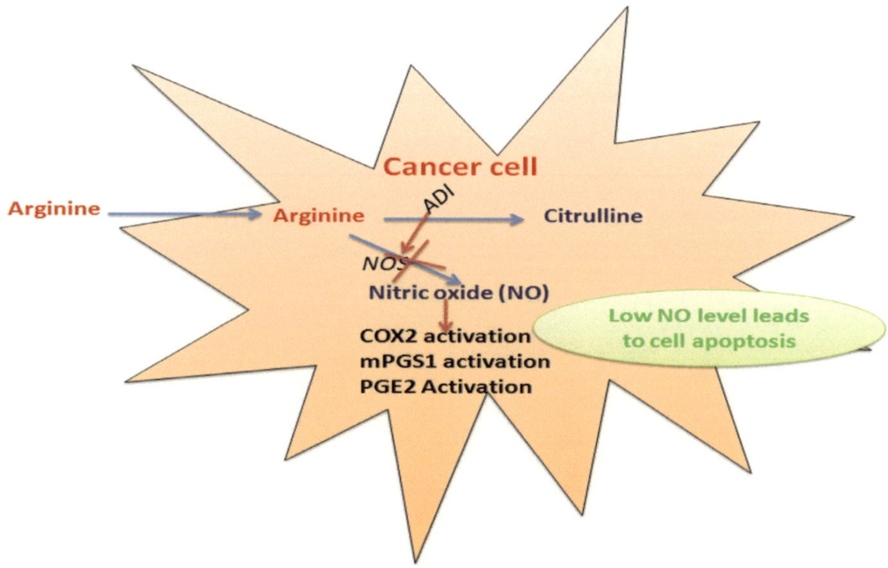

Figure 7.
Impact of ADI enzyme which leads the apoptosis in cancer cell.

mitochondrial inner membrane depolarization and Bcl-2-associated X protein (BAX) activation which is well known as program type II or caspase-independent cell death [55]. It was also reported that ASS1-deficient cancer cells, prolonged autophagy activated upon ADI treatment and impaired mitochondrial functions by inducing oxidative stress, chromatin autophagy and DNA leakage which finally causes cell death [61] as presented in **Figure 7**. In addition to this arginine limitation by pegylated (PEG) arginase causes cell death induced by necrosis as observed in acute myeloid leukemia (AML). In contrast, cell cycle arrest in AML cells did not induce cell apoptosis, autophagy, and rapid production of reactive oxygen species [62].

8. Role of arginine deprived agents

Deprivations of arginine from cancerous cells not only have a cytotoxic effect on cell but also induce specific cell cycle arrest. The cell cycle arrest analysis was done to check the surviving population of pancreatic and ovarian cancer cells to examine the consequence of arginine deprivation on cell cycle. The first reported arginine deprivation agent was ADI enzyme which degraded arginine and prevents cell growth s in culture from growing [63]. Human Arginase 1 (HuArgI) is second arginine deprivation agent and used to target arginine auxotrophic cancer cell lines and it is stable longer in serum, improved catalytic activity and less exposed with immune system [64]. Different types of cancerous cell lines undergo different mechanisms of cell death when deprived to arginine such as the process of autophagy, when cell degrades itself during nutrients limitation leads to starvation and cell death. Moreover, autophagy inhibited by HuArgI may indicate no caspase activation, no loss in membrane integrity and prevent the cell death caused by apoptosis [65].

9. In inhibition of cell migration

Cell migration is a well accepted attribute of the cancerous cell and arginine depletion majorly affect on cell viability and migration [66]. Low level of arginine

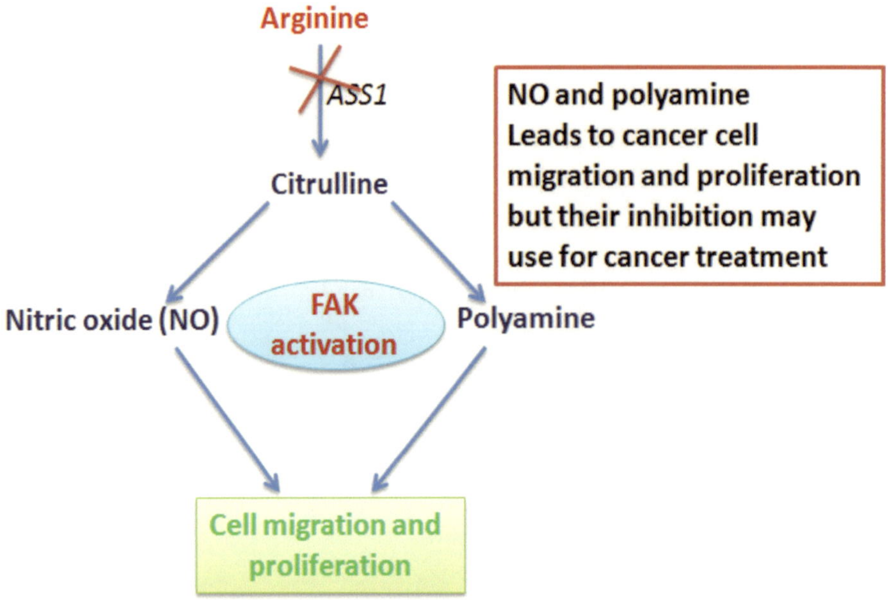

Figure 8.
Migration and proliferation of cancer.

accompany with the down-regulation of ASS1 expression which may leads the complete auxotrophic and mitigation of cancer cell migration [67]. The migration dependent on arginine requires optimum arginine to be catabolized two specific major enzymes nitric oxide synthase and arginase1 and produced citrulline and NO [68]. This increased level of promotes the cell viability, proliferation and migration during the process of wound healing [69]. Higher level of NO also activates signaling of focal adhesion kinase (FAK) cascade, which take parts in integrin assembly and disassembly as shown in **Figure 8**. Less arginine degradation to NO during wound healing showed a decrease in migration of colorectal cancer cells and added citrulline restored cell migration [70]. This higher nitric oxide synthase activity was recorded in intestinal epithelial cells in the presence of arginine and citrulline and NO production, which directly stimulate the cell migration [71]. The impact of arginine limitation and role of FAK was noticed when a study done on human intestinal epithelial cells which showed a significant role of NO production and cell migration [72]. Similar to this, other enzyme such as ornithine decarboxylase (ODC) used in polyamine biosynthesis also play important role in cell migration where polyamines increase the K^+ channel mediated Ca^{2+} influx and support to FAK activation [73]. The inhibition of ODC enzymes was majorly correlated with abnormal morphology of actin-cytoskeleton of metastatic cells migration [74]. Other signals including PI3K, Rho GTPases, microtubules and integrins always found to be interlinked and positive play important role in cell polarity by regulating intracellular junctions, cell adhesion, invasion and migration [75]. Arginine depletion also hamper the RhoA activation in colorectal cancer cells and during a report, the increased level of NO was majorly found to be involved in the RhoA activation in pancreatic cancer cells [76].

10. Conclusion

Several cancerous cells exhibited a higher metabolic requirement for specific amino acids to meet their rapid growth and migration. Therefore, specific amino acid limitation could be a novel therapy target to cure cancer. Arginine is a well known essential amino acid with the ability to regulate cellular activities and influence viability, proliferation, motility, migration, adhesion and invasion. There are several clinical studies have been reported which clearly explain the impact of arginine limitation as a therapy to cure arginine auxotrophic tumors and arginine converted in to polyamines and NO, majorly focused for cell proliferation and migration. The role of ADI enzyme and less expression of ASS1 gene was found to be directly correlated with the production of NO and polyamines and elimination of arginine auxotrophic tumors. Some tumors such as hepatocellular carcinoma and melanoma are found to be very sensitive for this treatment of arginine limitation because here arginine does not take parts in the urea cycle. Thus, the development of a new drug and drug resistance due to induction of ASS1 expression leads to a potential problem in tumors curing. Overall, the complete mechanism understanding of arginine limitation and inhibition of arginine auxotrophic cancer cell proliferation and migration is not clear and still further investigation is required to understand this cancer therapy.

Author details

Kapil Singh Narayan[1*] and Reenu Kashyap[2]

1 New Bolton Center, University of Pennsylvania, Kennett Square, PA, USA

2 Dairy Microbiology Division, National Dairy Research Institute, Karnal, Haryana, India

*Address all correspondence to: kapilnsnk@gmail.com

References

[1] Savaraj N, You M, Wu C, Wangpaichitr M, Kuo MT, Feun LG (2010) Arginine deprivation, autophagy, apoptosis (AAA) for the treatment of melanoma. Curr. Mol. Med. 10: 405-412.

[2] Fernandes H, Silva Teixeira C, Fernandes P, Ramos M, Cerqueira N (2017) Amino acid deprivation using enzymes as a targeted therapy for cancer and viral infections. Expert OpinTher Pat 27:283-297.

[3] Lu CD (2016) Pathways and regulation of bacterial arginine metabolism and perspectives for obtaining arginine overproducing strains. Appl. Microbiol. Biotechnol. 70:261-272

[4] Townsend DE, Kaenjak A, Jayaswal RK, Wilkinson BJ (1996) Proline is biosynthesized from arginine in *Staphylococcus aureus*. Microbiol. 142:1491-1497

[5] Lind DS (2004) Arginine and Cancer. J. Nutr. 134:2837-2841

[6] Nasreddine G, El-Sibai M, Abi-Habib RJ (2019) Cytotoxicity of [HuArgI (co)- PEG5000]-induced arginine deprivation to ovarian Cancer cells is autophagy dependent. Investig. New Drugs 38:10-19

[7] Wheatley DN, Campbell E (2002) Arginine catabolism, liver extracts and cancer. Pathol. Oncol. Res. 8:18-25

[8] Feun LG, Kuo MT, Savaraj N (2015) Arginine deprivation in cancer therapy. Curr. Opin. Clin. Nutr. Metab. Care. 18:78-82.

[9] Fung MKL, Chan GCF (2017) Drug-induced amino acid deprivation as strategy for cancer therapy. J. Hematol. Oncol. 10:144.

[10] Hellerstedt BA, Pienta KJ (2002) The current state of hormonal therapy for prostate cancer. CA Cancer J Clin. 52(3):154-179.

[11] Abdallah MS, Kennedy CR, Stephan JS, Khalil PA, Mroueh M, Eid AA, Faour W H (2018) Transforming growth factor-β1 and phosphatases modulate COX-2 protein expression and TAU phosphorylation in cultured immortalized podocytes. Infamm Res 67(2):191-201.

[12] Husson A, Brasse-Lagnel C, Fairand A, Renouf S, Lavoinne A (2003) Glutamine Stimulates Argininosuccinate Synthetase Gene Expression through Cytosolic O-Glycosylation of Sp1 in Caco-2 Cells. Eur. J. Biochem. 270: 1887-1899

[13] Ligthart-Melis GC, van de Poll MCG, Boelens PG, Dejong CHC, Deutz NEP, van Leeuwen PAM (2008) Glutamine is an important precursor for de novo synthesis of arginine in humans. Am. J. Clin. Nutr. 87:1282-1289

[14] Morris DR, Koffron KL (1967) Urea production and putrescine biosynthesis by *Escherichia coli*. J. Bacteriol. 94: 1516-1519

[15] Schneider B, Kiupakis AK, Reitzer LJ (1998a) Arginine Catabolism and the Arginine Succinyltransferase Pathway in *Escherichia coli*. J. Biotechnol. 80: 4278-4286.

[16] Marquis RE, Bender GR, Murray DR, Wong A. (1987) Arginine deiminase system and bacterial adaptation to acid environment. Appli. Environ. Microbiol. 53:198-200.

[17] Morris SM (2006) Arginine: beyond protein. Am. J. Clin. Nutr. 83:508-512.

[18] Zuniga M, Champomier-Verges M, Zagorec M, Perez-Martinez G (1998)

Structural and functional analysis of the gene cluster encoding the enzymes of the arginine deiminase pathway of *Lactobacillus sake*. J. Bacteriol. 180:4154-4159

[19] Rath M, Müller R, Kropf P, Closs E, Munder M (2014) Metabolism via arginase or nitric oxide Synthase: Two competing arginine pathways in macrophages. Front. Immunol. 5:532.

[20] JadeL X, Teng LL, Botelho MG, Lo RC, Lau KPS, Patrick CY (2016) Arginine metabolism in bacterial pathogenesis and cancer therapy. Int. J. Mol. Sci. 17:363.

[21] Wheatley DN, Campbell E (2003) Arginine deprivation, growth inhibition and tumor cell death: 3. Deficient utilization of citrulline by malignant cells. Br. J. Cancer. 89:573-576.

[22] Feun L, You M, Wu CJ, Kuo MT, Wangpaichitr M, Savaraj N (2008) Arginine deprivation as a targeted therapy for cancer. Curr. Pharm. Des. 14:1049-1057.

[23] Lortie MJ, William FN, Peterson OW, Vallon V, Malvey K, Mendonca M, Satriano J, Insel P, Thomson SC, Blantz RC (1996) Agmatine, a bioactive metabolite of arginine production, degradation, and functional effects in the kidney of the rat. Ameri. Socie. Cli. Investi. Inc. 9:413-420

[24] Wu WH, Morris DR (1973) Biosynthetic arginine decarboxylase from *Escherichia coli*. J. Biol. Chem. 248:1687-1695.

[25] Rastogi R, Dulson J, Rothstein SJ (1993) Cloning of tomato (*Lycopesiconesculentum* Mill.) arginine decarboxylase gene and its expression during fruit ripening. Plant Physiol. 103:829-834

[26] Morrissey J, McCracken R, Ishadoya S, Klahr S (1995) Partial cloning and characterization of an arginine decarboxylase in the kidney. Kidney Int. 47:1458-1461

[27] Sun MK, Regunathan S, Reis DJ (1995) Cardiovascular responses to agmatine, a clonidine displacing substance, in anesthetized rat. Clin. Exper. Hyperten. 17:115-128.

[28] Schneider BL, Kiupakis AK, Reitzer LJ (1998b) Arginine catabolism and the arginine succinyltransferase pathway in *Escherichia coli*. J. Bacteriol. 180:4278-4286.

[29] Shaibe E, Metzer E, HalpernYS (1985) Metabolic pathway for the utilization of L-arginine, L-ornithine, agmatine, and putrescine as nitrogen sources in *Escherichia coli*K-12. J. Bacteriol. 163:933-937.

[30] Caldwell RW, Rodriguez PC, Haroldo A, Toque S, Narayanan P, Ruth B (2018) Arginase: A multifaceted enzyme important in health and disease. Physiol Rev. 98: 641-665.

[31] Fuhrmann J, Clancy KW, Thompson PR (2015) Chemical biology of protein arginine modifications in epigenetic regulation. Chem. Rev. 115:5413-5461.

[32] Minois N, Gutierrez DC, Madeo F, (2011) Polyamines in aging and disease. Agi. 3:716-732.

[33] Bussiere FI, Chaturvedi R, Cheng Y (2005) Spermine causes loss of innate immune response to *Helicobacter pylori* by inhibition of inducible nitric-oxide synthase translation. J. Biol. Chem. 280:2409-2412.

[34] Merrell DS, Goodrich ML, Otto G, Tompkins LS, Falkow S (2003) pH-regulated gene expression of the gastric pathogen *Helicobacter pylori*. Infect. Immun. 71:3529– 3539.

[35] Wallace HM, Niiranen K (2007) Polyamine analogues-an update. Ami. Acid. 33:261-265.

[36] Cunin R, Glansdorff N, Pierard A, Stalon V (1986) Biosynthesis and metabolism of arginine in bacteria. Microbiol. 50:314-352.

[37] Verhoogt HJC, Abee HST, Gamper M, Driessen AJM, Haas D, Konings WN (1992) arcD, the first gene of the arc operon for anaerobic arginine catabolism in *Pseudomonas aeruginosa*, encodes an arginine-ornithine exchanger. J. Bacteriol. 174:1568-1573.

[38] Deibel RH (1963) utilization of arginine as energy source for the growth of *Streptococcus faecalis*. J. Bacteriol. 87:988-992.

[39] Marciniak BC, Pabijaniak M, De-Jong A, Dűhring R, Seidel G, Hillen W, Kuipers OP (2012) High- and low-affinity cre boxes for CcpA binding in *Bacillus subtilis* revealed by genome-wide analysis. BMC Genom. 13:401.

[40] Phillips MM, Sheaff MT, Szlosarek PW (2013) Targeting arginine-dependent cancers with arginine-degrading enzymes: Opportunities and challenges. Can. Res. Treat. 45:251-262.

[41] Nussler AK, Billiar TR, Liu ZZ, Morris SM (1994) Co-induction of nitric oxide synthase and argininosuccinate synthetase in a murine macrophage cell line. Implications for regulation of nitric oxide production. J. Biol. Chem. 269:1257-1261.

[42] Hattori Y, Campbell E B, Gross SS (1994) Argininosuccinatesynthetase mRNA and activity are induced by immune stimulants in vascular smooth muscle. Role in the regeneration of arginine for nitric oxide synthesis. J. Biol. Chem. 269:9405-9408.

[43] Nussler AK, Liu ZZ, Hatakeyama K, Geller DA, Billiar TR, Morris SM (1996) A cohort of supporting metabolic enzymes is co-induced with nitric oxide synthase in human tumor cell lines. Can. Let. 103:79 –84.

[44] Bruch-Gerharz D, Schnorr O, Suschek C, Beck KF, Pfeilschifter J, Ruzicka T, Kolb-Bachofen V (2003) Arginase1 over-expression in psoriasis: limitation of inducible nitric oxide synthase activity as a molecular mechanism for keratinocyte hyperproliferation. Am. J. Pathol. 162:203-211.

[45] Andrew PJ, Mayer B (1999) Enzymatic function of nitric oxide syntheses. Cardiovasc. Res. 43:521-531.

[46] Que LG, George SE, Gotoh T, Mori M (2002) Effects of arginase isoforms on NO production by nNOS. Nit. Oxi. 6(1):1-8.

[47] Mori M, Gotoh T (2004) Arginine Metabolic Enzymes, Nitric Oxide and Infection. J. Nutr. 134: 2743-2747.

[48] Fischer GM, Gopal YNV, McQuade JN,Peng W, Berardinis RJD, Davies MA (2018) Metabolic strategies of melanoma cells: mechanisms, interactions with the tumor microenvironment, and therapeutic implications. Pigm. Cell Melan. Res, 31:11-30.

[49] Dillon BJ, Prieto VG, Curley SA, Ensor CM, Holtsberg FW, Bomalaski JS, Clark MA (2004) Incidence and distribution of argininosuccinate synthetase deficiency in human cancers: A method for identifying cancers sensitive to arginine deprivation. Can. 100:826-833.

[50] Bowles TL, Kim R, Galante J, Parsons CM, rudachalam S, Kung HJ, Bold RJ (2008) Pancreatic cancer cell lines deficient in argininosuccinate synthetase are sensitive to arginine deprivation by arginine deiminase. Int. J. Can. 123:1950-1955.

[51] Qiu F, Chen YR, Liu X, Chu CY, Shen LJ, Xu J, Gaur S, Forman HJ,

Zhang H, Zheng S, Yen Y, Huang J, Kung HJ, Ann DK (2014) Arginine starvation impairs mitochondrial respiratory function in ASS1-deficient breast cancer cells. Sci. Signal. 7:ra31.

[52] Delage B, Luong P, Maharaj L, O'Riain C, Syed N, Crook T, Hatzimichael E, Papoudou-Bai A, Mitchell TJ, Whittaker SJ, Cerio R, Gribben J, Lemoine N, Bomalaski J, Li CF, Joel S, Fitzgibbon J, Chen LT, Szlosareket PW (2012) Promoter methylation of argininosuccinate synthetase-1 sensitises lymphomas to arginine deiminase treatment, autophagy and caspase-dependent apoptosis. Cell Death Dis. 3:e342.

[53] Kurlishchuk Y, Vynnytska-Myronovska, B, Grosse-Gehling P, Bobak Y, Manig F, Chen O, Merker SR, Henle T, Löck S, Stange DE, Stasyk O, Kunz-Schughart LA (2016) Co-application of canavanine and irradiation uncouples anticancer potential of arginine deprivation from citrulline availability. Oncotarget.7:73292-73308.

[54] Mörén L, Perryman R, Crook T, Langer JK, Oneill K, Syed N, Antti H (2018) Metabolomic profiling identifies distinct phenotypes for ASS1 positive and negative GBM. BMC Cancer volume, 18:167.

[55] Szlosarek PW, Klabatsa A, Pallaska A, Sheaff M, Smith P, Crook T, Grimshaw MJ, Steele JP, Rudd RM, Balkwill FR (2006) In vivo loss of expression of argininosuccinate synthetase in malignant pleural mesothelioma is a biomarker for susceptibility to arginine depletion. Clin. Can. Res. 12:7126-7131.

[56] Park IS, Kang SW, Shin YJ, Chae KY, Park MO, Kim MY, Wheatley DN, Min BH (2003) Arginine deiminase: a potential inhibitor of angiogenesis and tumor growth. Br. J. Cancer. 89:907-914.

[57] Beloussow K, Wang L, Wu J, Ann D, Shen WC (2002) Recombinant arginine deiminase as a potential anti-angiogenic agent. Can. Lett. 183:155-162.

[58] Huang Z, Hu H (2020) Arginine Deiminase Induces Immunogenic Cell Death and Is Enhanced by N-acetylcysteine in Murine MC38 Colorectal Cancer Cells and MDA-MB-231 Human Breast Cancer Cells *In Vitro*. Molecules. 26:511.

[59] Riess C, Shokraie F, Classen CF, Kreikemeyer B, Fiedler T, Junghanss C, Maletzki C (2018) Arginine-Depleting Enzymes – An Increasingly Recognized Treatment Strategy for Therapy-Refractory Malignancies. Cell Physiol. Biochem. 51:854-870.

[60] Zoua S, Wanga X, Liua P, Kea C, Xu S (2019) Arginine metabolism and deprivation in cancer therapy. Biomedi. Pharmaco. Therapy. 118:109210.

[61] Changou CA, Chen YR, Xing L, Yen Y, Frank YS, Chuang R, Cheng H, Bold RJ, Ann DK, Kung HJ (2014) Arginine starvation-associated atypical cellular death involves mitochondrial dysfunction, nuclear DNA leakage, and chromatin autophagy. PNAS.111: 14147-14152

[62] Mussai F, Egan S, Higginbotham-Jones J, Perry T, Beggs A, Odintsova E, Loke J, Pratt G, Pong KU, Lo A, Ng M, Kearns P, Cheng P, Santo CD (2015) Arginine dependence of acute myeloid leukemia blast proliferation: a novel therapeutic target. Blood. 125:2386-2396.

[63] North AK, Smith MC, Baumberg S (1989) Nucleotide sequence of a *Bacillus subtilis* arginine regulatory gene and homology of its product to the *Escherichia coli* arginine repressor. Gene. 80:29-38.

[64] Koussa HA, Mais NE, Maalouf H, Abi-Habib R, Sibai ME (2020) Arginine

deprivation: a potential therapeutic for cancer cell metastasis? A review. Can. Cell Int. 20:150.

[65] Al-Koussa H, Al-Haddad M, Abi-Habib R, El-Sibai M (2019) Human recombinant arginase I [HuArgI]-Induced arginine depletion inhibits colorectal cancer cell migration and invasion. Int J. Mol. Sci. 20:6018.

[66] Morris CR, Hamilton-Reeves J, Martindale RG, Sarav M, Ochoa Gautier JB (2017) Acquired amino acid defciencies: a focus on arginine and glutamine. Nutri. Clin. Pract. 32:30S–47S.

[67] Shan YS, Hsu HP, Lai MD, Yen MC, Chen WC, Fang JH, Weng TY, Chen YL (2015) Argininosuccinate synthetase 1 suppression and arginine restriction inhibit cell migration in gastric can. cell lines Sci. Rep. 5:9783.

[68] Spitler R, Schwappacher R, Wu T, Kong X, Yokomori K, Pilz RB, Boss GR, Berns MW (2013) Nitrosyl-cobinamide (NO-Cbi), a new nitric oxide donor, improves wound healing through cGMP/cGMP-dependent protein kinase. Cell Sign. 25:2374-2382.

[69] Zhan R, He W, Wang F, Yao Z, Tan J, Xu R, Zhou J, Wang Y, Li H, Wu J (2016) Nitric oxide promotes epidermal stem cell migration via cGMP-Rho GTPase signalling. Sci Rep. 6:30687.

[70] Sanders MA, Basson MD (2000) Collagen IV-dependent ERK activation in human Caco-2 intestinal epithelial cells requires focal adhesion kinase. J. Biol. Chem. 27538040-27538047.

[71] Rhoads J, Chen W, Gookin J, Wu G, Fu Q, Blikslager A, Rippe R, Argenzio R, Cance W, Weaver E (2004) Arginine stimulates intestinal cell migration through a focal adhesion kinase dependent mechanism. Gut. 53:514-522.

[72] Case LB, Waterman CM (2015) Integration of actin dynamics and cell adhesion by a three-dimensional, mechanosensitive molecular clutch. Nat. Cell Biol. 17:955.

[73] Vandenberg CA. Integrins step up the pace of cell migration through polyamines and potassium channels. ProcNatlAcad Sci. 2008;105(20):7109-7110.

[74] Riveline D, Thiagarajan R, Lehn JM, Carlier MF (2014) Synthetic polyamines: new compounds specific to actin dynamics for mammalian cell and fission yeast. Bio. Architec. 4:144-148.

[75] Sedgwick A, D'Souza-Schorey C (2016) Wnt signaling in cell motility and invasion: drawing parallels between development and cancer. Can. 8:80.

[76] Fujita M, Imadome K, Endo S, Shoji Y, Yamada S, Imai T (2014) Nitric oxide increases the invasion of pancreatic cancer cells via activation of the PI3K–AKT and RhoA pathways after carbon ion irradiation. FEBS Lett. 588:3240-3250.

Chapter 11

Electro-Spinning and Electro-Spraying as Innovative Approaches in Developing of a Suitable Food Vehicle for Polyphenols-Based Functional Ingredients

Mahmoud Ghorbani and Ricardo Santos Aleman

Abstract

With recent advances in medical and nutrition sciences, functional foods and nutraceuticals fortified with natural polyphenols have received a lot of attention from both health professionals and the common population in the last few years since their chemical structure allows them to exert various health effects (e.g., antioxidant, anti-inflammatory, immune, antitumor and prebiotic properties). Nonetheless, there are several hurdles to applications of polyphenols in the food system. The most critical hurdle includes polyphenols' tendency to lose their anti-oxidative properties or bioactive functionalities during food processing, as well as inclusion of poly-phenol compounds may impart an astringent or bitter taste, or introduce a degree of brown coloring causing serious sensorial impacts on food products. On this basis, interest has increased in understanding the development of new and efficient food vehicles as delivery systems for polyphenols-based functional ingredients. In this context, one approach that could augment the growth of polyphenols-based functional foods is electro-hydrodynamic processing, as the most versatile method to produce nanoscale fibers or particulates suitable for application in food technology by encapsulation to form nanoscale delivery systems.

Keywords: Polyphenols, Functional foods, Nutraceuticals, Electro-hydrodynamic processing, Electro-spinning, electro-spun fibers, electro-spraying, electro-sprayed particles and core-shell structured fibers or particles

1. Introduction

During the past few years, there has been increasing awareness about health-promoting effects of dietary polyphenols abundant in functional foods (*natural or processed foods that contain known or unknown biologically-active compounds; which in defined, effective and non-toxic amounts, provide a clinically proven and documented health benefit for prevention, management or treatment of chronic disease*) [1] and nutraceuticals (Natural bioactive or chemical compounds that besides offering a nutritional value provide health-promoting, and disease curing or prevention properties) [2] as functional ingredients to provide a health benefit beyond basic nutrition. Polyphenols are naturally occurring compounds in plants endowed with antioxidant and anti-inflammatory, immune, antitumor and prebiotic properties [3]; widely present in a wide variety of fruit, vegetables, seeds, herbs and beverages in particular in beer, red wine, fruit juice, coffee, tea, cocoa, chocolate and dry legumes and cereal [2, 4–8] and are therefore an integral part of the human diet [9]. From a chemical standpoint, this large family of secondary plant metabolites constitutes a large heterogeneous class of compounds characterized by hydroxylated phenyl moieties [2, 10–12] with more than 8000 identified compounds so far [2, 13–16]. Numerous food matrixes naturally enriched with dietary polyphenols are the most potent sources of plant-derived bioactive compounds eliciting many beneficial health effects in man. Despite their interesting biologic properties, their presence and abundance in nature, chemical instability of polyphenols during processing, handling and storage [17], the low oral bioavailability [3, 8] and rapid fast-pass metabolism of polyphenols might greatly restrict their biologic effects and applications in the functional foods and nutraceuticals [18]. Further, these extracts or their isolated individual compounds have the potential to interact with other compounds in the environment in particular proteins, resulting in formation of sensory characteristics and organoleptic properties in foods and beverages including High Molecular Weight (HMW) brown color [19], flavor and taste attributes like bitterness or astringency [4, 5, 20, 21].

In conclusion, the main drawback to using polyphenols as functional ingredients to develop functional food products and dietary supplements, nutraceuticals is their poor bioavailability and the variable bio-accessibility in the human body and variety of molecular interactions between polyphenols and other food components; however, in order to preserve the structural integrity, polyphenols need to be shielded by a finishing formulation that is, able to protect and to deliver them to the physiologic targets without losing any bioactivity [22]. Encapsulation system applied to polyphenols through the development of micro and nano-sized particle systems, as a reliable tool to overcome the problems related with the direct use of dietary polyphenols in their free form in food matrixes will ensure protection of these bioactive compounds and additionally, functional properties to the final product [23, 24]. The administration of encapsulated polyphenols instead of their free form can overcome the drawbacks of their instability; relieve unpleasant tastes or flavors in food matrixes, and as well improve their bioavailability in gastrointestinal tract (GIT). Numerous encapsulation processes have been developed to encapsulate polyphenols-based functional ingredients each with their own merits and demerits including Ionic gelation, layer-by-layer deposition, extrusion, coprecipitation, coacervation and phase separation, spray/freeze drying, emulsification/emulsion polymerization, inclusion complexation, liposome entrapment, fluidized bed coating, supercritical fluid, etc. [22, 25]. However, polyphenols are oxidized easily due to light, heat and oxidant; therefore, chemical instability of these compounds is the major constraint to encapsulation through the processes stated above since they mostly require heating and/or pressure, and the use of strong and toxic organic solvents or expensive equipment [26]. In this regard, electro-hydrodynamic processing referring to the dynamics of electrically charged fluids has emerged as an Innovative and environmentally friendly alternative technology for encapsulation that needs neither temperature nor expensive equipment; therefore, heat-sensitive compounds such as dietary polyphenols may be successfully processed and also, the use of organic solvents can be avoided by adjusting some processing conditions (i.e., use of molten polymer).

In this chapter the drawbacks related to the incorporation of dietary polyphenols as possible functional ingredients in food formulations and novel strategy to improve their efficiency is discussed; starting from bio-accessibility and bioavailability of polyphenols, continuing to the chemical structure of polyphenols, nature of food matrix as well as interaction with other food constituents and also food processing influencing their stability and, consequently their availability and concluding to consider electro-hydrodynamic processing as novel strategy to improve delivery efficiency and controlled release of polyphenols.

2. Bio-accessibility and bioavailability of dietary polyphenols

Polyphenols, a class of chemical compounds consisting of one or more hydroxyl groups (OH) attached directly to at least two phenyl rings lacking nitrogen-based functional group in their most basic structural expression are plant secondary natural metabolites, ubiquitous in all vascular plants arising biogenetically from either the shikimate derived phenylpropanoid and/or the polyketide pathway(s) [27–29]. Polyphenols are classified into diverse classes on the basis of their chemical structures and/or the attachment of hydroxyl groups to the aromatic rings structure while the main classes of polyphenols consist of flavonoids, phenolic acid, tannins (phenolic polymers), phenylethanoid [30], stilbenes and lignans [4, 27, 31–34].

Flavonoids are recognized as one of the largest and most abundant type of polyphenols in the diet that constitute approximately two-thirds of intake. The core structural unit of flavonoids encompasses a common carbon skeleton of diphenyl propane in which two benzene rings (A, B) are linked by a linear three-carbon chain, forming a closed pyran ring with the A benzene ring. Flavonoids (**Figure 1**) are then subdivided into several subclasses based on the central pyran ring's oxidation state that the most important of them follow as: flavonols (e.g. Quercetin, kaempferol), flavones (e.g. luteolin, apigenin), anthocyanins (e.g. cyaniding, pelargonidin), flavanones (e.g. naringenin, hesperetin), flavanols also known as flavan-3-ols (e.g. catechin, epicatechin), and isflavones (e.g. daidzein, genistein) [4, 7, 32, 35, 36].

Phenolic acids (**Figure 2**) ubiquitously found in plant materials at varying levels are divided into two sub-classes hydroxybenzoic (e.g., gallic, phydroxybenzoic, vanillic, syringic, and protocatechuic acids) that are often the component of a complex structure like lignins and hydrolyzable tannins, and hydroxycinnamic acids (e.g., p-coumaric, caffeic, ferulic, sinapic and cinnamic acids). Further, decarboxylation of benzoic acid and phenylpropanoid derivatives leads to the formation of simple phenols namely, phenol, o-cresol, 4-ethylphenol, guaiacol, 4-vinylguaiacol and eugenol [4]. Some phenolic acids are found in free form in red fruits and vegetables such as strawberries and blackberries, black radish, onions, and tea [35], but hull, bran, and seed contain phenolic acids that in bound form that are released by acid, alkali, and enzyme hydrolysis [7].

Tannins are compounds of intermediate to high molecular weight (500–20,000 Da) [37] and are more extensively hydroxylated [35]. Depending on their structures, tannins are classified into two major groups including hydrolyzable and non-hydrolyzable tannins, also called condensed tannins or proanthocyanidins (PAs). Hydrolyzable tannins (HTs) consist a center of glucose or a polyhydric alcohol partially or completely esterified with simple phenolic acids such as gallic acid (gallotannins) or hexahydroxydiphenic acid (ellagitannins) while condensed tannins are oligomers and polymers of flavonoids, specifically flavan-3-ols [4, 31, 34, 37, 38]. Ellagitannins (e.g., punicalagin [39]) (**Figure 3**) are esters of hexahydroxydiphenoic acid and monosaccharide (most commonly glucose) naturally occurred in some fruits (pomegranate, strawberry, blackberry, and raspberry), nuts (walnuts, almonds) and seeds. While ellagitannins are slowly hydrolyzed in

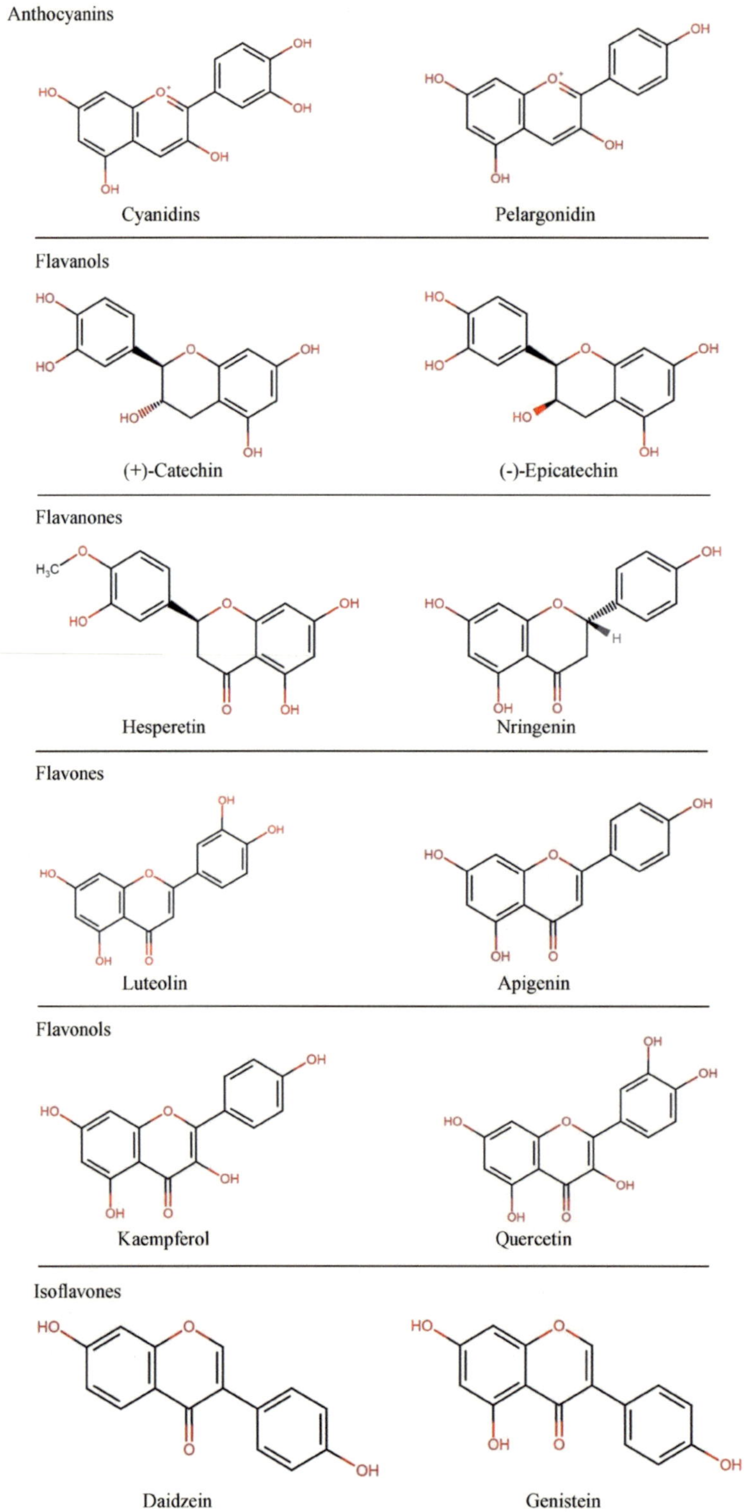

Figure 1.
Structures of flavonoids.

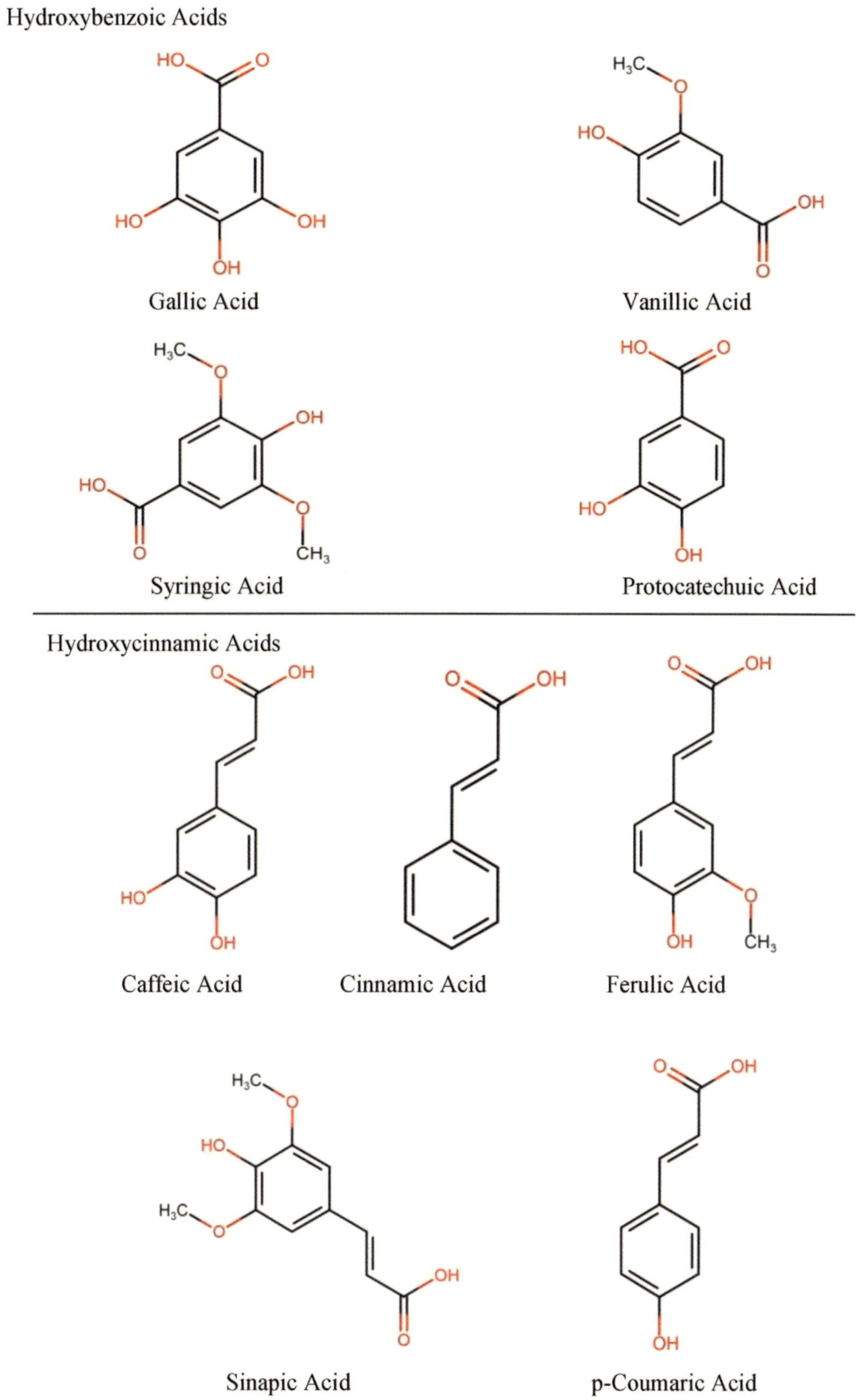

Figure 2.
Structures of phenolic acids.

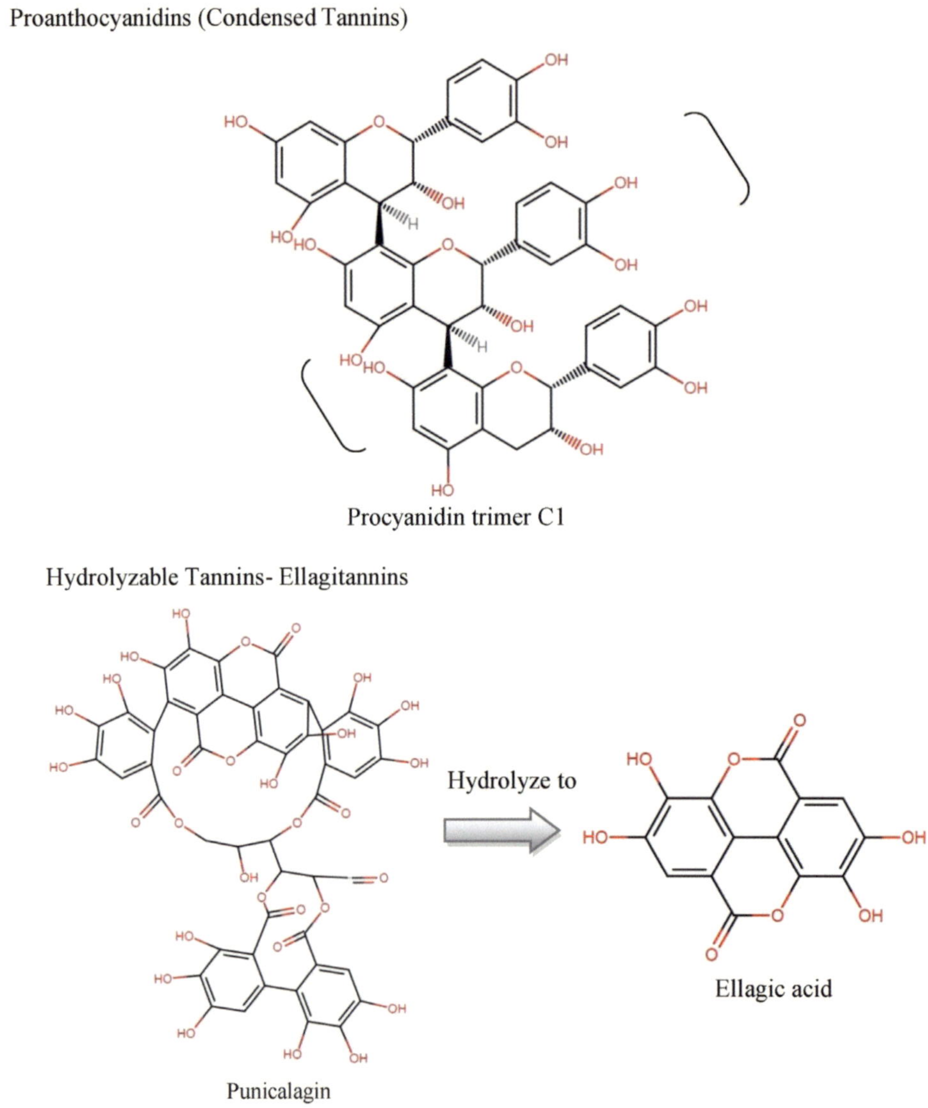

Figure 3.
Structures of tannins.

the digestive tract releasing the ellagic acid molecule, the main determinant of the physicochemical properties and biologic activity is their chemical structure [39]. Based on the hydroxylation pattern of A- and B-rings, non-hydrolyzable taanins or PAs can be divided into procyanidins (e.g. procyanidin trimer C1), propelargonidins and prodelphinidins [4, 37].

Tyrosol, hydroxytyrosol and oleuropein (**Figure 4**) are the prominent types in phenylethanoid class, found mainly in olive leaf and oil [30, 36]. Besides, Rueda et al. reported that the minor values of both tyrosol and hydroxytyrosol present in other edible virgin vegetable oils (argan, wheat germ and sesame) [40].

Stilbenes are a family of hydrocarbons that share with similar chemical structure to flavonoids consisting of two phenyl groups linked by a methylene group (or "methylidene") that occur naturally in either a cis or a trans configuration [32, 33]. Resveratrol, pterostilbene, and piceatannol are primary representatives [4, 41] while resveratrol (3,5,40-trihydroxy-trans-stilbene) (**Figure 5**) presented in the both cis and trans

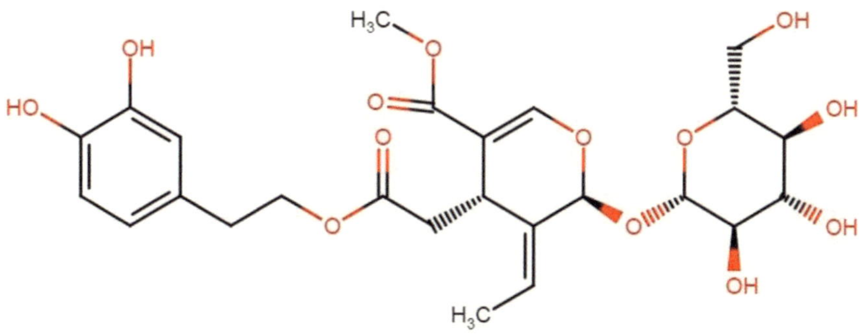

Oleuropein

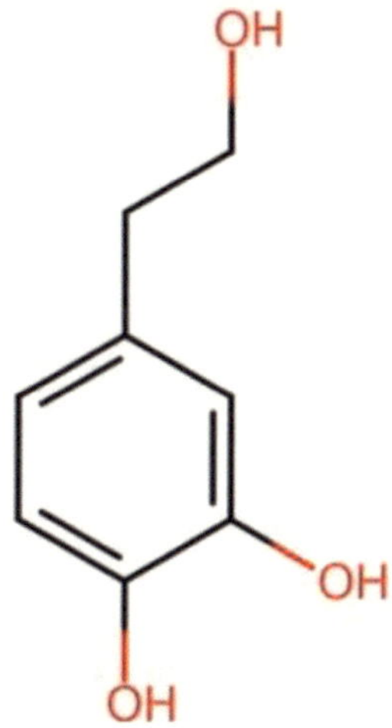

Hydroxytyrosol

Figure 4.
Structure of phenylethanoid.

isomeric forms as well as its derivatives including glucosylated, prenylated, methylated, and hydroxylated modifications are the most widely known stilbenes that do are important from a health perspective. Polydatin, also known as piceid (resveratrol-3-O-β-mono-D-glucoside, POLY) is a glucoside of resveratrol in which the glucoside group bonded in position C-3 substitutes a hydroxyl group. This substitution brings about conformational changes of the molecule resulting in changes in the biologic properties. There exist numerous reports suggesting trans-resveratrol to be the more stable form in nature and the most bioactive form of this molecule though upon exposure to UV light, trans-resveratrol (isomeric) can be readily converted to cis-resveratrol (isomeric) and also been unstable when exposed to high pH [42]. Stilbenes are reported to be present in grapes, berries, peanuts and red wine [4, 31, 32, 42].

Secoisolariciresinol and matairesinol (**Figure 6**) characterized by the union of two phenylpropanoid (C6C3) units with β-β or C8-C8 linkages are the major types of lignans [4, 28]; being present in the free form or glycosidically linked to different fiber-associated polyphenols [30]. Flaxseed [4], rye bran and flour, various berry fruits [43], legumes, vegetables, black and green tea [43] are sources of secoisolariciresinol and matairesinol [32].

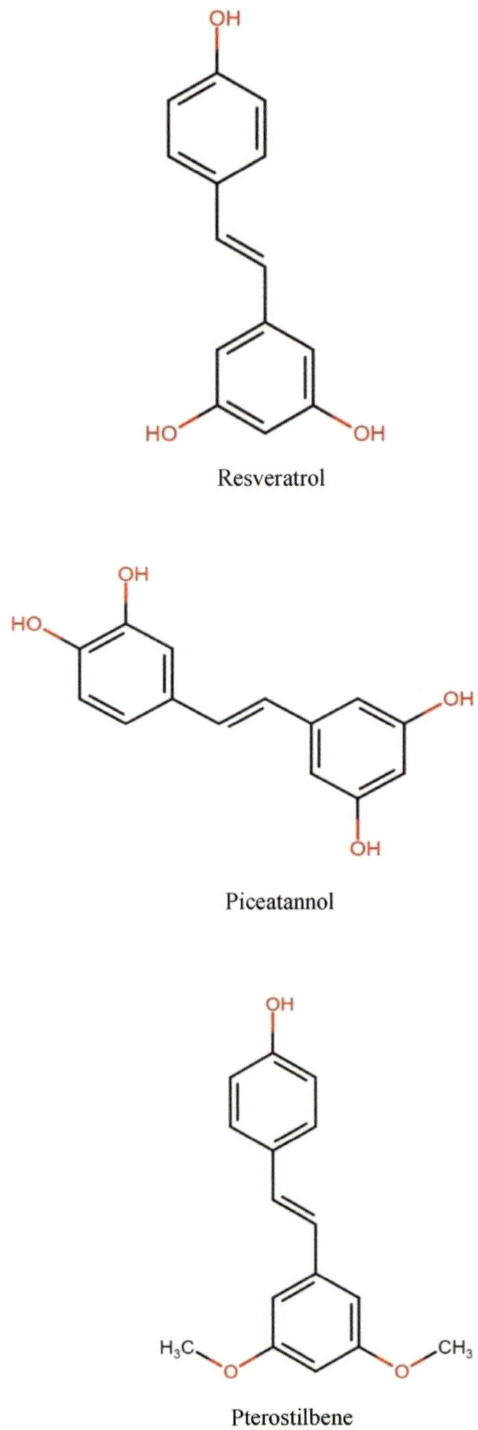

Figure 5.
Structures of stilbenes.

These compounds are potent antioxidant compounds able to counteract oxidative stress and chronic inflammation that could be separated or recovered from food waste and agricultural by-products [35, 44]; thus, this group of natural products could be used as additives and functional ingredients in the novel

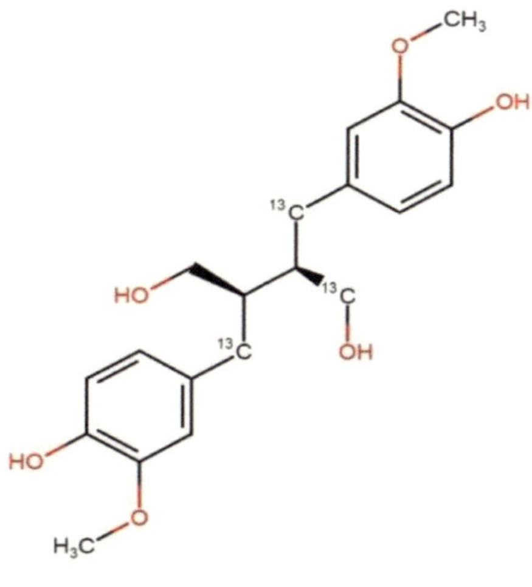

Secoisolariciresinol

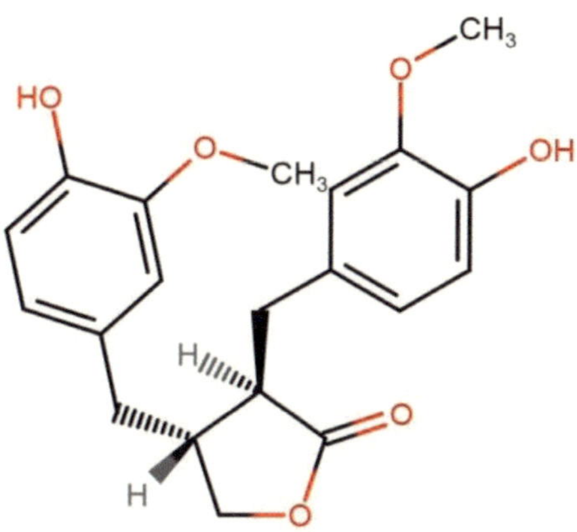

Matairesinol

Figure 6.
Structures of lignans.

functional foods and beverages [45]. There are numerous *in-vitro* and *in-vivo* studies that have proved, poly-phenol intake is associated with a wide spectrum of potential biologic activities related to health-promoting properties including; anti-inflammatory, antioxidant, pro-DNA repair, anti-helmintic, anti-microbial, anti-viral, insecticidal, anti-cancer, anti-aging, and anti-radiation effects [36]; it is known as a preventive for the certain cancers, cardiovascular diseases, type 2 diabetes, osteoporosis, pancreatitis, gastrointestinal problems, lung damage, and neurodegenerative diseases [11, 46–51]. Due to these healthy characteristics, there is increasing interest in the development of new product with enhanced potential health-promoting action while the effectiveness of polyphenols depends on preserving the stability, bioactivity, and bioavailability of the active ingredients.

Once ingested, polyphenols are metabolized by the human body as xenobiotic compounds which may undergo several biologic processes limiting their potential uptake by humans [10, 52, 53]. However, putative bioactivity and bio-efficacy [18, 54, 55] of dietary polyphenols are therefore strictly associated with related to the concepts of bio-accessibility and bioavailability. In this regard, to exploit the real biologic potential effect it is crucial to know the quantity of polyphenols properly released from food matrix during gastrointestinal digestion, which is the potential absorption of polyphenols to be available for subsequent metabolic pathways. This parameter is known as bio-accessibility, which can be defined as: "the fraction of a bioactive compound within the food matrix where it is contained, that has the potential to be bio-available and reach systemic circulation; this means that it has been released from the food matrix by the action of digestive enzymes [46, 52, 56, 57]. Potentially, bioavailability refers to the fraction of the bio-accessible ingested nutrient, either parent compound or active metabolite that reaches the systemic circulation and becomes available at the site of the action where it can exert biologic effects [46, 52, 58, 59]; in the case of polyphenols, this is the amount of polyphenols which can be absorbed and exert effects on specific tissues. As per this principle, bioavailability is a process that depends on the intrinsic and extrinsic factors of the host; this means that, the process depends on the food matrix ingested and the gastrointestinal conditions within the individual. Investigations have revealed which the bioavailability of polyphenols contains seven main digestive processes: (1) the release of polyphenols from food matrix (bio-accessible polyphenols); (2) changes in polyphenols during gastric and small intestine digestion conditions; (3) the cellular uptake of aglycones and conjugated forms of polyphenols; (4) microbial metabolism conjugated non-bio-accessible fraction polyphenols by the colonic micro-biota; (5) phase I (oxidation, reduction and hydrolysis) and particularly phase II (conjugation) bio-transformations in the enterocytes and then the hepatocytes, followed by generating methyl, glucuronide and sulfate derivatives; (6) transit to systemic circulation and tissue distribution; (7) urinary excretion or excretion back into the gut via bile and pancreatic juices [52, 57, 60]. Numerous studies have pointed out that the bioavailability of polyphenols is rather low and the magnitude of the relative urinary excretion of the intake fluctuates from 0.3% for anthocyanins to 43% for isoflavones that demonstrates the great variability in the bioavailability from one poly-phenol to another and the most abundant dietary poly-phenol was not necessarily the one leading to the highest levels of active metabolites in plasma [34, 51, 61]. Consequently, to explore and to determine the mechanisms of action of dietary polyphenols and their role in disease prevention, it is crucial to understand the factors that constrain bio-accessibility and bioavailability of polyphenols, some related to the food (e.g., chemical structure of the compound, food matrix, food processing and dose) while others depend on the individual (e.g., gastric emptying, intestinal transit time, composition of the micro-biota) [58, 62, 63]. However, the discussion of factors influencing the bio-accessibility and bioavailability of polyphenols will focus on food related factors such as the polyphenols' chemical structure, the nature of the food matrix and food processing, since these are the first hurdles that polyphenols face prior to absorption.

The interested reader may consult some of the accounts of the concentration and bio-accessibility of poly-phenol compounds with potential antioxidant activity as affected by simulated *in vitro* digestion; for a more detailed description, see the references [64–68].

2.1 Factors leading to degradation of polyphenols and low bioavailability

Dietary polyphenols to exert their health-promoting effect need to endure the food processing conditions; second, could be released from the food matrix and become bio-accessible in the gastrointestinal tract, and then undergo metabolism and reach the target tissue of interest. As a result, chemical structure of polyphenols, nature of food matrix as well as interaction with other food constituents in particular proteins, lipids and carbohydrates and food processing play a significant role upon coming to bio-accessibility and bioavailability of polyphenols, since they represent the first step in the challenging journey of well-known dietary polyphenols reaching the target tissues.

2.1.1 Chemical composition and structure of polyphenols

Dietary polyphenols has been a most exotic topic in modern food chemistry not only as structural diversity and major plant secondary metabolites, but also as compounds that express a wide range of applications in various aspects of commercial as well as general public interests [27]. The importance of their molecular structure lies in the fact that the molecular size, the parent structure, degree of polymerization or glycosylation, solubility, hydrophobicity, isomer configuration and conjugation with other phenolics [5, 34, 60, 69, 70] have a strong impact on their bio-accessibility and bioavailability. Most of the polyphenols, especially those containing adjacent dihydroxyl groups (e.g. catechins and procyanidins) are especially prone to polymerization and loss through oxidation [71]. Relatively, simple phenolic derivatives such as phenolic acids (e.g., gallic acid, caffeic acid, vanillin, and coumaric acid) and flavonoids including isoflavones are readily absorbed through the gut tract that are followed by catechins, flavanones, and quercetin glucosides [34]. On the contrary, proanthocyanindins which are compounds of high molecular weight are very poorly absorbed as well as galloylated tea catechins and the anthocyanins [34, 60]. Among the various poly-phenol compounds, reported bioavailability is so highly variable that the highest bioavailability has been reported for isoflavones, followed by flavanols, flavanones and flavonol glycosides, while the proanthocyanidins, flavanol gallates and anthocyanidins are the most poorly absorbed [34, 49, 51, 61, 72].

2.1.2 Food matrix

Food products fortified with dietary poly-phenol rich extracts may lead to changes in the nutritional, chemical and rheological properties of the fortified food. Apart from potential biologic activities related to health-promoting properties, when included in a food product depending on the type of extract, the poly-phenol compounds may impart an astringent and/or bitter taste, or introduce a degree of brown coloring [5, 21, 53, 73, 74]. Concerning taste, PAs resulting from oxidative reactions are mostly responsible for some unpleasant organoleptic properties such as astringency and bitterness [51, 70, 71]. "*Astringency is a tactile sensation defined as dryness, tightening and puckering sensations perceived in the oral cavity during the ingestion of astringent molecules, mainly tannins, alums and some metal ions*" [74]. Concerning color, it is worth to note that anthocyanins are one of the most important natural pigments though they represent a problem owing to their high instability [74].

Polyphenols possess the ability to interact, both with food matrix constituents in particular carbohydrates, lipids and proteins, as well as with biologic compounds, namely proteins. All these interactions can affect the accessibility and availability both of polyphenols and other compounds as well as organoleptic properties of fortified food products and consumer acceptance. Polyphenols interact mostly to components of food matrix through non-covalent hydrophobic interactions but in the cases of interactions between polyphenols and proteins or/and carbohydrates, hydrogen bonds also contribute significantly. Nonetheless, some covalent bonds may also occur under certain food processing conditions [74]. Polyphenols form complexes with proteins that can be occurred by non-covalent interaction (reversible), primarily driven by hydrogen bonds and hydrophobic interactions and covalent interaction (mostly irreversible) after poly-phenol activation either by oxidation, i.e. as quinones, or as carbocations resulting from proanthocyanidin cleavage under hot acidic conditions [38]. As a whole, polyphenols with elevated molecular weight and a more abundance of hydroxyl group, which provide more than one site for interaction reveal a higher affinity to interact with proteins [19, 75]. However, tannins are polyphenols capable of precipitating proteins from aqueous solutions, which are synthesized via the shikimic acid pathway [4]. In terms of non-covalent associations, amino acids (e.g., alanine, valine, isoleucine, leucine, methionine, phenylalanine, tyrosine, tryptophan, cysteine and glycine) may react to tannins through hydrophobic interactions and hydrogen bonds. From a mechanistic point of view, the hydrogen bindings with the carboxyl group of proteins are associated with capability of the hydroxyl groups of polyphenols to donate a hydrogen atom to the nitrogen or oxygen molecule of amino acids (e.g., lysine, arginine, histidine, asparagine, glutamine, serine, threonine, aspartic acid, glutamic acid, tyrosine, cysteine and tryptophan) [19, 38]. In terms of covalent interactions, polyphenols namely tannins can be oxidized under alkaline conditions and reactive oxygen species through enzymatic and non-enzymatic oxidation reactions causing the generation of highly reactive quinone radicals. Following the oxidation step, resultant quinone radical reacts to another quinone radical that is named condensation reaction to form a dimer; a high molecular weight brown color brown color pigment named as tannin that can further interact with amino acids in a polypeptide chain through covalent binding. At the end, these dimers remain highly reactive, in turn they are re-oxidized and cross linked to another polypeptide chain [19, 73]. In this sense, research studies carried-out by Rodríguez-Roque et al., assessed impact of food matrix (water-, milk- and soymilk-fruit juice beverages) on the *in vitro* bio-accessibility of phenolic compounds and hydrophilic antioxidant activity from fruit juice-based beverages and observed that the combination of a blended fruit juice with milk or soymilk could decrease the bio-accessibility of dietary polyphenols due to the formation of complexes among these compounds and proteins of milk and soymilk though the protein precipitation could mask the poly-phenol astringent or bitter taste [67, 68, 76].

Similar to polyphenols strongly associated with proteins, evidence reveals that polyphenols can also form complexes with Carbohydrates (digestible and non-digestible) that are, highly dependent on the molecular weights of the polyphenols, the hydrophilicity of the poly-phenol, and the structure of the carbohydrate (high molecular weight, low solubility, and conformational flexibility) [75, 77]. The associations between carbohydrates and polyphenols can also affect the organoleptic properties [78] but depending on the compounds, these interactions could have positive and negative effects [74]. Besides influencing astringency perception by tannins-proteins interactions, PAs-carbohydrates associations also can lead to an astringency taste and bitterness modulation into fortified foods while tannins have less affinity to carbohydrates than to proteins due to the strong hydrogen bond formation with protein's carboxyl group. Apart from tannins, anthocyanins have the capability to interact with carbohydrates [78], in turn the association with carbohydrates could lead for, on one hand, to a lower extraction yield, and therefore lower color intensity on the final product, and on the other hand to stabilization and enhancement of anthocyanins color [74]. Interestingly, some studies reported that bioavailability of polyphenols could be reduced due to the interaction with polysaccharides [46, 52, 63, 79] while other studies revealed that polysaccharides from human diet could enhance the polyphenols' uptake [78]; however, polyphenols interactions with dietary fibers (non-digestible) are of particularly significant since non-digestible polysaccharides may play role of "ploy-phenol carrier" as an "essential physiologic function" of polysaccharides contributing to the overall health effects of fiber-rich diets [38, 78].

Concerning polyphenols interactions with lipids, only a few studies have investigated the effect of dietary lipid–poly-phenol interactions on taste that have not been of special importance, except in case of plant oils—primarily the one made from olives [46, 74, 78]. In contrary, it should be highlighted that polyphenols can decrease the synthesis of fats and fatty acids in the liver, or delay their absorption in intestines [74, 78]. As reviewed, dietary polyphenols are known to form complexes with macromolecules and to affect on antioxidant values and bio-accessibility that can impair bioavailability of both polyphenols and macromolecules.

Apart from interactions between polyphenols and macromolecules, polyphenols are also known for their strong metal-chelating capabilities. A number of polyphenols (e.g., phenolic acids, flavonoids [79] and also tannins [80]) efficiently chelate trace metal ions, such as Al^{3+}, Fe^{3+}, and Cu^{+} [26, 81, 82] that undergo redox cycling reactions and possess the ability to produce reactive radicals such as superoxide anion radical and nitric oxide in biologic systems [83]. This action attributed to the galloyl and catechol groups of poly-phenol compounds results in the diminution intestinal absorption of minerals and trace elements [79, 80].

2.1.3 Food processing

Processing of plant foods and fortified foods with polyphenols-based functional ingredients exert a main impact on bio-accessibility of polyphenols and consequently bioavailability as well as their content and potential antioxidant activity, which depends on the type poly-phenol-rich food, the nature and location of poly-phenol compounds in the food matrix, the intensity and duration of treatment, as well as presence of components that affect absorption efficiency [46, 48, 49, 63, 69, 76, 84–88]. In overall, the bioavailability of polyphenols is determined by their bio-accessibility [89]; this means that the availability as well as accessibility of polyphenols is likely to be affected by the processing methods since food processing brings about changes in chemical composition and structure of polyphenols and in molecular interactions that have an influence on the capacity of a given compound to be extracted during digestion [69] and thus, it may also increase or diminish the bio-accessibility of such health-promoting components [76, 90] that these components may be those that are added to functional foods or are found naturally in foods such as dietary polyphenols. In other words, food processing can give rise to the degradation of polyphenols; thus, reducing their bio-accessible and non-bio-accessible fractions as well can result to chemical or physical modifications in food in such a way that fosters the release and absorption of polyphenols during digestion [69, 70]. This could be explained different effects found according to the intensity and duration of treatment depict two different scenarios including; (1) increase in the bio-accessible and non-bio-accessible contents but no effects on compounds bio-accessibility; (2) decrease in the bio-accessible and non-bio-accessible contents but a modified (increased or decreased) bio-accessibility [69, 70, 76, 90, 91]. In this sense, precise discernment of the concentration of dietary polyphenols reaching the bio-accessible fraction is much more important than the concentration of these compounds in the corresponding food products [70, 76]. Research study carried-out by Rodríguez-Roque et al. [76], assessed impact of processing [high-intensity pulsed electric fields (HIPEF); high-pressure processing (HPP); and thermal treatment (TT)] on the *in vitro* bio-accessibility of phenolic compounds and hydrophilic antioxidant activity from fruit juice-based beverages and observed an improvement up to 38% in the bio-accessibility of individual polyphenols (caffeic and p-coumaric acids from both water-fruit juice beverage (WB) and milk-fruit juice beverage (MB); chlorogenic and ferulic acids from MB; hesperidin and rutin from all beverages after treatments), mainly by non-thermal methods (HIPEF and HPP). On the contrary, all treatments did not change the bio-accessibility of caffeic and chlorogenic acids from soymilk-fruit juice beverage (SB), as well as naringenin from both WB and MB but diminished the bio-accessibility of ferulic acid from WB. Besides, bio-accessibility of chlorogenic and p-hydroxybenzoic acids from WB were also significantly reduced by HIPEF (between 10 and 11%) and TT (between 11 and 24%) [76]. In another research study by Ribas-Agustí et al., results clearly showed the overall decrease in bio-accessible polyphenols after pulsed electric fields treatments (1.8 and 7.3 kJ kg^{-1}) can be linked to decreased contents in undigested apple, which was probably consequence of their degradation due to process-induced oxidative reactions [91].

Apart from bio-accessibility, polyphenols may lose their antioxidant activities or bioactivity during processing since they are oxidized easily when exposed to high temperature, oxygen and enzymes [46, 92–96], which should be taken into account when processing poly-phenol-rich food matrixes [54, 97, 98]. Dietary polyphenols are degraded at high temperature; thus, thermal treatments diminish the poly-phenol content in polyphenols-rich fortified food and jeopardize the amount of bio-available polyphenols (referring to bio-accessible fraction) due to the loss of thermo-labile phenols or their polymerization [99–101]; however, it has been shown that based on thermal processing technique used, high temperature also gives rise to other modifications turning into positive for the dietary polyphenols bioavailability such as degradation or modification of cell wall polysaccharides, proteins and other matrix factors that may lead to compounds more accessible to absorption [69, 70, 92, 95, 102, 103]. In these cases, the effect of processing can be accounted for by multiplying raw food poly-phenol content by a retention factor (RF), which describes the change in poly-phenol content for a given food due to a given process and was then calculated from the poly-phenol contents of corresponding raw and processed foods and the yield factor value [97, 104]. RFs are calculated according to Eqs. (1) and (2) as follows:

$$\text{Retention factor}\,(RF) = \frac{\textit{concentration of polyphenol in processed food}}{\textit{Concentration of polyphenol in raw food}} \times \text{Yield Factor} \tag{1}$$

$$\text{Yield Factor} = \frac{\textit{weight of food after processing}}{\textit{wieght of food before processing}} \tag{2}$$

Thus, the calculated values of RF < 1 indicate a reduced poly-phenol content in the processed food whereas RF = 1 and RF > 1 indicate full retention or an increase, respectively [97]. Most studies evaluating the impact of thermal and non-thermal processing on poly-phenol compounds in terms of quantity and availability have focused on traditional processing technologies such as heat-related thermal treatment (TT) and on novel emerging non-thermal techniques such as HPP, HIPEF and have used to preserve manufactured food or cooking, as well as during the food preparation, i.e. pretreatments on raw material to obtain food. Some interesting studies evaluating the effect of thermal and non-thermal treatments on retention of polyphenols and their antioxidant capacity as well as bio-availability have been conducted [84, 87, 92, 94, 98, 105–108]. The main finding concerning the impact of thermal and non-thermal processing on retention of individual polyphenols is summarized below and in **Table 1**.

Poly-phenol components	Food before processing	Food after processing	Retention Factors					Ref
			Mean RF Value	Min	Max	SD	N	
Hesperetin	Orange juice	Orange pure juice, pasteurized 70°C,30 s	1.03	1.00	1.07	0.03	2	[110, 111]
	Orange juice	Orange pure juice, high-pressure proccessed (400 MPa/40°C/1 min)	1.27	1.16	1.39	0.11	2	[110, 111]
	Orange juice	Orange pure juice, high intensity pulsed electric fields (35 kV cm-1/750 μs)	1.00	0.96	1.04	0.04	2	[110, 111]
Naringenin	Orange juice	Orange pure juice, pasteurized 70°C,30 s	0.91	0.84	0.99	0.07	2	[110, 111]
	Orange juice	Orange pure juice, high-pressure processed (400 MPa/40°C/1 min)	1.16	1.13	1.20	0.03	2	[110, 111]
	Orange juice	Orange pure juice, high intensity pulsed electric fields (35 kV cm-1750 μs)	0.9	0.87	0.93	0.03	2	[110, 111]
Myricetin	Strawberry juice	Strawberry pure juice, pasteurized 90°C, 30 & 60 s	0.93	0.90	0.95	0.03	2	[112]
	Strawberry juice	Strawbery pure juice, high-intensity pulsed electric fields(35 kV/cm for 1700 μs)	1.00	1.00	1.00	0	1	[112]
Kaempferol	Strawberry juice	Strawberry pure juice, pasteurized 90°C, 30 & 60 s	1.03	1.01	1.05	0.02	2	[112]
	Strawberry juice	Strawbery pure juice, high-intensity pulsed electric fields (35 kV/cm for 1700 μs)	1.05	1.05	1.05	0.00	1	[112]
Ferulic acid	Milk- fruit juice beverage	Milk-fruit juice beverage, high-intensity pulsed electric fields (35 kV cm-1800 μs)	0.89	0.89	0.89	0.00	1	[76]
	Milk- fruit juice beverage	Milk-fruit juice beverage, high-pressure processed (400 MPa/40°C/5 min)	0.89	0.89	0.89	0.00	1	[76]
	Milk- fruit juice beverage	Milk-fruit juice beverage, pasteurized 90°C,60 s	0.81	0.81	0.81	0.00	1	[76]
	Soymilk- fruit juice beverage	Soymilk-fruit juice beverage, high-intensity pulsed electric fields fields (35 kV cm-1800 μs)	0.82	0.82	0.82	0.00	1	[76]
	Soymilk- fruit juice beverage	Soymilk-fruit juice beverage, high-pressure processed (400 MPa/40°C/5 min)	0.81	0.81	0.81	0.00	1	[76]
	Soymilk- fruit juice beverage	Soymilk-fruit juice beverage, pasteurized 90°C,60 s	0.66	0.66	0.66	0.00	1	[76]

Poly-phenol components	Food before processing	Food after processing	Retention Factors					Ref
			Mean RF Value	Min	Max	SD	N	
p-coumaric acid	Water-fruit juice beverage	Water-fruit juice beverage, high-intensity pulsed electric fields fields (35 kV cm-1800 μs)	0.87	0.87	0.87	0.00	1	[76]
	Water-fruit juice beverage	Water-fruit juice beverage, high-pressure processed(400 MPa/40°C/5 min)	0.91	0.91	0.91	0.00	1	[76]
	Water-fruit juice beverage	Water-fruit juice beverage, pasteurized 90°C,60 s	0.78	0.78	0.78	0.00	1	[76]
p-hydroxybenzoic acid	Water-fruit juice beverage	Water-fruit juice beverage, high-intensity pulsed electric fields fields (35 kV cm-1800 μs)	0.69	0.69	0.69	0.00	1	[76]
	Water-fruit juice beverage	Water-fruit juice beverage, high-pressure processed(400 MPa/40°C/5 min)	0.82	0.82	0.82	0.00	1	[76]
	Water-fruit juice beverage	Water-fruit juice beverage, pasteurized 90°C,60 s	0.64	0.64	0.64	0.00	1	[76]
Chlorogenic acid	Fruit juice-soymilk beverage	Fruit juice-soymilk beverage, high-intensity pulsed electric fields fields (35 kV cm-800 & 1400 μs)	0.91	0.85	0.98	0.06	2	[113]
	Fruit juice-soymilk beverage	Fruit juice-soymilk beverage,, pasteurized 90°C,60 s	0.81	0.81	0.81	0.00	1	[113]
Sinapic acid	Fruit juice-soymilk beverage	Fruit juice-soymilk beverage, high-intensity pulsed electric fields fields (35 kV cm-800 & 1400 μs)	0.99	0.99	1.00	0.00	2	[113]
	Fruit juice-soymilk beverage	Fruit juice-soymilk beverage,, pasteurized 90°C,60 s	0.87	0.87	0.87	0.00	1	[113]
Coumaric acid	Fruit juice-soymilk beverage	Fruit juice-soymilk beverage,, pasteurized 90°C,60 s	0.97	0.97	0.97	0.00	1	[113]
Apigenin	Fruit juice-soymilk beverage	Fruit juice-soymilk beverage, high-intensity pulsed electric fields fields (35 kV cm-800 & 1400 μs)	0.89	0.79	1.00	0.1	2	[113]
	Fruit juice-soymilk beverage	Fruit juice-soymilk beverage,, pasteurized 90°C,60 s	0.69	0.69	0.69	0.00	1	[113]

DOI: http://dx.doi.org/10.5772/intechopen.98626

Poly-phenol components	Food before processing	Food after processing	Retention Factors					Ref
			Mean RF Value	Min	Max	SD	N	
Clorogenic acid	Apple, Whole unpeeled	apple, high-intensity pulsed electric fields fields (0.4 kV cm^{-1}, 5 pulses (0.01 kJ kg^{-1}, 20 μs total treatment time)	0.88	0.75	1.02	0.13	2	[92, 114]
	Apple, Whole unpeeled	Apple, high-intensity pulsed electric fields fields (2.0 kV cm^{-1}, 35 pulses (1.8 kJ kg^{-1}, 140 μs total treatment time)	0.58	0.58	0.58	0.00	1	[114]
	Apple, Whole unpeeled	Apple, high-intensity pulsed electric fields fields (3.0 kV cm^{-1}, 65 pulses (7.3 kJ kg^{-1}, 260 μs total treatment time)	0.33	0.33	0.33	0.00	1	[114]

Table 1.
Showing retention factors for individual polyphenols in foods and beverages with the processing technologies as following: Pasteurization, High-intensity pulsed electric fields and high-pressure processed- Analysis by chromatography after hydrolysis [56] & Yield factor values collected based on USDA National Nutrient Database for Standard Reference, Release 23 [109].

3. Encapsulation of polyphenols with electro-hydrodynamic techniques

The incorporation of polyphenols as possible functional ingredients in fortified food products, in particular due to their antioxidant capacity is becoming a growing area of research for the food industry. Nevertheless, their application as bioactive components is often hindered by their poor stability and solubility during food processing and storage, burst release and their low bioavailability or chemical instability when exposed to the conditions of the upper GIT which can significantly compromise their envisioned biologic benefits [115–117]; however, in order to preserve the structural integrity, polyphenols need to be sheltered by a finishing formulation that is, able to protect and to deliver them to the physiologic targets without losing any bioactivity [22, 26, 118].

An attractive approach to avoid the degradation of polyphenols is the process of encapsulation, which is referring to a process that involves the entrapment of an active ingredient through a polymeric matrix that seeks to increase stability and to improve bioavailability and the controlled release of bioactive agents through protecting these compounds from adverse environmental conditions or from the GIT (e.g., stomach acid) [117, 119]. Even though recent trends in the field of encapsulation have been focused on the development of techniques to encapsulate food bioactive ingredients such as polyphenols each with their own merits and demerits including Ionic gelation, layer-by-layer deposition, extrusion, coprecipitation, coacervation and phase separation, spray/freeze drying, emulsification/emulsion polymerization, inclusion complexation, liposome entrapment, fluidized bed coating, supercritical fluid, etc. [22, 25, 117], the application of these techniques is restricted owing to the mostly require heating and/or pressure, and the use of strong organic/non-polar solvents or expensive equipment, which result in degradation of heat-sensitive ingredients as well as associated toxicity concerns [120, 121]. In this regard, electro-hydrodynamic (EHD) processing, which refers to the dynamics of electrically charged fluids [120, 122] has emerged as an attractive alternative technology for encapsulation that needs neither temperature nor expensive equipment; therefore, heat-sensitive compounds may be successfully processed and also, the use of organic solvents can be avoided by adjusting some processing conditions (i.e., use of molten polymer) [120, 123, 124]. Principally, electro-spinning also called "electrostatic spinning" [120, 125] and electro-spraying also known as electro-hydrodynamic atomization (EHDA) [126–128] processes considered as kindred EHD processes [129, 130] are very cost effective, highly flexible and robust techniques, where use a uniform electro-hydrodynamic force to break the liquids into fine jets [121, 124, 128, 130–132]. However, electro-spinning is a drawing process based on electrostatic interactions [133] for papering fibers while electro-spraying is a process of liquid atomization by electrical forces [130] for papering particles. These approaches are promising techniques to fabricate delivery vehicles presenting structural and functional benefits for encapsulation of bioactive ingredients while their use in the field food processing and preservation is considerably less explored. Hence, in following section, focus will be on the current work aim to recognize the prospective of both electro-spinning and electro-spraying techniques for one-step encapsulation of dietary polyphenols, respectively into polymeric fibers and particles of micro-and nano-meter diameters.

3.1 Fundamental of electro-hydrodynamic process encapsulation

The incorporation of dietary polyphenols within polymeric particles and fibers of micro-and nanometer diameters is a promising technique to enhance the performance polyphenols-based functional ingredients in food industry [115, 117]. Micro and nano-sized particles and fibers could enhance stability, encapsulation efficiency (≥80%) [134] and oral bioavailability of polyphenols, as well develop controlled delivery or release [126]; thus, facilitating the development of innovative functional foods. Therefore, a clear and precise understanding of electro-hydrodynamic processes is essential to optimize the operating conditions for the nano-encapsulation of various polyphenols and thus, broadening the potential industrial application in food science. The basic setup of uniaxial electro-hydrodynamic technique generally consists of a high voltage power supply, a spinneret with a metallic needle or capillary tube of small diameter (up to 1 mm) [135], a pumping system, and an electrically conductive collector screen connected to an electrical earth, either can be a flat plate or rotating drum [115–117, 120, 123, 130, 131, 134, 136, 137]. In a typical EHD process, the bioactive agent dispersed in a carrier polymer solution or polymer melt is delivered at a fixed solution flow rate [124, 127, 134] to a capillary spinneret connected to the voltage supply [120, 126] by a pump, which forms a droplet at the spinneret apex. Once the droplet is charged under an applied electrical field to the spinneret, the hemispherical surface of droplet is deformed into a conical shape known as the Taylor cone [132, 138, 139] through the action of two major electrostatic forces including internal electrostatic repulsion of similar charges and the coulombic force of external electric field, which is applied between the spinneret apex and the collector [122–125]. With the increase of electric field strength, more electrical charges accumulate on the surface of suspended droplet, especially until a critical point is reached, where internal electrostatic repulsion eventually overcomes the intrinsic molecular tension forces present at the surface of the droplet at the tip of the Taylor cone; an electrically charged jet of the polymer is then ejected from the tip of the Taylor cone and is driven towards the conductive collector (counter electrode [140]) that is, usually held at earth potential to encourage fibers and particles capture. As the jet takes flight between the spinneret and the collector, it experiences a range of competing instabilities including the surface tension driven Rayleigh-Plateau instability [141, 142] and the electrically driven axisymmetric conducting instability and whipping/bending instability [136, 142] (more correctly described as an expanding helix) [122, 123, 129]. Electro-spun fibers are formed if the degree of molecular chains entanglement in the polymer solution and the solution concentration (directly proportional to viscosity and surface-tention) are high enough, the polymer jet from Taylor cone is stabilized, and elongation occurs in flight in initial linear trajectory and continues at an increased rate after the onset of the so-called "whipping instability" (actually a consistent expanding helix) [123], thereby inhibiting the formation of a filament of discrete droplets while electro-sprayed particles are formed providing the solution concentration is low, the polymer jet is destabilized due to varicose instability and hence, fine particulates are formed. These highly charged aerosols self-disperse in space, thereby preventing droplet aggregation and coagulation as well deposited on the collector as micro- or nano-particles [122, 130, 143, 144]. One important advantage of electro-spinning and electro-spraying is that due to whipping/bending instability of the jet and high surface to volume ratio [143], the evaporation of the solvent occurs at an increased rate during jet flight [123] or by blowing hot air on the extruded filament [120] and no heating is needed, which makes these technologies suitable for dealing with thermally sensitive materials (e.g., polyphenols, probiotic bacteria) [145] as well as the use of organic solvents can be avoided by adjusting some processing conditions [124].

In EHD process, molecular weight of the polymer reflecting the entanglement of polymer chains in solutions and solution concentration (directly proportional to viscosity and surface-tention) have the most effect on the formation bead or fiber morphology from the electro-spinning or spraying process [122, 123, 146]. Depending on the viscoelasticity of the polymer solution, the dominating instability leads to either electro-spray or electro-spinning [123, 127]. Both techniques work on the same physical principles of the ejection of a continuous jet; however, if the degree of molecular cohesion is below a critical level, particulates are formed from the ejecta and not a continuous fiber. This phenomenon is known Rayleigh-Plateau instability as characteristic of the electro-spraying process which is more commonly achieved with low-viscosity, low-molecular weight or low-concentration polymer solutions [125]. In this context, if Rayleigh-Plateau instability [141] dominates the process and manifests varicose waves on the surface of an EHD jet, the jet breaks up to form highly charged fine particles/beads, dispersed in a radial fashion due to coulomb repulsion. In other word, electro-spray transits to electro-spinning when the viscoelasticity of the polymer solution partially or completely suppresses Rayleigh-Plateau instability resulting in necklace-like beaded fibers or long continuous fibers [129]. The most effective parameters, which affect the fabrication of electro-spun fibers or electro-sprayed particles are divided into parameters related to polymer solvent properties (e.g, conductivity, viscosity, and surface tension), parameters related to the process (e.g., the applied electrical field, solution flow rate, and the distance between the tip of the needle and the collector) and ambient parameters (e.g., temperature, humidity and air flow) [124, 147]. Therefore, by manipulating these parameters, multiple morphologies can be attained and continuous polymeric fibers and beads with diameters ranging from a few nanometers to a few microns can be obtained.

The interested reader may consult some of the accounts of the effect of processing parameters on the properties of electro-spun or electro-sprayed materials; for a more detailed description, see the references [122, 123, 148].

3.2 Methods of electro-spinning/spraying encapsulation

Various strategies are available for encapsulation purposes using electro-spinning and electro-spraying. Direct incorporation of the bioactive food compounds such as dietary polyphenols into the polymeric/bio-polymeric carrier is the most common approach to encapsulate these compounds. Using this path, the bioactive component is randomly distributed throughout the fibers or the particles [144]. In this sense, a number of challenges are available that need to be overcome when developing this type of structures. First of all, many natural biopolymers are polyelectrolytes; having strong intermolecular interactions which need to be overcome for the subsequent formation of electro-spun/electro-sprayed structures [149, 150], as well as a certain fraction of the dispersed component is distributed nearby or on the surface of both electro-spun fibers or electro-sprayed particles which these unprotected species are susceptible to degradation owing to exposure of undesirable external environmental factors [134]. Aceituno-Medina et al. [151] encapsulated quercetin within hybrid amaranth protein isolate (API):pullulan (Pul) ultrathin fibers by using the electro-spinning technique. Their finding revealed that the thermal stability of quercetin decreased upon encapsulation, probably due to the dispersion of this antioxidant. However, a sustained-release of quercetin with a rate of ~ 52% from the API:Pul electro-spun fibers was reported during *in vitro* digestion, which probably corresponded to the amount of bioactive molecules distributed nearby to the fiber surface. Similarly, Blanco-Padilla et al. [152] encapsulated two different concentrations of curcumin (0.05 % & 0.075%) within API and Pul ultrafine fibers using the electro-spinning technique. Their finding revealed the release behavior of curcumin from the electro-spun fibers during an in-vitro digestion process (under simulated gastrointestinal conditions) (pH = 2). The burst release of curcumin from electro-spun fibers was reported 14.5-28.6% during the first 10 min, followed by a more gradual increase up to 28.6-55.8% released at 120 min. Third, the blend formulations often give rise to burst release of some encapsulated compounds [117, 134]. Fuenmayor et al. [118] investigated two types of highly antioxidant phenolic compounds of very different hydrophobicity included gallic acid (GA) (phenolic acid, water-solubility: ~1.4 x 104 mg/kg at 23°C) and naringenin (NAR) (flavanone, poorly water solubility: ~1.6 x 101 mg/kg at 23°C) that were homogeneously incorporated by conventional electro-spinning in ultrafine fibers made of zein (Z) a hydrophobic protein extracted from corn maize. It was reported that release of the loaded polyphenols into aqueous environments is pH-dependent. In the sense, release studies revealed a burst release trend with accumulative release threshold minimum for pH 2 and maximum for pH 7, probably due to pH-dependent differences in the molecular cargo-carrier interactions. Forth, conventional EHD also faces enormous challenges for the encapsulation of hydrophilic bioactive molecules into hydrophobic polymers or the hydrophobic bioactive molecules into hydrophilic polymers [117, 127, 134]. Besides, water is not the ideal solvent for electro-hydrodynamic processing since in comparison with organic solvents, it has a high evaporation temperature and high surface tension [127, 150], as well as the presence of organic solvents can result in the inactivation or denaturation of some hydrophilic bioactive substances [117, 127, 134]. However, other novel approaches such as emulsion or coaxial electro-hydrodynamic process have attracted a great deal of attention due to the fabrication core-sheath structures for encapsulation purpose.

3.2.1 Coaxial electro-hydrodynamic process

Coaxial electro-hydrodynamic is a controlled and one-step technique for encapsulation of fragile compounds such as extracts-rich poly-phenol into core-shell structured nano-fibers/particles using a couple of capillary tube where a smaller one is inserted concentrically inside the larger capillary [124, 144]. Coaxial electro-spinning/electro-spraying overcomes technical limitations of direct incorporation of the polyphenols into the polymer solution by its core-shell design [120, 123, 153] resulted in suppressing the initial burst release [115, 154] and thereby, delivering compounds in a controlled manner [117]. In particular, in the coaxial technology, the active component (core) is fed through the inner capillary spinneret while the polymer solution is extruded through the outer capillary spinneret simultaneously in order to acquire core–sheath structures; thus, the component immiscibility problem is alleviated [149]. One important advantage of coaxial configuration is that coaxial structures can be used to generate multiple core-shell structures [155] which involve one or more additional layers for the bioactive ingredients and the potential to adjust the release kinetics of active component by adjusting the number of layers of the protective shell [121, 127, 150, 156]. In the coaxial configuration, the core liquid containing the food bioactive compounds is pumped through inner capillary spinneret and simultaneously, the shell liquid containing polymeric material is extruded through outer capillary spinneret allowing the formation of a charged compound jet consisting of concentrically co-flowing liquids. And then, core-shell structured particles are formed during the charged compound or coaxial jet with appropriate parameters that is generally known as coaxial electro-spraying [121]; however, core-shell particles transit to fibers with an encapsulated core if the outermost shell polymer solution has sufficient viscoelasticity [129]. This technique is known as co-electro-spinning or coaxial electro-spinning. Compared to uniaxial electro-spun fibers and electro-sprayed particles, the coaxial electro-spun fibers and electro-sprayed particles obtain higher encapsulation efficiency [130], enhanced bioactive protection [141], controlled and tunable release of functional compounds and encapsulation of different compounds in the same structure allowing their release at different stages [153–155]. Torkamani et al. [157] studied encapsulation of poly-phenolic antioxidants obtained from *Momordica charantia* fruit within zein/gelatin shell core fibers via coaxial electro-spinning. Bitter gourd (*Momordica charantia* L) (BG) fruit is rich in flavonoids and polyphenols making it of certain potential value for use in food and nutraceutical industries. This study dealt with encapsulation of bitter gourd extract within bi-layer zein/gelatin fiber nano-structure as alternative polymer geometry, different than spherical configurations achieved by conventional methods. Their finding revealed that produced coaxial fibers showed higher thermal properties than their zein and gelatin uniaxial fiber counterparts; high encapsulation efficiency and sufficient shelf stability demonstrated the suitability of the coaxial electro-spinning process and the robustness of fabricated fibers which could replace conventional methods such as spray drying or freeze drying, as well as coacervation encapsulation method; coaxial electro-spun encapsulated fibers possessed the potential to be used as stand-alone nutraceutical supplement products or as an ingredient (e.g., filling or edible wrapper) in various food products [157]. Similarly, Yang Mao, et al. [158] investigated ferrulic acid/zein composite fibers prepared using a modified coaxial electro-spinning process to improve drug release profiles. Clearly, results of *in vitro* dissolution tests demonstrated that the fibers from the modified coaxial electro-spinning process exhibited a better drug release performance than those from the single-fluid electro-spinning process in terms of initial burst effect, release period, and tailing-off period compared with those from the blend process [158]. In another study, Yuan Shuai et al. [153] encapsulated curcumin in poly (lactic-co-glycolic acid) (PLGA) micro-particles by an improved coaxial electro-spray process and obtained Core-shell structured micro-particles with designated morphologic characteristics and high drug encapsulation efficiency are obtained in the stable cone-jet mode. Their results demonstrate that coaxial electro-spraying process yields micro-particles with improved drug release profiles in comparison with traditional microencapsulation methods.

3.2.2 Emulsion electro-hydrodynamic process

The emulsion electro-hydrodynamic techniques have been also explored to fabricate core-shell structured fibers or particles using water in oil (W/O) or oil in water (O/W) emulsions which can be developed to encapsulate hydrophilic and hydrophobic compounds, e.g., vitamins, carotenoids, polyphenols, enzymes, peptides, oils, flavors, and probiotics respectively. In this approach, an immiscible liquid containing food bioactive compounds (core material) is firstly stabilized by an emulsifier consist of the original emulsions until a stable emulsion is formed and then electro-hydrodynamic solution is prepared by adding shell polymer into emulsion [128]. The core-shell structured electro-spun fibers or electro-sprayed particles obtain by adjusting the operating parameters (voltage, flow rate, receiving distance, etc). Also, the properties of emulsion (viscosity, droplet size, emulsion stability, etc.) play important roles to ensure the success of emulsion electro-hydrodynamic process [121, 159]. Different from coaxial electro-hydrodynamic that utilize a couple of capillary tube where an inner one is inserted concentrically inside the outer capillary to fabricate core-shell structures, emulsion electro-hydrodynamic processing is utilized to fabricate core-shell structures using a single feeding capillary [121, 144] that the formation of electro-spun fibers and electro-sprayed particles is due to the solidification of polymer and coating on emulsion minimizing the amount of organic solvents used in food systems [121, 159]. Referring to recent studies reveal that the application of such a system can prevent the primary release of ingredients and can achieve targeted delivery and controlled release since the encapsulated bioactive components need to pass through the core-shell structure matrix prior to entering the surrounding medium during the release process and enhance the encapsulation efficiency, solubility, stability, bioavailability and bioactivity of bioactive compounds [118, 159]. Paximada et al. [160] used emulsion-electro-spraying technique to prepare epigallocatechin-3-gallate (EGCG) as well as a modified lipophilic version of EGCG loaded micro and sub-micron structured bacterial cellulose–whey protein isolate (BC-WPI) particles. Two different catechin, hydrophilic (H-EGCG) or lipophilized (L-EGCG), were encapsulated either on the aqueous or the oily phase of the emulsions and then emulsion was electro-sprayed. Particle size and encapsulation efficiency were highly dependent on the type of EGCG (hydrophilic versus lipophilic) and emulsification method and whether the bioactive compound was added to the oily or aqueous phase. The highest encapsulation efficiency was obtained with lipophilic EGCG, which had been added to the oily phase of the emulsion and emulsified by ultrasound (USLO capsules). The stability of EGCG in USLO capsules was tested under different storage conditions. Overall, capsules prepared with WPI and bacterial cellulose protected EGCG from moisture, heat, and dissolution conditions leading to their potential use to enhance EGCG shelf life when incorporated into foods. However, testing of the capsules in food systems remains to be investigated annual report [160].

4. Conclusions

As reviewed, chemical integrity, retention during processing and matrix interactions are some food-related factors hindering polyphenols bio-accessibility and consequently bioavailability that is, a prerequisite for their bioactivity in humans; however, it is possible to overcome it by entrapping these health-promoting components within polymeric particles and fibers of micro-and nanometer diameters through encapsulation process that entail an enhanced release of dietary polyphenols and/or higher absorption in the gastrointestinal tract, but choosing the most adequate encapsulation matrix, optimal core-to-carrier ratio, and operational parameters need to be performed in order to yield a high-quality product. In the case of dietary polyphenols, electro-spun/electro-sprayed structures can be used as the delivery system in foods to protect them during the processing and storage and to transfer these health-promoting components to the target site in the body as well enhance their bioactive functionalities and mask unpleasant taste, such as astringency of some polyphenols. The key advantage of electro-spinning/spraying process is the absence of heat that is, important for preserving the structure and achieving high loadings of polyphenols upon processing storage and thus, as a novel delivery approach for bioactive compounds, it opens a new horizon in food technology with the possibility of commercialization in the near future.

Author details

Mahmoud Ghorbani[1*] and Ricardo Santos Aleman[2]

1 Functional Food Group, Esfahan, Iran

2 School of Nutrition and Food Science, Louisiana State University, Baton Rouge, Louisiana

*Address all correspondence to: mahmoud.ghorbani@functionalfoodgroup.ir; mahmoud.ghorbani@live.com

References

[1] Westmark C. J. Healthy, Functional, and Medical Foods. Similarities and Differences between these Categories: Bioactive Food Compounds. In: Martirosyan D. M. editor. Introduction to Functional Food Science.3rd ed. Textbook. Functional Food Institute; Food Science 2015.p.35-45.

[2] Gupta ch, Prakash D, Gupta S. Relationships between Bioactive Food Components and Their Health Benefit. In: Martirosyan D. M. editor. Introduction to Functional Food Science.3rd ed. Textbook. Functional Food Institute;Food Science; 2015.p.113 - 128.

[3] Kawabata K, Yoshioka Y, Terao J. Role of Intestinal Microbiota in the Bioavailability and Physiological Functions of Dietary Polyphenols-Review. Molecules. 2019; 24, 370. DOI: 10.3390/molecules24020370.

[4] Shahidi F, Naczk M. Phenolics in Food and Nutraceuticals. Boca Raton: CRC Press; 2004.576p.

[5] Wang L, Bohn T. Health-Promoting Food Ingredients and Functional Food Processing. In Jaouad Bouayed. editor. Nutrition, Well-Being and Health. IntechOpen. 2012;p.201-224. DOI:10.5772/25862.

[6] Vergara-Salinas J.R, Cuevas-Valenzuela J, Pérez-Correa J.R. Pressurized hot water extraction of polyphenols from plant material. In: Gupta V.K, Tuohy M.G. editors. Biotechnology of Bioactive Compounds: Sources and applications. Chichester: Wiley; 2015.p.63-101.

[7] Munawar A, Farhan S, Faqir M.A, Muhammad A, Tabussam T, Muhammad S.B, Adnan I, Shahzad H, Hafiz Ansar R.S. Natural polyphenols: An overview. International Journal of Food Properties. 2017;20:8,1689-1699, DOI: 10.1080/10942912.2016.1220393.

[8] Dall'Asta M, Calani L, Tedeschi M, Jechiu L, Brighenti F, Del Rio D. Identification of microbial metabolites derived from in vitro fecal fermentation of different polyphenolic food sources. Nutrition. 2012;28:197-203. DOI: 10.1016/j.nut.2011.06.005.

[9] Passon M. Dietary Phenolic Compounds in Biological Samples: Current Challenges in Analytical Chemistry. In: Mérillon J.M, Ramawat K.G. editors. Bioactive Molecules in Food. Springer Nature Switzerland; 2019.p.1921-1958. DOI: 10.1007/978-3-319-78030-6_19.

[10] Cardona F, Andrés-Lacueva C, Tulipani S, Tinahones F.J, Queipo-Ortuño M.I. Benefits of polyphenols on gut microbiota and implications in human health-REVIEWS: CURRENT TOPICS. Journal of Nutritional Biochemistry.2013;24:1415-1422. DOI: 10.1016/j.jnutbio.2013.05.001.

[11] Quirós Sauceda A.E, Pacheco-Ordaz R, Ayala-Zavala J.F, Mendoza A.H, González-Córdova A.F, Vallejo-Galland B, González-Aguilar G.A. Impact of Fruit Dietary Fibers and Polyphenols on Modulation of the Human Gut Microbiota. In: Yahia E.M. editor. Fruit and Vegetable Phytochemicals: Chemistry and Human Health. 2nd ed. Chichester: Wiley; 2018.p.405-421.

[12] Edwards C.A, Havlik J, Cong W, Mullen W, Preston T, Morrison D.J, Combet E. Polyphenols and health: Interactions between fibre, plant polyphenols and the gut microbiota-NEWS FROM THE RESEARCH COUNCILS. Nutrition Bulletin. 2017;42:356-360. DOI: 10.1111/nbu.12296.

[13] Shahidi F, Senadheera R. Encyclopedia of Food Chemistry:

Protein–Phenol Interactions. In: Melton L, Shahidi F, Varelis P. editors. ENCYCLOPEDIA OF FOOD CHEMISTRY. Volume2. Elsevier; 2019.p.532-538. DOI:10.1016/B978-0-08-100596-5.21485-6.

[14] Wan M.L.Y, Anna Co V & El-Nezami H. Dietary poly-phenol impact on gut health and microbiota. Critical Reviews in Food Science and Nutrition. 2020. DOI: 10.1080/10408398.2020.1744512.

[15] Tsao R. Chemistry and Biochemistry of Dietary Polyphenols-Review. Nutrients. 2010;2:1231-1246. DOI: 10.3390/nu2121231.

[16] Bhuyan D.J, Basu A. Phenolic Compounds Potential Health Benefits and Toxicity. In: Vuong Q.V. editor. Utilisation of Bioactive Compounds from Agricultural and Food Waste. Boca Raton, FL: CRC Press, Taylor & Francis Group; 2017.p.27-59.

[17] Dey G. Developing a Suitable Food Vehicle for Bioactive Ingredients. In: Martirosyan D. M. editor. Introduction to Functional Food Science.3rd ed. Textbook. Functional Food Institute; Food Science 2015.p.96-112.

[18] Chidambara Murthy K.N, Shivapriya M, Monika P, Tejashree B. Challenges in Optimal Utilization of Bioactive Molecules Clinically. In: Mérillon J.M, Ramawat K.G. editors. Bioactive Molecules in Food. Springer Nature Switzerland; 2019.p.109-136. DOI: /10.1007/978-3-319-78030-6_49.

[19] Ozdal T, Yalcinkaya I.E, Toydemir G, Capanoglu E. Poly-phenol-Protein Interactions and Changes in Functional Properties and Digestibility. In: Melton L, Shahidi F, Varelis P. editors. ENCYCLOPEDIA OF FOOD CHEMISTRY. Elsevier; 2019.p.566-577. DOI: 10.1016/B978-0-08-100596-5.21490-X.

[20] Sun-Waterhouse D. Stability and Bio-accessibility of Fruit Bio-actives in Foods: Food Component Interactions and Matrix Effect. In: Skinner M, Hunter D. editors. Bioactives in Fruit: Health Benefits and Functional Foods. 1st ed. Chichester: Wiley; 2013.p.467-507.

[21] Buchwald-Werner S, Gärtner C, Horlacher P, Schwarz G. Fortification with substances other than vitamins and minerals (polyphenols, carotenoids, fatty acids and phytosterols). In: Ottaway P.B, editor. Food fortification and supplementation: Technological, safety and regulatory aspects. Cambridge: Woodhead Publishing Limited; 2008.p.41-59.

[22] Fang, Z, Bhandari B. Encapsulation of polyphenols– A review. Trends in Food Science and Technology. 2010;21(10),510-523. DOI:10.1016/j.tifs.2010.08.003.

[23] Inês Dias M, Caleja C, Ferreira I.C.F.R, Filomena Barreiro M. The use of encapsulation to guarantee the stability of phenolic compounds. In: Ruiz Ruiz J.C, Campos M.R.S. editors. New polymers for encapsulation of nutraceuticals compounds. Chichester: Wiley; 2017.p.121-143.

[24] Botelho G, Canas S, Lameiras J. Development of Phenolic Compounds Encapsulation Techniques as a Major Challenge for Food Industry and for Health and Nutrition Fields. In: Grumezescu A.M. editor. Nutrient Delivery: Nanotechnology in the Agri-Food Industry. Volume5. London: Academic Press;2017.p.535-586. DOI: 10.1016/B978-0-12-804304-2.00014-7.

[25] Mishra M.k. Overview of Encapsulation and Controlled Release. In: Mishra M. editor. Handbook of Encapsulation and Controlled Release. Boca Raton, FL: CRC Press, Taylor & Francis Group; 2016.p.3-19.

[26] Munin A, Edwards-Lévy F. Encapsulation of natural poly-phenolic compounds: A review. Pharmaceutics. 2011;3,793-829. DOI:10.3390/pharmaceutics3040793.

[27] Ziaullah and Vasantha Rupasinghe H.P. Application of NMR Spectroscopy in Plant Polyphenols Associated with Human Health. In: Rahman A, FRS. Choudhary M.I, editors. Applications of NMR Spectroscoppy. Volume2. Bentham Science: Elsevier; 2015.p.3-92. DOI:10.1016/B978-1-60805-999-7.50001-X.

[28] Lattanzio V. Phenolic Compounds: introduction. In: Ramawat K.G, Me´rillon J.M. editors. Natural Products: Phytochemistry, Botany and Metabolism of Alkaloids, Phenolics and Terpenes. Springer-Verlag Berlin Heidelberg; 2013.p.1543-1580. DOI: 10.1007/978-3-642-22144-6_57.

[29] González-Laredo R.F, Rocha-Guzmán N.E, Gallegos-Infante J.A, Moreno- Jiménez M.R, Gamboa-Gómez C.I. Phenolic Compounds in Nature. In: Nollet L.M.L, Gutierrez-Uribe J. A. editors. Phenolic Compounds in Food (Characterization and Analysis). Boca Raton: CRC Press; 2018.p.21- 31.

[30] NA Aryee A, Agyei D, Akanbi T.O. Food for Oxidative Stress Relief: Polyphenols. In: Melton L, Shahidi F, Varelis P. editors. ENCYCLOPEDIA OF FOOD CHEMISTRY. Elsevier; 2019.p.392-398. DOI: 10.1016/B978-0-08-1005965.21779-4.

[31] Durazzo A, Lucarini M, Souto E.B, Cicala C, Izzo A.A, Novellino E, Santini A. Polyphenols: A concise overview on the chemistry, occurrence, and human health. Phytotherapy Research. 2019;33:2221-2243. DOI: 10.1002/ptr.6419.

[32] Zhao Y, Wu Y, Wang M. Bioactive Substances of Plant Origin. In: Cheung P.C.K, Mehta Bh.M. editors. Handbook of Food Chemistry. Heidelberg: Springer; 2015.p.967-1008. DOI: 10.1007/978-3-642-36605-5_13.

[33] Craft BD, Kerrihard A.L, Amarowicz R, Pegg R.B. Phenol-based antioxidants and the in vitro methods used for their assessment. Compr Rev Food Sci Food Saf. 2012;11:148-173. DOI: 10.1111/j.1541-4337.2011.00173.x.

[34] Martin K.R, Appel C.L. Polyphenols as dietary supplements: a double-edged sword. Nutr Diet Sup. 2010;2:1-12.DOI: 10.2147/NDS.S6422.

[35] Catalkaya G, Venema K, Lucini L, Rocchetti G, Delmas D, Daglia M, De Filippis A, Xiao H, Quiles J.L, Xiao J, Capanoglu E. Interaction of dietary polyphenols and gut microbiota: Microbial metabolism of polyphenols, influence on the gut microbiota, and implications on host health-Review. FOOD FRONTIERS. 2020;1:109-133. DOI: 10.1002/fft2.2.

[36] Speisky H, Fuentes J, Dorta E, Lopez-Alarcón C. Polyphenols: Sources and Main Characteristics. In: Cuevas-Valenzuela J, Vergara-Salinas J.R, Pérez-Correa J.R. editors. Advances in technologies for producing food-relevant polyphenols. Boca Raton, FL: CRC Press, Taylor & Francis Group; 2017.p.1-32.

[37] Santana-Gálvez J, Jacobo-Velázquez D. A. Classification of Phenolic Compounds. In: Nollet L.M.L, Gutierrez-Uribe J. A. editors. Phenolic Compounds in Food (Characterization and Analysis). Boca Raton : CRC Press, Taylor & Francis Group; 2018.p.3 – 20.

[38] Le Bourvellec C, MGC Renard C. Interactions Between Polyphenols and Macromolecules: Effect of Tannin Structure. In: Melton L, Shahidi F, Varelis P. editors. ENCYCLOPEDIA OF FOOD CHEMISTRY. Elsevier; 2019.p.515-521. DOI: 10.1016/B978-0-08-100596-5.21486-8.

[39] Lipińska L, Klewicka E, Sójka M. Structure, Occurrence and Biological Activity of Ellagitannins: A General Review*. Acta Sci. Pol., Technol. Aliment. 2014;13(3):289-299. DOI: 10.17306/j.afs.2014.3.7.

[40] Rueda A, Samaniego-Sánchez C, Olalla M, Giménez R, Cabrera-Vique C, Seiquer I, Lara L. Combination of analytical and chemometric methods as a useful tool for the characterization of extra virgin argan oil and other edible virgin oils: Role of polyphenols and tocopherols. J. AOAC Int. 2016;99:489-494. DOI:10.5740/jaoacint.15-0121.

[41] Leopoldini M, Russo N, Toscano M. The molecular basis of working mechanism of natural polyphenolic antioxidants - Review. Food Chem. 2011;125:288-306. DOI: 10.1016/j.foodchem.2010.08.012.

[42] Lanzilli G, Cottarelli A, Nicotera G, Guida S, Ravagnan G, Fuggetta M.P. Anti-inflammatory effect of resveratrol and polydatin by in vitro IL-17 modulation. Inflammation. 2012;35(1):240-248. DOI: 10.1007/s10753-011-9310-z.

[43] Davies K.M. Metabolic Engineering of Bioactive Phenylpropanoids in Crops. In: Bagchi D, Lau F.C, Gosh D.K. editors. Boca Raton: CRC Press, Taylor & Francis Group; 2010.p.181-195.

[44] Mourtzions I, Goula A. Polyphenols in Agricultural Byproducts and Food Waste. In: Watson R. R. editors. Polyphenols in Plants: Isolation, Purification and Extract Preparation. 2nd ed. Academic Press; 2019. p.23-44. DOI:10.1016/B9780-12-813768-0.00002-5.

[45] Tian Sh, Sun Y, Chen Zh, Yang Y, Wang Y. Functional Properties of Polyphenols in Grains and Effects of Physicochemical Processing on Polyphenols-Review. Journal of Food Quality. 2019; Article ID 2793973.8p. DOI: 10.1155/2019/2793973.

[46] Dueik V, Bouchon P. Inclusion of Food of Polyphenols into Food Matrices. In: Cuevas-Valenzuela J, Vergara-Salinas J.R, Pérez-Correa J.R. editors. Advances in technologies for producing food-relevant polyphenols. Boca Raton, FL: CRC Press, Taylor & Francis Group; 2017.p.303-325.

[47] Gómez-Maqueo A, Escobedo-Avellaneda Z, Cano M.P, Welti-Chanes J. Phenolic Compounds in Food. In: Nollet L.M.L, Gutierrez-Uribe J.A. editors. Phenolic Compounds in Food (Characterization and Analysis). Boca Raton: CRC Press, Taylor & Francis Group; 2018.p.33- 57.

[48] Cory H, Passarelli S, Szeto J, Tamez M, Mattei J. The Role of Polyphenols in Human Health and Food Systems: A Mini-Review. Frontiers in Nutrition. 2018;5.Article 87. DOI: 10.3389/fnut.2018.00087.

[49] Rubió L, Macià A, Motilva M-J. Impact of Various Factors on Pharmacokinetics of Bioactive Polyphenols: An Overview. Current Drug Metabolism. 2014;15:62-76.

[50] Benmeddour Z, Mehinagic E, Le Meurlay D, Louaileche H. Phenolic composition and antioxidant capacities of ten Algerian date (Phoenix dactylifera L.) cultivars: A comparative study. Journal of Functional Foods. 2013;5:346-354. DOI: 10.1016/j.jff.2012.11.005.

[51] Selma M.V, Espi´ n J.C, Toma´ s-Barbera´ n F.A. Interaction between Phenolics and Gut Microbiota: Role in Human Health-Review. J. Agric. Food Chem. 2009;57: 6485-6501. DOI: 10.1021/jf902107d.

[52] Velderrain Rodríguez G.R, Blancas-Benítez F.J, Wall-Medrano A, Sáyago-Ayerdi S.G,

González-Aguilar G.A. Bioaccessibility and Bioavailability of Phenolic Compounds from Tropical Fruits. In: Yahia E.M. editor. Fruit and Vegetable Phytochemicals: Chemistry and Human Health. 2nd ed. Chichester: Wiley; 2018.p.155-164.

[53] Singh A.K, Cabral C, Kumar R, Ganguly R, Rana H.K, Gupta A, Rosaria Lauro M, Carbone C, Reis F, Pandey A.K. Beneficial Effects of Dietary Polyphenols on Gut Microbiota and Strategies to Improve Delivery Efficiency-Review. Nutrients; 2019:11,2216. DOI: 10.3390/nu11092216.

[54] Ifie I, Marshall L.J. Food processing and its impact on phenolic constituents in food-review. Cogent Food & Agriculture, 2018;4:1,1507782, DOI: 10.1080/23311932.2018.1507782.

[55] Aschoff J.K, Kaufmann S, Kalkan O, Neidhart S, Carle R, Schweiggert R.M. In vitro bioaccessibility of carotenoids, flavonoids, and vitamin C from differently processed oranges and orange juices [Citrus sinensis (L.) osbeck]. Journal of Agricultural and Food Chemistry. 2015;63(2):578-587. DOI: 10.1021/jf505297t.

[56] Ludwig I.A, Rubio L, Macia A, Romero M.P. Hydroxycinnamates. In: Tomas- Barberan F.A, González-Sarrías A, García-Villalba, editors. Dietary Polyphenols: Metabolism and Health Effects. 1st ed. Hoboken, NJ : Wiley-Blackwell; 2021.p.129-164.

[57] Macagnan F.T, Bender A.B.B, Speroni C.S, Dietary Fiber and Non-extractable Polyphenols. In: Saura-Calixto F, Pérez-Jiménez J. editors. Non-extractable Polyphenols and Carotenoids: Importance in Human Nutrition and Health. Royal Society of Chemistry; 2018. p.17-36.

[58] Olmo-García L, Monasterio R.P, Bajoub A, Carrasco-Pancorbo A. Analytical Strategies for Determining Polyphenols in Foods and Biological Samples. In: Tomas- Barberan F.A, González-Sarrías A, García-Villalba, editors. Dietary Polyphenols: Metabolism and Health Effects. 1st ed. Hoboken, NJ : Wiley-Blackwell; 2021.p.85-128.

[59] Bhuyan D.J, Basu A. Phenolic Compounds Potential Health Benefits and Toxicity. In: Vuong Q.V. editor. Utilisation of Bioactive Compounds from Agricultural and Food Waste. Boca Raton, FL: CRC Press, Taylor & Francis Group; 2017.p.27-59.

[60] Carbonell-Capella J.M, Buniowska M, Barba F.J, Esteve M.J, Frígola A. Analytical methods for determining bioavailability and bioaccessibility of bioactive compounds from fruits and vegetables: A review. Compr. Rev. Food Sci. Food Saf. 2014;13:155-171. DOI: 10.1111/1541-4337.12049.

[61] Chen L, Cao H, Xiao J. Polyphenols: absorption, bioavailability, and metabolomics. In: Galanakis Ch. M. editor. Polyphenols: Properties, Recovery, and Applications. Woodhead; 2018.p.45-67. DOI: 10.1016/B978-0-12-813572-3.00002-6.

[62] Renard C, Watrelot A, Le Bourvellec C. Interactions between polyphenols and polysaccharides: Mechanisms and consequences in food processing and digestion. Trends in Food Science and Technology. 2017;60:43-51. DOI:10.1016/j.tifs.2016.10.022.

[63] Manach C, Scalbert A, Morand Ch, Rémésy Ch, Jime´nez L. Polyphenols: food sources and bioavailability. Am J Clin Nutr. 2004;79:727-747.

[64] Rodríguez-Roque M.J. In vitro bioaccessibility of health-related compounds from beverages based on fruit juice, milk or soymilk: Influence of

food matrix and processing [Doctoral Thesis]. Escola Tècnica Superior d'Enginyeria Agrària Departament de Tecnologia d'Aliments: UNIVERSITAT DE LLEIDA; 2014.

[65] Rodríguez-Roque M.J, Rojas-Graü M.A, Elez-Martínez P, Martín-Belloso O. Changes in vitamin C, phenolic, and carotenoid profiles throughout in vitro gastrointestinal digestion of a blended fruit juice. Journal of Agricultural and Food Chemistry. 2013;61(8):1859-1867.

[66] Rodríguez-Roque M.J, Rojas-Graü M.A, Elez-Martínez P, Martín-Belloso O. Soymilk phenolic compounds, isoflavones and antioxidant activity as affected by in vitro gastrointestinal digestion. Food Chemistry. 2013;136(1):206-212.

[67] Rodríguez-Roque M.J, Rojas-Graü M.A, Elez-Martínez P, Martín-Belloso O. In vitro bioaccessibility of health-related compounds as affected by the formulation of fruit juice- and milk-based beverages. Food Research International. 2014;62:771-778.

[68] Rodríguez-Roque M.J, Rojas-Graü M.A, Elez-Martínez P, Martín-Belloso O. In vitro bioaccessibility of health-related compounds from a blended fruit juice-soymilk beverage: Influence of the food matrix. Journal of Functional Foods. 2014;7:161-169.

[69] Ribas-Agustí A, Martín-Belloso O, Soliva-Fortuny R, Elez-Martínez P. Food processing strategies to enhance phenolic compounds bioaccessibility and bioavailability in plant-based foods. Critical Reviews in Food Science and Nutrition. 2018;58(15):2531-2548. DOI: 10.1080/10408398.2017.1331200.

[70] Barba F.J, Mariutti L.R.B, Bragagnolo N, Mercadante A.Z, Barbosa-Cánovas G.V, Orlien V. Bioaccessibility of bioactive compounds from fruits and vegetables after thermal and non-thermal processing. Trends in Food Science & Technology. 2017;67:195-206. DOI: 10.1016/j.tifs.2017.07.006.

[71] Lea A. Analysis of polyphenol antioxidants in fortified foods and supplements. In: Ottaway P.B, editor. Food fortification and supplementation: Technological, safety and regulatory aspects. Cambridge: Woodhead Publishing Limited; 2008.p.175-194.

[72] Parada J, Aguilera J.M. Food microstructure affects the bioavailability of several nutrients. Journal of Food Science. 2007;72:R21-R32. DOI: 10.1111/j.1750-3841.2007.00274.x.

[73] Lindsay R.C. Flavors. In: Damodaran S. Parkin K.L. (Kirk Lindsay).editors. Fennema's food chemistry.5th ed. Boca Raton: CRC Pres; 2017.p.753-802.

[74] Soares S, Mateus N, de Freitas V. Polyphenol Interactions and Food Organoleptic Properties. In: Melton L, Shahidi F, Varelis P. editors. ENCYCLOPEDIA OF FOOD CHEMISTRY. Elsevier; 2019.p.650-655. DOI: 10.1016/B978-0-08-100596-5.21865-9.

[75] Scholz S, Williamson G. Interactions Affecting the Bioavailability of Dietary Polyphenols in Vivo. Int. J. Vitam. Nutr. Res. 2007;77(3):224-235. DOI: 10.1024/0300-9831.77.3.224.

[76] Rodríguez-Roque M.J, de Ancos B, Sánchez-Moreno C, Cano M.P, Elez-Martínez P, Martín-Belloso O. Impact of food matrix and processing on the in vitro bioaccessibility of vitamin C, phenolic compounds, and hydrophilic antioxidant activity from fruit juice-based beverages. Journal of Functional Foods. 2015;14:33-43. DOI:10.1016/j.jff.2015.01.020.

[77] Campos-Vega R, Oomah B.D, Hernández-Arriaga A.M, Salazar-López N.J, Vázquez-Sánchez K. Tannins. In: Nollet L.M.L, Gutierrez-Uribe J.A. editors. Phenolic Compounds in Food: Characterization and Analysis. Boca Raton: CRC Press, Taylor & Francis Group; 2018.p.211-258.

[78] Kardum N, Glibetic M. Polyphenols and Their Interactions with Other Dietary Compounds: Implications for Human Health. In: Toldra F. Serial editor. Advances in Food and Nutrition Research. Volume 84. Academic Press; 2018. p.103-144. DOI: 10.1016/bs.afnr.2017.12.001.

[79] Bohn T. Dietary factors affecting polyphenol bioavailability. Nutrition Reviews. 2014;77(2):429-452. DOI: 10.1111/nure.12114.

[80] Bravo L. Polyphenols: chemistry, dietary sources, metabolism, and nutritional significance. Nutrition Reviews. 1998;56(11):317-333. DOI:10.1111/j.17534887.1998.tb01670.x.

[81] Fraga C.G. Plant Polyphenols: How to Translate their in vitro Antioxidant Actions to in vivo Conditions- Critical Review. Informa Healthcare. 2007;59(4-5):308-315. DOI: 10.1080/15216540701230529.

[82] Belščak-Cvitanović A, Durgo K, Huđek A, Bačun-Družina V, Komes D. Overview of polyphenols and their properties.: In: Galanakis Ch.M. editor. Polyphenols: Properties, Recovery, and Applications. Woodhead; 2018.p.3-44. DOI: 10.1016/B978-0-12-813572-3.00001-4.

[83] Sánchez-Vioque R, Polissiou M, Astraka K, de los Mozos-Pascual M, Tarantilis P, Herraiz-Penalver D, Santana-Méridas O. Polyphenol composition and antioxidant and metal chelating activities of the solid residues from the essential oil industry. Industrial Crops and Products. 2013;49:150-159. DOI: 10.1016/j.indcrop.2013.04.053.

[84] Rawson A, Patras A, Tiwari B.K, Noci F, Koutchma T, Brunton N. Effect of thermal and non thermal processing technologies on the bioactive content of exotic fruits and their products: Review of recent advances. Food Research International. 2011;44:1874-1887. DOI: 10.1016/j.foodres.2011.02.053.

[85] Liang Sh, Ling K. Millet grain as a candidate antioxidant food resource: a review. International Journal of Food Properties. 2019;22(1):1652-1661. DOI:10.1080/10942912.2019.1668406

[86] Shahidi F. Nutraceutical and functional foods: Whole versus processed foods, Review. Trends in Food Science and Technology. 2009;20:376-387. DOI: 10.1016/j.tifs.2008.08.004.

[87] Toydemir G, Capanoglu E, Kamiloglu S, Boyacioglu D, de Vos R.C.H, Hall R.D, Beekwilder J. Changes in sour cherry (Prunus cerasus L.) antioxidants during nectar processing and in vitro gastrointestinal digestion. Journal of Functional Foods. 2013;5:1402-1413. DOI: 10.1016/j.jff.2013.05.008.

[88] Colle I, Van Buggenhout S, Van Loey A, Hendrickx M. High pressure homogenization followed by thermal processing of tomato pulp: influence on microstructure and lycopene in vitro bioaccessibility. Food Research International. 2010;43:2193-2200. DOI: 10.1016/j.foodres.2010.07.029.

[89] Anson N.M, Selinheimo E, Havenaar R, Aura A-M, Mattila I, Lehtinen P, Bast A, Poutanen K, Haenen G.R.M.M. Bioprocessing of wheat bran improves in vitro bioaccessibility and colonic metabolism of phenolic compounds. Journal of Agricultural and Food Chemistry. 2009;57:6148-6155. DOI: 10.1021/jf900492h.

[90] Cilla A, Bosch L, Barberá R, Alegría A. Effect of processing on the bioaccessibility of bioactive compounds – A review focusing on carotenoids, minerals, ascorbic acid, tocopherols and polyphenols. Journal of Food Composition and Analysis. 2018;68:3-15. DOI: 10.1016/j.jfca.2017.01.009.

[91] Ribas-Agustí A, Martín-Belloso O, Soliva-Fortuny R, Elez-Martínez P. Influence of pulsed electric fields processing on the bioaccessible and nonbioaccessible fractions of apple phenolic compounds. Journal of Functional Foods. 2019;59:206-214. DOI: 10.1016/j.jff.2019.05.041.

[92] Otavio Minatel I, Vanz Borges C, Ferreira M.I, Gomez Gomez H.A, Oliver Chen Ch-Y, Pace Pereira Lima G. Phenolic Compounds: Functional Properties, Impact of Processing and Bioavailability. In: Soto-Hernández M. editor. Phenolic Compounds: Biological Activity. London: IntechOpen; 2017.p.1-24. DOI: 10.5772/66368.

[93] Abbès F, Kchaou W, Blecker Ch, Ongena M, Lognay G, Attia H, Besbes S. Effect of processing conditions on phenolic compounds and antioxidant properties of date syrup. Industrial Crops and Products. 2013;44:634-642. DOI: 10.1016/j.indcrop.2012.09.008.

[94] Georgé S, Tourniaire F, Gautier H, Goupy P, Rock E, Caris-Veyrat C. Changes in the contents of carotenoids, phenolic compounds and vitamin C during technical processing and lyophilisation of red and yellow tomatoes. Food Chemistry. 2011;124:1603-1611. DOI: 10.1016/j.foodchem.2010.08.024.

[95] Terefe N.S, Kleintschek T, Gamage Th, Fanning K.J, Netzel G, Versteeg C, Netzel M. Comparative effects of thermal and high pressure processing on phenolic phytochemicals in different strawberry cultivars. Innovative Food Science and Emerging Technologies. 2013;19:57-65. DOI: 10.1016/j.ifset.2013.05.003.

[96] Kamiloglu S, Ayfer Pasli A, Ozcelik B, Van Camp J, Capanoglu E. Influence of different processing and storage conditions on in vitro bio-accessibility of polyphenols in black carrot jams and marmalades. Food chemistry. 2015;186:74-82. DOI: 10.1016/j.foodchem.2014.12.046.

[97] Rothwell J.A, Medina-Remon A, Perez-Jim´enez J, Neveu V, Knaze V, Slimani N, Scalbert A. Effects of food processing on polyphenol contents: A systematic analysis using Phenol-Explorer data. Mol. Nutr. Food Res. 2014;00:1-11. DOI:10.1002/mnfr.201400494.

[98] Jolic S.M, Radojčic´ Redovnikovic I, Markovic K, Šipušic D.I, Delonga K. Changes of phenolic compounds and antioxidant capacity in cocoa beans processing. International Journal of Food Science and Technology. 2011;46:1793-1800. DOI: 10.1111/j.1365-2621.2011.02670.x.

[99] Chandrasekara A, Naczk M., Shahidi F. Effect of processing on the antioxidant activity of millet grains. Food Chemistry. 2012;133(1):1-9. DOI: 10.1016/j.foodchem.2011.09.043.

[100] Chandrasekara A, Shahidi F. Bioaccessibility and antioxidant potential of millet grain phenolics as affected by simulated in vitro digestion and microbial fermentation. Journal of Functional Foods. 2012;4:226-237. DOI: 10.1016/j.jff.2011.11.001.

[101] Wang T, He F, Chen G. Improving bioaccessibility and bioavailability of phenolic compounds in cereal grains through processing technologies: A concise review. Journal of Functional Foods. 2014;7(1):101-111. DOI: 10.1016/j.jff.2014.01.033

[102] Hostetler G.L, Riedl K.M, Schwartz S.J. Effects of food

formulation and thermal processing on flavones in celery and chamomile. Food Chemistry. 2013;41:1406-1411. DOI:10.1016/j.foodchem.2013.04.051.

[103] White B.L, Howard L.R, Prior R.L. Impact of Different Stages of Juice Processing on the Anthocyanin, Flavonol, and Procyanidin Contents of Cranberries. J. Agric. Food Chem. 2011;59:4692-4698. DOI: 10.1021/jf200149a.

[104] Rothwell J.A, Perez-Jimenez J, Neveu V, Medina-Remón A, M'Hiri N, García-Lobato P, Manach C, Knox C, Eisner R, Wishart D.S, Scalbert A. PhenolExplorer 3.0: a major update of the Phenol-Explorer database to incorporate data on the effects of food processing on polyphenol content. Database. 2013;2013:1-8. DOI: 10.1093/database/bat070.

[105] Deng J, Yang H, Capanoglu E, Cao H, Xiao J. Technological aspects and stability of polyphenols. In: Galanakis Ch.M. editor. Polyphenols: Properties, Recovery, and Applications. Woodhead; 2018.p.295-323. DOI: 10.1016/B978-0-12-813572-3.00009-9.

[106] Oracz J, Zyzelewicz D, Nebesny E. The content of polyphenolic compounds in cocoa beans (Theobroma cacao L.), depending on variety, growing region and processing operations: A review. Critical Reviews in Food Science and Nutrition. 2015;55(9):1176-1192. DOI: 10.1080/10408398.2012.686934.

[107] Capanoglu E, de Vos R.C.H, Hall R.D, Boyacioglu D, Beekwilder J. Changes in polyphenol content during production of grape juice concentrate. Food Chemistry. 2013;139:521-526. DOI: 10.1016/j.foodchem.2013.01.023.

[108] Gancel A.L, Feneuil A, Acosta O, Pérez A.M, Vaillant F. Impact of industrial processing and storage on major polyphenols and the antioxidant capacity of tropical highland blackberry (Rubus adenotrichus). Food Research International. 2011;44:2243-2251. DOI: 10.1016/j.foodres.2010.06.013.

[109] U.S. Department of Agriculture, Agricultural Research Service, USDA Nutrient Data Laboratory. 2010. USDA National Nutrient Database for Standard Reference, Release 23.

[110] Plaza L, Sanchez-Moreno C, De Ancos B, Elez-Martinez P, Martin-Belloso O, Cano M.P. Carotenoid and flavanone content during refrigerated storage of orange juice processed by high-pressure, pulsed electric fields and low pasteurization. Lwt-Food Science and Technology. 2011;44:834-839. DOI: 10.1016/j.lwt.2010.12.013.

[111] Sanchez-Moreno C, Plaza L, Elez-Martinez P, De Ancos B, Martin-Belloso O, Pilar Cano M. Impact of High Pressure and Pulsed Electric Fields on Bioactive Compounds and Antioxidant Activity of Orange Juice in Comparison with Traditional Thermal Processing. J. Agric. Food Chem. 2005;53:4403–4409. DOI: 10.1021/jf048839b.

[112] Odriozola-Serrano I, Soliva-Fortuny R, Martin-Belloso O. Phenolic acids, flavonoids, vitamin C and antioxidant capacity of strawberry juices processed by high-intensity pulsed electric fields or heat treatments. European Food Research and Technology. 2008;228:239-248.

[113] Morales-de la Peña M, Salvia-Trujillo L, Rojas-Graü M.A, Martín-Belloso O. Changes on phenolic and carotenoid composition of high intensity pulsed electric field and thermally treated fruit juice–soymilk beverages during refrigerated storage. Food Chemistry. 2011;129:982-990. DOI: 10.1016/j.foodchem.2011.05.058.

[114] Ribas-Agustí A, Martín-Belloso O, Soliva-Fortuny R, Elez-Martínez P.

Enhancing hydroxycinnamic acids and flavan-3-ol contents by pulsed electric fields without affecting quality attributes of apple. Food research International. 2019;121:433-440. DOI: 10.1016/j.foodres.2018.11.057.

[115] Wen P, Zong M.H, Linhardt R.J, Feng K, Wu H. Electro-spinning: A novel nano-encapsulation approach for bioactive compounds- Review. Trends in Food Science & Technology. 2017;70:56-68. DOI: 10.1016/j.tifs.2017.10.009.

[116] Faki R, Gursoy O, Yilmaz Y. Effect of Electro-spinning Process on Total Antioxidant Activity of Electro-spun Nano-fibers Containing Grape Seed Extract. Open Chem. 2019;17:912-918. DOI: 10.1515/chem-2019-0098.

[117] Wen P, Wen Y, Zong M.H, Linhardt R.J, Wu H. Encapsulation of Bioactive Compound in Electro-spun Fibers and Its Potential Application. J. Agric. Food Chem. 2017;65:9161-9179. DOI: 10.1021/acs.jafc.7b02956.

[118] Fuenmayor C. A, Cosio M.S. Encapsulation of Antioxidant Phenolic Compounds in Zein Ultra-thin Fibers via Electro-spinning. Edición especial N.3. 2016;12:E13-E26. DOI: 10.14508/reia.2016.12.e3.13-26.

[119] González-Cruz E.M, Calderón-Santoyo M, Barros-Castillo J.C, Ragazzo-Sánchez J.A. Evaluation of biopolymers in the encapsulation by electro-spraying of polyphenolic compounds extracted from blueberry (Vaccinium corymbosum L.) variety Biloxi. Polymer Bulletin. 2020. DOI: 10.1007/s00289-020-03292-3.

[120] Busolo M.A, Castro S, Lagaron J.M. Electro-Hydrodynamic Processes (Electro-spinning and Electro-spraying): Non-thermal Processes for Micro- and Nano-encapsulation. In: Krokida M.K. editor. Thermal and Non-thermal Encapsulation Methods. Boca Raton, FL: CRC Press, Taylor & Francis Group; 2018.p.115-136.

[121] Niu B, Shao P, Luo Y, Sun P. Recent advances of electro-sprayed particles as encapsulation systems of bio-actives for food application. Food Hydrocolloids.2020;99:105376. DOI: 10.1016/j.foodhyd.2019.105376.

[122] Ghorani B, Alehosseini A, Tucker N. Electro-spinning as a Novel Delivery Vehicle for Bioactive Compounds in Food Nanotechnology. In: Moreno J.J. editor. Innovative Processing Technologies for Foods with Bioactive Compounds. Boca Raton, FL: CRC Press, Taylor & Francis Group; 2017.p.287-325.

[123] Ghorani B, Alehosseini A, Tucker N. Nano-capsule formation by electro-spinning. In: Jafari S.M. editor. Nanoencapsulation Technologies for the Food and Nutraceutical Industries. London: Academic Press; 2017.p.264- 319.

[124] López-Córdoba A, Duca C, Cimadoro J, Goyanes S. Electro-spinning and Electro-spraying Technologies and Their Potential Applications in the Food Industry. In: Rai V.R, Bai J.A. editors. Nanotechnology Applications in the Food Industry. Boca Raton, FL: CRC Press, Taylor & Francis Group;2018.p.461-479.

[125] Fuenmayor C.A, Espitia P.J.P. Electro-spun Nanofibers Development and Potential in Food Packaging Applications. In: Rai V.R, Bai J.A. editors. Nanotechnology Applications in the Food Industry. Boca Raton, FL:CRC Press, Taylor & Francis Group;2018.p.141-180.

[126] Coelho S.C, Estevinho B.N, Rocha F. Encapsulation in food industry with emerging electro-hydrodynamic techniques: Electro-spinning and electro-spraying – A review. Food

Chemistry.2021;339:127850. DOI: 10.1016/j.foodchem.2020.127850.

[127] Gómez-Mascaraque L.G, Gómez-Mascaraque A. Production of food bioactive loaded nano-particles by electro-spraying. In: Jafari S.M. editor. Nanoencapsulation of Food Ingredients by Specialized Equipment. Volume 3. London: Academic Press;2019.p.107-149. DIO: 10.1016/B978-0-12-815671-1.00003-2.

[128] Echegoyen Y, Fabra M.J, Castro-Mayorga J.L, Cherpinski A, Lagaron J.M. High Throughput Electro-hydrodynamic Processing in Food Encapsulation and Food Packaging Applications: Viewpoint. Trends in Food Science & Technology.2016;60:71-79. DOI: 10.1016/j.tifs.2016.10.019.

[129] Luo C.J, Edirisinghe M. Core-Liquid-Induced Transition from Coaxial Electro-spray to Electro-spinning of Low-Viscosity Poly(lactide-co-glycolide) Sheath Solution. Macromolecules. 2014;47:7930-7938. DOI: 10.1021/ma5016616.

[130] Bhushani J.A, Anandharamakrishnan C. Electro-spinning and electro-spraying techniques: potential food based applications-Review. Trends Food Sci. Technol.2014;38(1):21-33. DOI: 10.1016/j.tifs.2014.03.004.

[131] Pirez-Masia R, Fabra M.J, Lagaron J.M, Lopez-Rubio A. Use of Electro-spinning for Encapsulation. In: Mittal V. editor.Encapsulation Nantechnologies. Massachusetts: Scrivener Publishing; 2013.p.107-135.

[132] Lim L.T. Encapsulation of bioactive compounds using electro-spinning and electro-spraying technologies. In: Sabliov C.M, Chen H, Yada R.Y. editors. Nanotechnology and Functional Foods: Effective Delivery of Bioactive Ingredients.1st ed. Chichester: Wiley; 2015.p.297-317.

[133] Li F, Zhao Y, Song Y. Core-Shell Nanofibers: Nano Channel and Capsule by Coaxial Electrospinning. In: Kumar A. editor. Nano-fibers. London: InTechOpen;2010.p.219-438.

[134] Shahiri Tabarestani H, Jafari S.M. Production of food bioactive-loaded nano-fibers by electro-spinning. In: Jafari S.M. editor. Nanoencapsulation of Food Ingredients by Specialized Equipment. Volume 3. London: Academic Press;2019.p.31-105. DOI: 10.1016/B978-0-12-815671-1.00002-0.

[135] Akkurt S, Shu Liu L, Tomasula P. Electrospinning of Edible, Food-Based Polymers. In: Rai V.R, Bai J.A. editors. Nanotechnology Applications in the Food Industry. Boca Raton, FL:CRC Press, Taylor & Francis Group;2018.p.293-314.

[136] Faridi Esfanjani A, Jafari S.M. Biopolymer Nano-particles and Natural Nano-carriers for Nano-encapsulation of Phenolic Compounds. Colloids and Surfaces B: Biointerfaces. 2016; 146:532-543. DOI:10.1016/j.colsurfb.2016.06.053.

[137] Aman Mohammadi M, Rostami M.R, Raeisi M, Tabibi Azar M. Production of Electrospun Nanofibers from Food Proteins and Polysaccharides and Their Applications in Food and Drug Sciences. Jorjani Biomedicine Journal.2018;6(4): 62-77. DOI: 10.29252/jorjanibiomedj.6.4.62.

[138] Taylor G. The force exerted by an electric field on a long cylinder conductor. Proceedings of the Royal Society of London, Series A: Mathematical and Physical Sciences.1966;291:145-158.

[139] Yarin A.L, Koobhongse S. Reneker D.H. Taylor cone and jetting from liquid droplets in electrospinning of nanofibers. Journal of Applied Physics. 2001;90:4836-4846.

[140] Da Silva J.A.L. Functional nanofibers in food processing. In: Wei Q. editor. Functional Nano-fibers and their Applications. Woodhead Publishing;2012.p.262-304. DOI: 10.1533/9780857095640.2.262.

[141] Wang P, Ding M, Zhang T, Wu T, Qiao R, Zhang F, Wang X, Zhong J. Electrospraying Technique and Its Recent Application Advances for Biological Macromolecule Encapsulation of Food Bioactive Substances. Food Reviews International. 2020. DOI: 10.1080/87559129.2020.1738455.

[142] Rajendran R, Gudimalla A, Mishra R, Bajpai SH, Kalarikkal N, Thomas S. Electro-spinning as a Novel Delivery Vehicle for Bioactive Compounds in Food. In: Rajendran R, George A, Kalarikkal N, Thomas S. editors. Innovative food science and emerging technologies. Oakville: Apple Academic Press;2019.p.426-453.

[143] Lim L.T, Mendes A.C, Chronakis I.S. Electro-spinning and electro-spraying technologies for food applications. In: Lim L.T, Rogers M. editors. Advances in Food and Nutrition Research. Volume 88. Academic Press;2019.p.167-234. DOI: 10.1016/bs.afnr.2019.02.005.

[144] López-Córdoba A, Lagarón J.M, Goyanes S. Fabrication of electrospun and electrosprayed carriers for the delivery of bioactive food ingredients. Issue December. In: L. Melton, F. Shahidi, & P. Varelis editors. Encyclopedia of food chemistry.2016.p.733– 739. DOI: 10.1016/b978-0-08-100596-5.03088-2.

[145] Neo Y.P, Ray S, Jin J, Gizdavic-Nikolaidis M, Nieuwoudt M.K, Liu D, Quek S.Y. Encapsulation of food grade antioxidant in natural biopolymer by electrospinning technique: A physicochemical study based on zein-gallic acid system. Food Chemistry.2013;136(2):1013-1021. DOI: 10.1016/j.foodchem.2012.09.010.

[146] Khoshnoudi-Ni S, Sharif N, Jafari S.M. Loading of phenolic compounds into electrospun nanofibers and electrosprayed nanoparticles. Trends in Food Science & Technology.2020;95:59-74. DIO: 10.1016/j.tifs.2019.11.013.

[147] Davis F.J, Mohan S.D, Ibraheem M.A. Introduction. In: Mitchell G.R. editor. Electro-spinning: Principles, Practice and Possibilities. Cambridge: The Royal Society of Chemistry;2015.p.1-21. DOI:10.1039/9781849735575-00001.

[148] Tapia-Hernández J.A, Rodríguez-Félix F, Katouzian I. Nano-capsule formation by electro-spraying. In: Jafari S.M. editor. Nanoencapsulation Technologies for the Food and Nutraceutical Industries. London:Academic Press;2017.p.320-345.

[149] Drosou C.G, Krokida M.K, Biliaderis C.G. Encapsulation of bioactive compounds through electrospinning/electrospraying and spray drying: A comparative assessment of food-related applications. Drying Technology.2017;35(2):139-162. DOI:10.1080/07373937.2016.1162797.

[150] Gómez-Mascaraque L.G, Tordera F, Fabra M.J, Martínez-Sanz M, Lopez-Rubio A. Coaxial electrospraying of biopolymers as a strategy to improve protection of bioactive food ingredients. Innovative Food Science and Emerging Technologies.2019;51:2-11. DOI: 10.1016/j.ifset.2018.03.023.

[151] Aceituno-Medina M., Mendoza S, Rodríguez B.A, Lagaron J.M, López-Rubio A. Improved antioxidant capacity of quercetin and ferulic acid during in-vitro digestion through encapsulation within food-grade electrospun fibers. J. Funct. Foods.2015;12:332-341.

[152] Blanco-Padilla A, López-Rubio A, Loarca-Piña G, Gómez-Mascaraque L.G, Mendoza S. Characterization, release and antioxidant activity of curcumin loaded amaranth-pullulan electrospun fibers. LWT - Food Science and Technology.2015;63(2):1137-1144. DOI: /10.1016/j.lwt.2015.03.081.

[153] Yuan S, Lei F, Liu Z, Tong Q, Si T, Xu R.X. Coaxial electrospray of curcumin-loaded microparticles for sustained drug release. PLoS One. 2015;10(7):1-15. DOI: 10.1371/journal.pone.0132609.

[154] Isik B.S, Altay F, Capanoglu E. The uniaxial and coaxial encapsulations of sour cherry (*Prunus cerasus L.*) concentrate by electrospinning and their *in vitro* bioaccessibility. *Food Chemistry.* 2018;265:260-273. DOI:10.1016/j.foodchem.2018.05.064.

[155] Costa M.J, Ramos Ph.E, Fuciños P, Teixeira J.A, Pastrana L.M, Cerqueira M.Â. Development of Bio-Based Nanostructured Systems by Electrohydrodynamic Processes. In: Rai V.R, Bai J.A. editors. Nanotechnology Applications in the Food Industry. Boca Raton, FL:CRC Press, Taylor & Francis Group; 2018.p.3-20.

[156] Han D, Steckl A.J. Coaxial Electrospinning Formation of Complex Polymer Fibers and their Applications. ChemPlusChem.2019;84:1453 –1497. DOI: 10.1002/cplu.201900281.

[157] Torkamani A.E, Syahariza Z.A, Norziah A.H, Wan A.K.M, Juliano P. Encapsulation of polyphenolic antioxidants obtained from Momordica charantia fruit within zein/gelatin shell core fibers via coaxial electro-spinning. Food Bioscience.2018;21:60-71. DOI: 10.1016/j.fbio.2017.12.001.

[158] Yang J.M, Zha L.Sh, Yu D.G, Liu J. Coaxial electro-spinning with acetic acid for preparing ferulic acid/zein composite fibers with improved drug release profiles. Colloids and Surfaces B: Biointerfaces.2013;102:737-743. DOI: 10.1016/j.colsurfb.2012.09.039.

[159] Zhang C, Feng F, Zhang H. Emulsion electrospinning: Fundamentals, food applications and prospects. Trends in Food Science & Technology. 2018;80:175-186. DOI: 10.1016/ j.tifs.2018.08.005.

[160] Paximada P, EchegoyenY, Koutinas A.A, Mandala I.G, Lagaron J.M. Encapsulation of hydrophilic and lipophilized catechin into nanoparticles through emulsion electrospraying. Food Hydrocolloids. 2017;64:123-132. DOI: 10.1016/j.foodhyd.2016.11.003.

Chapter 12

Scope, Nutritional Importance and Value Addition in Palmyrah (*Borassus flabellifer L.*): An Under Exploited Crop

Merugu Chandra Surya Rao, Dokka Venkata Swami, P. Ashok, Satya Prakash Nanda and Banavath Babu Rao

Abstract

Palmyrah palm has great economic potential and every part of the palm is useful in one way or the other is considered as 'kalpaga tharu'. The palm is found growing widely in southern states of India. As the value addition in palmyrah is not standardized, the palmyrah products *viz.* tender fruit endosperm (*nungu*), *neera*, jaggery and tuber flour are not commercialized so far. Even though palmyrah is an economically important palm for its nutritional aspects, it has not received proper attention from the agricultural research workers, probably on account of the fact that it is very slow growing palm and mostly found in the wild state. In this context, knowing the physico-chemical properties and development of value added products and popularizing the same is essential.

Keywords: *nungu*, *neera*, jaggery, tuber flour, value addition

1. Introduction

Palmyrah (*Borassus flabellifer* L.) belongs to very ancient family of trees *i.e* Arecaceae and order Arecales. Palmyrah is also known as toddy palm and sugar palm. It is a tropical palm tree which is easily cultivated and also found to grow wild. It is native to Indian sub-continent and South-East Asia. It is widely cultivated from Western India through Indo-china to the lesser Sunda Islands of Indonesia including Bangladesh, Cambodia, China South-Central, Jawa, Laos, Malaya, Myanmar, Socotra, Sri Lanka, Sulawesi, Thailand and Vietnam.

Palmyrah (*Borassus flabellifer* L.) belongs to family Arecaceae and order Arecales. Palmyrah belongs to very ancient family of trees. Palmyrah is also known as toddy palm and sugar palm. It is a tropical palm tree which is easily cultivated and also found to grow wild. It is native to Indian sub-continent and South-East Asia. It is widely cultivated from Western India through Indo-china to the lesser Sunda Islands of Indonesia including Bangladesh, Cambodia, China South-Central, Jawa, Laos, Malaya, Myanmar, Socotra, Sri Lanka, Sulawesi, Thailand and Vietnam.

The name borassus was derived from a Greek word means leathery covering of the fruit and the word flabellifer means Fan bearer. *Borassus flabellifer* is a robust tree that can live more than 100 years and reach the height of 50 to 60 meters. These can be grown in waste lands, farm filed boundaries, sea costs, parks, industrial estates and house colonies. The trunk is grey, robust and old leaves remain attached to the trunk for several years before falling cleanly. The leaves are look like fan-shaped and it grow up to 3 meters long with robust black teeth on the petiole margins. The palmyrah palm throws out spathes during the flowering season and on tapping the young inflorescence a clear, transparent, sweet, pleasant smelling and refreshing and popular drink called *neera* is obtained with high nutritive value, delicious taste and agreeable flavor. The tapping of *neera* and making it into sugar candy was observed by Chinese traveler Magestanes. The different parts of the plant such as roots, leaves, seeds and fruits are used for various purposes. Now a day's palm trees are being cut by people because of not knowing the medical and commercial values.

In India, palmyrah adorns the dry landscape of the semi arid regions of Tamil Nadu, Andhra Pradesh, Gujarat, Odisha, West Bengal, Bihar, Karnataka and Maharashtra. Currently, palmyrah palm wealth of India is estimated as 102 million palms and half of them are in Tamil Nadu. Out of 51.90 million palms in Tamil Nadu, more than 50% of palms are concentrated in the Southern district of Thoothukudi [1]. Government of Tamil Nadu in the year 1978 recognized Palmyrah as State Tree.

Mccurrah [2] enlisted the following 7 species under the genus Borassus (**Table 1**).

Based on the pigmentation of fruit skin Palmyrah palm can be broadly classified into two varieties.

1. Black skin fruits

2. Red skin fruits

1.1 Black skin fruits

Less red pigment is found on the fruit skin. Yield is less but superior seedlings with more starch content and less fibre content noticed. Pulp extraction process is easier. Alkaloids, minerals and free amino acids are lesser than red coloured fruits.

1.	*Borassus flabellifer*	Indian and Malayan spp.
2.	*Borassus aethiopicum*	African spp.
3.	*Borassus deleg*	Sudan
4.	*Borassus heiniana*	New Guinea
5.	*Borassus madagascariensis*	Madagascar
6.	*Borassus sambiranensis*	Madagascar
7.	*Borassus machadonis*	Malaya

Table 1.
Species of Borassus.

1.2 Red skin fruits

Variable amount of black pigment observed on the fruit skin. Fruit yield per tree is significantly high. Pulp, sugar and starch content are less when compared to black skin fruits.

Both the black skin and red skin fruits are recorded for essential amino acids, lysine and methionine. In view of the fruit characters and sap yield the red skinned fruit varieties are seemed to favour for selection for commercial exploitation (**Figure 1**).

In order to prevent huge quantitative as well as qualitative losses in horticultural crops like plantation crops (cashew nut, areca nut, tea, coffee, oil palm, coconut and Palmyrah palm - as these are consumed mostly after the processing) all steps of improved postharvest technology must be carefully implemented from harvesting and ending with consumption and utilization of their products (value added products). In spite of adequate food production, there is existence of hunger and malnutrition. That might be due to the result of uneven distribution of food, losses and deterioration of available food produce. Hence, maximum utilization of available

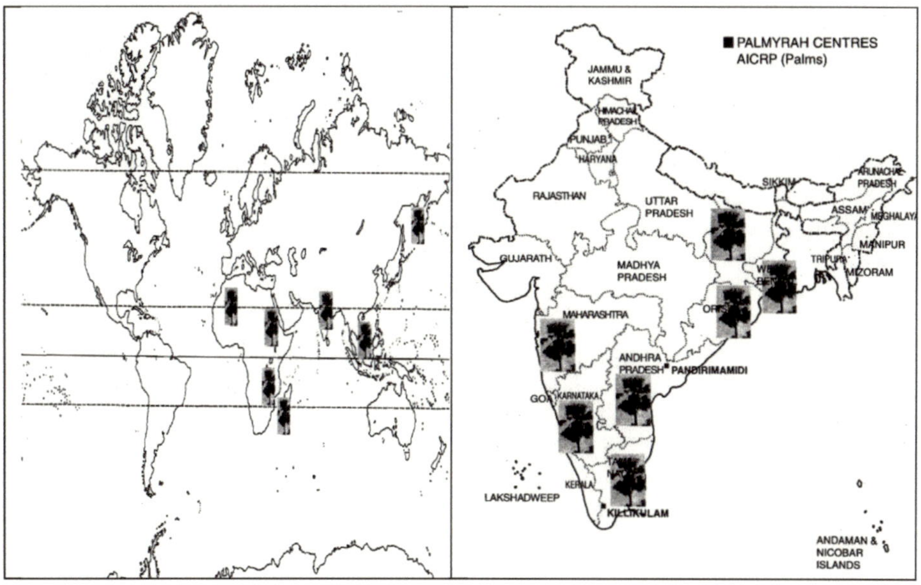

Figure 1.
Palmyrah palm distribution in Worldwide and India.

food and minimization of postharvest losses are absolutely essential in the current scenario.

Food material are subjected to spoilage. The aim of food processing or value addition is to protect food against deterioration or spoilage. The rate of spoilage of raw or fresh food commodities may be very high. The spoilage of food is due to three main causes: (1) microbial, (2) enzymatic and (3) chemical. Foods during storage are more or less infected with microbes that lead to decomposition of the food constituents, sometimes produce evil-smelling and toxic substances. Hence, prevention of microbiological spoilage is essential in any preservation method.

2. Need for post harvest management in palmyrah palm

Palmyrah nungu, neera, tuber flour are highly prone to post harvest losses due to spoilage because they are composed of living tissues. These tissues must be kept alive and healthy throughout the process of marketing. Several factors influence the post harvest losses in commodity. These includes primary factors like mechanical injuries, physiological changes, pathological infections or environmental factors such as temperature, humidity *etc.* Secondary factors which are responsible for deterioration of produces are inadequate facilities of proper harvesting, proper handling, post harvest treatments, packaging and storage. Palmyrah is one such crop where post harvest handling protocols and value addition was not standardized so far.

Post harvest life is also governed by moisture content, rate of respiration, ethylene production and external factors like temperature, relative humidity and atmospheric compositions. Post harvest losses can be considerably minimized and the shelf life can be greatly increased by careful manipulation of the above factors. Deterioration of palmyrah products like nungu, neera and tuber flour during storage depends largely on temperature. One way to slow down these changes can be achieved by lowering the temperature to an desirable level.

3. Scope of palmyrah natural and value added products

Value addition involves change in the physical form of the agricultural produce which leads to its greater acceptability, extended availability, enhanced market viability and increased cost to benefit ratio for the grower of the agricultural produce. The spatial and temporal availability of the produce is enhanced and it becomes less sensitive to price fluctuations in the market. Value addition involves commoditization of the agricultural produce. Therefore, value addition is desirable from both the producer's as well as the consumer's point of view and has aptly been termed as secondary agriculture as stated by United States Department of Agriculture.

However, with the increasing population, industrialization and urbanization, India is facing serious challenges in food security. In order to sustain the agricultural production and address the challenges of food and livelihood security, agricultural diversification has to be adopted through the concept of value addition by minimizing the dependency on main staple crops.

Introduction of new species in the agricultural production system in India is the need of the hour to increase the resiliency of agriculture. In this context, palmyrah palm is one such under exploited crop which have received less attention from agricultural research workers, probably on account of the fact that it is very slow growing palm found mostly in the wild state inspite of having a good number of produce in fresh form (palm neera, nungu) as well as in value added form with a capacity to provide high nutritional value which is having the potential to overcome the problem of malnutrition in developing countries like India. Palmyrah palms are suitable for popularization through value addition (tuber flour, jaggery) which helps in income generation and thus it will improve food security to the poor and livelihood security of the marginal farmers. They can potentially reduce the dependency on few major species (wheat, sugar cane) while diversifying the agricultural production system and making it sustainable.

4. Importance of value addition in palmyrah

Fresh palmyrah tender fruit endosperm (*Nungu*), sap (*Neera*) and tuber flour are perishable and highly prone to post harvest losses due to spoilage. After removing from husk, outer skin of palmyrah tender fruit endosperm starts browning and looses appearance and will be fermented which cause sour odour. It is main factor for consumer to judge its freshness. In normal conditions, *nungu* will have very short shelf life of 2–3 days. Palmyrah sap is naturally prone to fermentation within few hours of extraction and becomes alcoholic beverage (Toddy). Under these circumstances, the processing of tender fruit endosperm and sap into value added products with sufficient shelf life is most important to utilize the products further. Thus the shelf life, quality and availability of the products can be improved by concept of value addition and the value added products have to be commercialized.

Hence, there is a need to study the scope of postharvest techniques for value added products and their shelf life in palmyrah.

Postharvest treatments, packing material and storage conditions significantly increase shelf life, reduce postharvest losses and maintain nutritional quality of palmyrah tender fruit endosperm as well as sap. Hence, it becomes necessary to find out suitable postharvest treatments such as packing material and storage conditions to extend the shelf life and reduce losses of PTFE, neera and other value added products *viz.,* palm jaggery and palm tuber flour.

5. Uses and nutritional importance of palmyrah natural and value added products

Palmyrah is referred as tree of life with nearly 800 uses including food, beverage, fiber, fodder, medicinal and timber. Among the various uses of the palm, the sweet sap from the inflorescence for making *neera* obtained by tapping the tip of the inflorescence either male or female is traditionally collected in hanging earthen pots and used to quench thirst. The sweet sap collected early in the morning is refreshing and light drink called *neera* in telugu and marathi and "*pathaneer*" in tamil. *Neera* has sugary sweet in taste, oyster white in colour, translucent with high nutritive value but susceptible to natural fermentation at ambient temperature within a few hours of extraction [3]. The sap collected in the evening or after fermentation becomes sour which is called *kallu* in telugu and *tadi* in marathi. *Tadi* is mostly consumed by villagers as raw alcoholic beverage.

When the fruit is very young, the kernel is hollow, soft as jelly and translucent like ice and is accompanied by a watery liquid, sweetish and potable. The jelly part of the fruit is covered with a thin, yellowish-brown skin. These are known to contain watery fluid inside the fleshy white body. Palmyrah tender fruit endosperm (PTFE) contains 43 kcal of energy, 87.6 g of water, 0.8 g of protein, 0.1 g of fat and 10.9 g of carbohydrates per 100 g fresh weight of palmyrah tender fruit endosperm [4].

Palmyrah tuber is an important edible shoot grown in loose soil from the seed of ripe fruit. Tuber is eaten by many people directly by cooking in open fire after peeling off outer layer. Roasted, dried tubers are ground to make flour which is blended with wheat flour for baking. The flour can be made into a number of food items which are used traditionally. It is used in preparation of odiyal consumed as porridge called khool and a steamed product called pittu. *Odiyal* made from palmyrah tuber flour contains 1423 kcal of energy, 10.8 g of moisture, 3.1 g of protein, 77.1 g of carbohydrates and 5.6 g of crude fiber per 100 g [5].

Ripened fruit is used in preparation of various foods at home level. The soft orange-yellow pulp (mesocarp) of the ripe fruit is sugary, dense and edible, rich in vitamin A and C. Palmyrah also contains bitter compounds called flabelliferrins, which are steroidal saponins. Ripe fruit pulp can be processed into soft beverages, jam, spread, toffee, delicious food items and sweets [6].

The palm jaggery is processed from the unfermented palmyrah tree sap called *neera*. Palm jaggery contains of 65–68% sucrose and 5–15% reducing sugars which is directly used in ayurvedic preparations and believes to reduce the lung cancer. Hundred grams of palmyrah jaggery contains 0.35% of protein, 0.17% of fat, 90.6% of carbohydrates, 24 mg of vitamin B-1, 11.0 mg of vitamin-C and 0.74% of minerals [7].

6. Palmyrah value added products

See **Table 2**.

Plant part	**Value added products**
Edible value added products	
Inflorescence sap	Toddy, Jaggery, Sugar, Honey, Wine
Fruit	Toffee, Spread, Jam, Pickle, Sweets (Burfi), Beverages (RTS, Squash, Nectar)
Kernel	Canned products
Non-edible value added products	
Leaf	Mats, Baskets, Fans, Hats, Umbrellas, Buckets, Writing Material, Fence, Fibre extracted is used to make brushes and handicrafts
Fruits	Fibre extracted is used to make toys and fancy items
Stem	As poles for sheds construction and as timber source

Table 2.
Value added products of palmyrah palm (edible and non-edible products).

7. Value addition in palmyrah (edible products)

7.1 Neera

Increasing health consciousness among the population in India has boosted the growth of health drinks industry in India. *Neera*, also called palm nectar (phloem sap extracted with zero percent alcohol) extracted from the inflorescence of toddy palms which is used as a nutritious health drink. *Neera* is called sweet toddy since it contains zero percent alcohol and known as *padaneer* in Tamil Nadu and *kallu* in

Telugu. Toddy and *neera* can be called as fermented sap and non-fermented sap respectively. *Neera* is susceptible to natural fermentation at ambient temperature within few hours of extraction due to enzymatic and microbial fermentation since it is rich in sugars, vitamins, proteins and minerals [8]. Once fermented, *neera* becomes toddy which is unsuitable as health drink or as value added product.

Neera is rich source of sugars, minerals and vitamins. It contains substantial amounts of iron, phosphorus and ascorbic acid. It is more nutritious than any of the commercially marketed fruit juices available in the country. The nutritional composition of *neera* is given in **Table 3**. Palmyrah palm *neera* contain very low Glycemic Index (GI) of 35 (normal table sugar has GI 70), *neera* can be used by diabetic patients also.

The high nutritive value of *neera* makes it an emerging health drink. It is cool and good for improving the general health. It can be served to supplement the iron and vitamin deficiency. Clinical studies proved neera has potential medical applications for treating asthma, tuberculosis, bronchial suffocation and piles. It is believed to facilitate clear urination and prevent jaundice. High amount of glutamic acid is present in neera. It is the amino acid which is used by the body to build proteins. Neera can be used to treat eye abnormalities and eczema as it contain high amount of inositol. Neera can be used as stimulant and antiphalegmatic which is considered to be useful in treating inflammatory infections. Women suffering from anemia due repeated pregnensis are advised to take neera to ameliorate health.

8. Season and stage for tapping

The extraction of sap from the inflorescence is called tapping and tapping vary according to the sex of the palm and age of the inflorescence. Long tapping duration is noticed in female palms (April to December) when compared to male palms (December to February). Dry season tapping is done mostly in the low lying lands where palms do not suffer due to moisture stress during drought period. The spathe is considered ready for tapping when the inflorescence opens or is just about to

Physico chemical parameter	Neera	Endosperm/Nungu	Jaggery	Tuber flour
Moisture (%)	—	—	09.02	09.32
TSS (°Brix)	10.00	08.50	07.50	05.10
Ph	04.28	06.44	05.51	05.54
Titrable acidity (%)	00.57	00.06	—	00.40
Total sugars (%)	14.85	08.83	09.30	14.39
Reducing sugars (%)	05.16	05.11	05.10	08.50
Non reducing sugars (%)	09.69	03.72	04.20	05.89
Starch (%)	—	—	—	32.96
Fiber (%)	—	—	—	10.20
Protein (%)	—	—	—	02.96
Phenols (mg)	00.28	—	00.16	10.43
Browning (%)	—	00.01	—	—
Alcohol content (%)	02.00	—	—	—

Table 3.
Proximate physico chemical composition of palmyrah natural and value added products on initial day of storage [9].

burst. The female flowers within the unopened spadix causes a swelling at the base and this indicates the appropriate stage for tapping.

Neera is widely consumed in countries like India, Sri Lanka, Africa, Malaysia, Indonesia, Thailand and Myanmar. It is a potential health and therapeutic drink since it is rich source of Vitamin C and vitamin B complex, having more calories which fight against diabetes, cancer, electrolyte deficiency and even hair fall. *Neera* is rich source of sugars, essential elements micronutrients and minerals. It contains acids like nicotinic acid (Vit.B3) and riboflavin (Vit.B2) and also can be consumed by people suffering from diabetes since it has a low glycemic index (GI) [10]. Consumption of *neera* prevents jaundice and also facilitates clear urination. It keeps the human biological system cool and helps to improve digestion. In a large scale *neera* production is noticed in an un-organized manner with major consumption by rural population. Hence, there is a wide scope for commercializing the product.

9. Tapping in male palm

9.1 Aripanai

Sap extracted from 2 weeks old inflorescence by removing the sheet covering the inflorescence and the inflorescence is kept to dry for three days. Later a new surface is made by cutting and pot is tied directly to the inflorescence to collect the sap. In this method no pressing or stroking to the inflorescence is made as like other methods of tapping.

9.2 Vallupanai

Sap is extracted from one month old inflorescence. In this tapping method male spikes bearing sessile flowers are subjected to pressing and stroking and such three to six spikes are brought together wrapped with palmyrah leaves kept in a pot for collection of the sap.

10. Tapping in female inflorescence

10.1 Tattupalai

Sap is extracted from the young inflorescence by softening the tissues. The inflorescence main axis is hitted with iron rod and fork is used to give a small press in the region where the fruits used to develop.

10.2 Kaivetty

The sap is extracted from the inflorescence at an age of two to three months where the matured fruits on the inflorescence are sliced for collection of the sap (July to November).

10.3 Nungu (palmyrah tender fruit endosperm)

The tender fruit endosperm is a summer delicacy consumed in the southern and eastern parts of India. On the hot summer day *nungu* acts as a coolant for the parching throat. It provides a proportionate balance of minerals and sugar which is required for the body during the summer season to attain cooling. The tender fruit endosperm, which is available in abundance during the summer season, is rich in vitamin B, iron, calcium niacin and riboflavin. It is used to cure ulcers, urinary infections and heat rashes which mainly occur during summer months [11].

The palmyra tender fruit has the outer fibrous layer containing the sugary gelatinous endosperm. Fresh tender endosperm is perishable and highly prone to postharvest losses due to spoilage as it is composed of living tissues. These tissues must be kept alive and healthy throughout the process of marketing. Several factors influence the postharvest losses of most commodities. These include primary factors *viz.*, mechanical injuries, physiological changes and environmental factors such as temperature, humidity *etc.* Secondary factors which are responsible for deterioration of produce is due to inadequate facilities for harvesting, handling, postharvest treatments, packaging, transportation, storage.

10.4 Jaggery

Palmyrah jaggery is superior to cane jaggery. Cane jaggery is sweet, but Palm jaggery is sweet and delicious it can be produced worth crores of rupees. Palm jaggery gives mineral salts too. Doctors have told me to eat jaggery and I always eat Palm jaggery. Nature has made this product in such a way that it cannot be manufactured in the Mills, it is produced in the Cottages. Where there are Palm trees, this jaggery can be easily produced. Andhra Desha has thousands of Palm trees, there jaggery is produced in every hamlet. This is the way to banish poverty from the land. This also is an antidote to poverty." (From a speech delivered at the opening of the village industry exhibition in Brindawan Bihar on 3 May 1939 by Mahatma Gandhi) which speaks about the potentiality of palmyrah jaggery.

Jaggery is a natural sweetener made by concentrating the palmyrah fresh neera with clarification to remove impurities and uniform heating in open pan. As the jaggery is made up of longer chains of sucrose it is complex in nature when compared to sugar that makes the jaggery to digest slowly than sugar and releases energy slowly and not spontaneously which provides the energy for a longer period of time so it is not harmful for the body. It is a sensitive product, getting affected by number of factors right from collection of neera to processing and storage of jaggery. The jaggery industry is still at cottage level because of some technological drawbacks in its export quality processing and storage. The keeping quality of jaggery largely depends on the atmospheric temperature and relative humidity [12].

Jaggery is mostly spoiled during the monsoon period because of the presence of higher humidity in the atmosphere. The major problem associated with jaggery storage is the presence of invert sugars and mineral salts which are hygroscopic in nature. In the coastal region of the country, where atmospheric humidity and rainfall very high as it is very difficult to store jaggery. The study showed that about 5–10% of stored jaggery get spoiled every year leading to a huge loss [12].

In India, the jaggery storage facilities at producer/farmer level are very poor as it is stored in godowns, household kitchens and cheap storage systems where hygienic conditions are not strictly maintained, which attract several pathogenic and non-pathogenic microorganisms. Cold storage godown is being used in west Godavari and Vishakhapatnam districts of Andhra Pradesh, Kolhapur district of Maharashtra and Muzaffarnager area of Uttar Pradesh in India. But for small farmers storing jaggery in the cold storage is very difficult due to cost and energy consumption factor [13].

11. Preparation of palmyrah jaggery

Palmyrah sap (neera) was collected in slacked lime treated earthen pots for experimental purpose. The cleared sap after lime sedimentation and filtration was transferred into the galvanized iron pan and boiled to 110°C. Few castor beans were crushed and put into iron pan to prevent over boiling. During boiling, a white scum arises on the surface which was skimmed off (removed with a ladle). Neera gets transformed into viscous fluid at 110°C. The fluid was stirred continuously to avoid charring at the bottom of vessel. Placing a few drops of fluid into cool water, the correct stage of formation of jaggery was judged. The hardening of fluid in the cold water is the indication of right stage of conversion of neera into jaggery. At this stage, jaggery fluid was poured into moulds and allowed to cool, then after some-time, fluid jaggery solidified in the moulds. The solid jaggery cubes were removed from the moulds and used in the experiment.

12. Palmyrah tuber flour

Tuber is an edible shoot grown in loose soil from the seed of ripe fruit. Tuber is eaten by many people directly by cooking in open fire after peeling off outer layer. Palmyrah tuber is rich in starch and fibre, which is helpful in controlling various diseases especially diabetic. Regular consumption of palmyrah tuber flour increase the body strength, reduce hunger and mixing of palmyrah tuber flour with other foods would positively reduce the malnutrition [14].

13. Bio active components in palmyrah tuber flour

Starch is the main carbohydrate present in palmyrah tuber flour. The starch has low viscosity and gelatinization temperature but exhibit good settling property as such food starch. The palmyrah tuber flour starch is devoid of bitterness as it is inherent property of palmyrah tuber and has a grain size of 40 μm similar to potato starch.

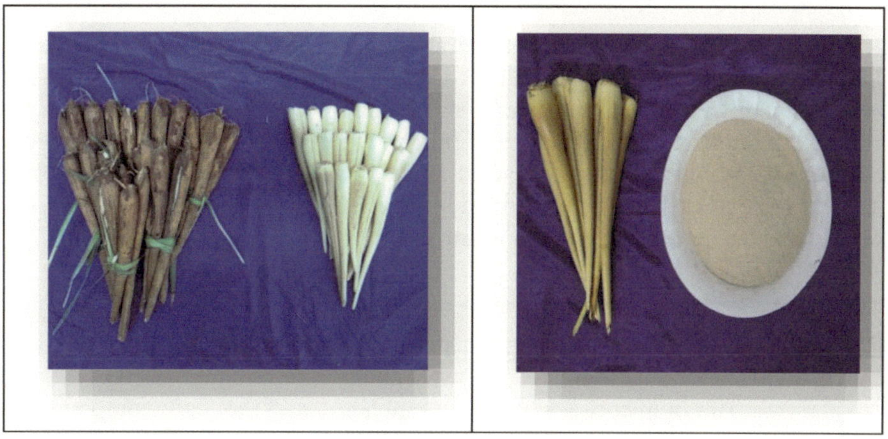

14. Palmyrah fruit

Palmyrah fruit pulp can be utilized to prepare food items and animal feed. About 40% pulp is obtained from the fruit which is dark yellow in colour with a characteristic taste, flavor and bitterness. Palmyrah pulp is mixed with other fruits to making jam, cordial, cream etc. As the palmyrah pulp is bitter in taste, it is better to prepare mixed fruit jam instead of palmyrah jam separately.

15. Extraction of pulp from palmyra palm

Fully ripened fruits obtained from the palm are washed, peeled and pulp is to be extracted manually by rubbing with the traditional Palmyra extractor. Additional water was used in the proportion of 1:1 ratio to extract the pulp which is adhered to the seeds. Heat treatment (70°C for a period of 10 min) was given to the pulp to obtain maximum pulp recovery [15]. Then the pulp is sieved to remove the fibrous material.

16. Preparation of palm spread

The best recipe for preparing palmyrah palm spread (pulp-1 kg, sugar-1 kg, skimmed milk powder-100 g, small cardamom-4 number citric acid-5 g) for the preparation of palm with good Total Soluble Solid (TSS) and acidity. For the preparation of palm spread, extracted pulp is mixed with other ingredients, heated at low flame with continuous stirring till the TSS reaches 65–68°Brix. Then cooked material was removed from heat, filled into broad mouth sterilized glass bottles, capped, labeled and stored at both room and refrigerated temperature.

17. Bio active components in palmyrah fruit pulp

Palmyrah fruit pulp consists of 75–80% moisture and the main components are carbohydrates. It also contains free amino acids like lysine, aspartate and glutamate. The main digestible carbohydrates found in palmyrah fruit pulp are sucrose, glucose and fructose. The content of carotenoids (beta carotenes) found but they varied.

Flabelliferrins (steroidal saponins) are reported to be the compounds which are responsible for the cause of bitterness in palmyrah fruits. A wide range of flabelliferrins were reported which act as anti-microbial properties. The term flabelliferrin was coined from specific word flabellifer (**Figure 2**).

17.1 Palm sugar

Palm sugar can be used as a substitute for cane sugar. To prepare palm sugar *neera* is strained through wire mesh to make the *neera* free from debris and it is boiled in an alloy vessel. *Neera* is boiled uniformly and the liquid is allowed to cool all the sediments have to be removed. Clarification is carried out by adding triple super phosphate to form insoluble calcium calcium phosphate as it react with the lime which is already present. Later it is heated to a temperature of 110°C for 2 hours until it reach honey like consistency then allowed to cool and passed through a crystallizer. After forming sugar crystals, it is centrifuged to collect sugar and dried and powdered to store.

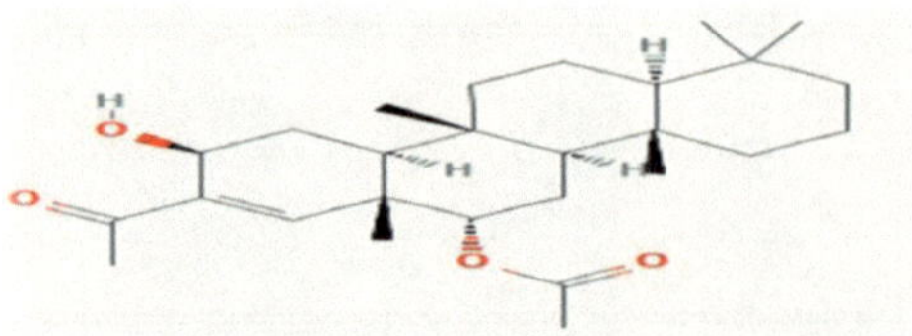

Figure 2.
Structure of flabelliferrin.

17.2 Palmyrah honey

Fresh *neera* is heated for 2 hours to obtain honey like consistency and it is transferred to mud pots. Ripe, dry and shelled tamarind fruits devoid of seeds are added to the boiled syrup. To prepare 10 liters of syrup about 1 Kg of tamarind fruits are required. The pot is closed with cloth and containers are kept in a shock proof, cool and dry place for a period of 130–180 days.

17.3 Toddy

Toddy is formed as a result of fermentation of neera/sap by wild yeasts and bacteria, which come into contact with the sap after tapping. This is an uncontrolled natural fermentation by number of different strains of yeast and bacteria. The alcohol content in naturally fermented toddy is reported to be 5%. But fermentation of palmyrah sap by using pure yeast culture gives about 7.8% alcohol content under laboratory conditions. The major sugars that are present in partly fermented toddy is sucrose, glucose and fructose but these will be gradually converted into ethyl alcohol during the process of fermentation [16].

17.4 Wine

Unfermented sap (*neera*) is sterilized which can be fermented with suitable strains of yeast to obtain palmyrah wine. Sweet toddy having a pH of 6–7, is sterilized and inoculated with good yeast that produces a very clear straw coloured wine. The alcohol strength increases by adding extra sugar to the sap. The wine prepared by this method is pleasant to drink which mask the specific characteristic toddy flavour and distinctive sour taste of the acids present in toddy [17].

18. Value addition from non edible products

18.1 Palmyrah leaves

The tender leaves which are in ivory colour are harvested from the palm are sized into narrow strips that can be utilized for making toys, flowers, garlands and fancy goods. Whereas, the matured leaves are used for making of containers. The harvested leaves have mid ribs that can be utilized for making of the brooms.

18.2 Palmyrah fibre

Fibre is extracted from the bifurcated base of leaf stalk. The palmyrah fibre is the product that having good export market in countries like Australia and United Kingdom and Japan. The fibre obtained from palmyrah palm leaf stalks is mainly valued for its high tensile strength leading itself for many industrial applications.

Naar is the fibrous material obtained from stalk of the fronds. Karukku are the longitudinal splits which are obtained by soaking the stalks after removing the sharp serrated margins of the petiole which can be used for tieing purpose.

18.3 Palmyrah timber

Palmyrah timber is most valued for the construction of houses in the villages and sometimes the timber is also utilized as rafters and beams. The timber obtained from palmyrah palm is also used as fire wood.

Palmyrah leaves	**Palmyrah fibre**	**Palmyrah wood**

19. Research and development organizations on Palmyrah

All India Coordinated Research Project on Palmyrah (AICRP ON Palms (Palmyrah).

Palmyrah is a mandatory crop under All India Coordinated Research Project on Palms (AICRP). Horticultural Research Station, Pandirimamidi under Dr. YSR Horticultural University, Andhra Pradesh and Horticulture College and Research Institute, Killikulam in Tamil Nadu are the two research centers under AICRP on palmyrah where the collection, conservation and evalution of existing germplasm in palmyrah and hybridization for developing dwarf types are focused.

Palmyrah Development Board (PDB), 244, Gallie Road, Bambaalapitiya, Colombo, Sri Lanka.

20. Conclusion

Palmyra tree plays an important role in human life. Every part of the tree is used for preparation of various types of products and it gives more health benefits. But everyone is not aware of this tree. So, it is necessary to create awareness regarding palmyrah natural (neera, nungu) and value added products (palm sugar, honey, toddy, wine, jaggery and flour. Post harvest losses can be considerably minimized and their storage life can be greatly increased by careful manipulation of moisture content, rate of respiration and atmospheric composition.

Value addition involves change in the physical form of the agricultural produce which leads to its greater acceptability, extended availability, enhanced market viability and increased cost to benefit ratio for the grower of the agricultural produce. However, with the increasing population, industrialization and urbanization, India is facing serious challenges in food security. In order to sustain the agricultural production and address the challenges of food and livelihood security, agricultural diversification has to be adopted through the concept of value addition by minimizing the dependency on main staple crops. Introduction of new species in the agricultural production system in India is the need of the hour to increase the resiliency of agriculture. In this context, palmyrah palm is one such under exploited crop having a good number of produce in fresh form (palm neera, nungu) as well as in value added form with a capacity to provide high nutritional value and having the potentiality to overcome the problem of malnutrition in developing countries like India. Palmyrah palms are suitable for popularization through value addition (tuber flour, jaggery) which helps in income generation and thus it will improve food security to the poor and livelihood security of the marginal farmers.

Author details

Merugu Chandra Surya Rao[1]*, Dokka Venkata Swami[2], P. Ashok[2], Satya Prakash Nanda[1] and Banavath Babu Rao[2]

1 MS Swaminathan School of Agriculture, Centurion University of Technology and Management, Odisha, India

2 Department of Horticulture, Dr. Y.S.R. Horticultural University, Andhra Pradesh, India

*Address all correspondence to: chandra.surya@cutm.ac.in

References

[1] Anonymous, 2015. Preliminary trial on vacuum packaging of nungu. AICRP (Palms). Annual Report 2014–15, *ICAR - All India Co-ordinated Research Project on Palms.* ICAR - CPCRI, Kasaragod. 108.

[2] Mccurrah, J.C. (1960). Palms in the world. Harper and Brothers, New York, pp.290.

[3] Vengaiah, P.C, Murthy, G.N, Sattiraju, M. and Maheswarappa. 2017. Value added food products from palmyrah palm (*Borassus Flabellifer* L.) *HP Journal of Nutrition and Health Sciences.* 4.

[4] Piyush, S.V. 2016. Standardization of packaging and storage technology for tender fruit endosperm and sap of palmyra palm (*Borassus flabellifer* L.) Ph. D. Thesis submitted to *Navsari Agricultural University*, Navsari.

[5] Jansz, E.R, Wickremasekara, N.T. and Sumuduni, K.A.V. 2002.A review of the chemistry and biochemistry of seed shoot flour and fruit pulp of the palmyrah palm (*Borassus flabellifer* L.). *Journal of National Science Foundation of Sri Lanka*. 30(1, 2): 61-87.

[6] Das, B.C. and Das, S.N. 2003. Cultivation of minor fruits. *Kalyani Publishers*, India.

[7] Vengaiah, P.C, Murthy, G.N, Prasad, K.R. and Kumari, K.U. 2013. Post-harvest technology of Palmyrah (*Borassus flabellifer* L.) Present practises and scope. *International Conference on Food Processing*. 22-24.

[8] Naveen, J, Gajanan, P, Deshmukh and Menon, R.R. 2018. Neera-A Potential Natural Health Drink. *Biomedical Journal of Scientific and Technical Research*. 11(3).

[9] Chandra Surya Rao, M. Swami, D. V. Ashok, P. Salomi Suneetha, D. R. Sujatha, R. V. and Sekhar V. 2020. Current Journal of Applied Science And Technology. 38(48): 396-401.

[10] Hebbar, K.B, Mathew, A.C, Manikantan, M. R. and Kukkamgai. 2013. Value added products from neera. *Indian Coconut Journal*. 28-33.

[11] Ramya, V. 2014. FouramazingbenefitsofPalmfruit. h ttp://www.wildturmeric.net/2014/12/4-amazing health benefits of palm fruit nungutadgolamunjal fruit.html.

[12] Shweta, D, Bhotmange, M.G, Prabodh. H. and Chitale, M. 2019. Study of traditional Indian sweetener 'Jaggery' and its storage behaviour. *International Journal of Chemical Studies.* 7(3): 410-416.

[13] Said, P.P. and Pradhan, R.C. 2013. Preservation and value addition of jaggery. *International Journal of Agricultural Engineering*. 6(2): 569-574.

[14] Thivya, P, Manoharan D. and Jaganmohan, R. 2018. Effect of debittering on the physical and chemical properties of palmyrah young shoots flour.

[15] Susanta K. Roy and R. N. Singh, 1979. Bael fruit (Aegle marmelos)—A potential fruit for processing. Economic botany, (33), 203-212.

[16] Savarimuthu, R., S.Ratnam, K. Subramaniam, L. Kandasamy and S. Kandiah. (1980). Proceedings 36th annual sessions. pp. 59,60.

[17] Theivendirarajah, K (1983). Presented at the workshop on palmyrah at Jaffna on 21st – 25th Feb.

Chapter 13

Phenolic Compounds

Mohd Abdul Gani and Shama M

Abstract

Phenolic compounds represent a group of molecules and its functions in the growth and development with a defense mechanism in plant. It includes pigments, signaling molecules, and flavors which will protect the plant against insects, fungi, bacteria, and viruses and plays a role to attract or repulse them. This current chapter includes different aspect of phenolic compounds were discussed such as the definition, chemical properties, classification, the biosynthesis process of phenolic compounds, extraction technologies in plants also includes the shikimate, pentose phosphate and phenylpropanoid pathways. They were having many health benefits like UV screens, attractants, signal compounds, and other response chemicals from different types. As per the human physiology, they are vital in protection and plays an important role in prevention and treatment of many chronic diseases. It also acts as antioxidant, antiseptic, anti-proliferative activities, antidiabetic, anti-inflammatory and anti-aging. They are useful to eat such plant foods that contains high antioxidant content, which can hamper the incidence of certain chronic diseases, such as cardiovascular diseases, diabetes and cancers, through the management of oxidative stress. Overall the phenolic compounds are a gift of god in our day to day lives.

Keywords: phenolic compound, chemical properties, biosynthesis, extraction technologies, health benefit

1. Introduction

Grains, mainly cereals and legumes, are important in every diets of human in any part of the world. They are rich in diverse nutrients and phytochemicals, and possess manifold bioactivities, such as antioxidant, antidiabetic, and anticancer effects [1–3]. Phenolic compounds [PC] are distributed everywhere in most of the plant tissues which includes the parts such as roots, stems, fruits, seeds, leaves, etc. [4]. There are more than 8000 individual plant with great chemicals isolated, structural variability and nearly 200000 were identified with diverse structures [5] and classes from higher plants around the planet. They are classified as primary metabolite and secondary metabolite [6]. The primary metabolite is required for cell nourishment, such as carbohydrates, proteins, fatty acids and nucleic acids. The secondary metabolite is essential to plant survival which directly involved in photosynthetic or respiratory metabolism. As differentiated from primary metabolite, the chemicals and structures of secondary metabolite are responsible for plant defense. They also protect the plant from oxidants and ultraviolet radiation and also act as attracting pollinators or animals for seed dispersion and signal compounds [6–8].

The secondary metabolite is classified according to their biosynthetic routes and structure; they are divided into three major groups: (1) flavonoids, allied phenolic, and polyphenolic compounds; (2) terpenoids, and (3) nitrogen-containing alkaloids and sulfur-containing compounds. These compounds are linked to primary metabolite by biosynthetic enzymes and building blocks [8]. Phenolic compounds (flavonoids, allied phenolic, and polyphenolic compounds) are one among the secondary metabolites more cosmopolitan in plants. The shikimate, pentose phosphate and phenylpropanoid pathways are extract from plants. These compounds perform an important role in the growth and reproduction of plants, giving protection against pathogens and predators. In vegetables and fruits, PC contribute to color and sensory characteristics [8, 9].

2. Phenolic compounds: definition, chemical properties, classification, biosynthesis, extraction technologies and medical importance

2.1 Definition

The compounds that have one or more hydroxyl groups connected straightway to the ring of an aromatic. The whole category is based on the arrangement of phenol (**Figure 1**).

In phenols, the hydroxyl group is linked to a chain of carbons which are alike to alcohols of aliphatic structures. Due to the existence of the aromatic ring, the phenolic hydroxyl group is affect. The hydrogen of the phenolic hydroxyl is unstable caused by the aromatic ring, that build the phenols as a weak acid [10].

Its structure consists of an aromatic ring that contain 1 or more hydroxyl substituents. It may be classified into simple phenolic molecule and extremely polymerized compounds. The PC occur naturally is associated with one or more phenolic groups when combine with mono- and polysaccharides. In addition, they also can be linked to esters and methyl esters. They have a wide range in structure diversity that occurs in nature. More than 8000 structures of phenolic compound are studied till now [8].

2.2 Chemical properties of phenolic compounds

2.2.1 Benzene ring

The carbon's atomic number is 6, i.e. it has 6 electrons and 6 protons. Electrons are around the atom's nucleus in orbitals. The benzene ring is representing one of two mesmeric complex. The double arrow indicates, the two drawn structures and the true structure of the molecule lies in between. Hence, as the six C-C bonds of the ring are identical, with the π-electrons over the entire ring which is more accurate to use structure. The affects of reactivity of aromatic compounds is due

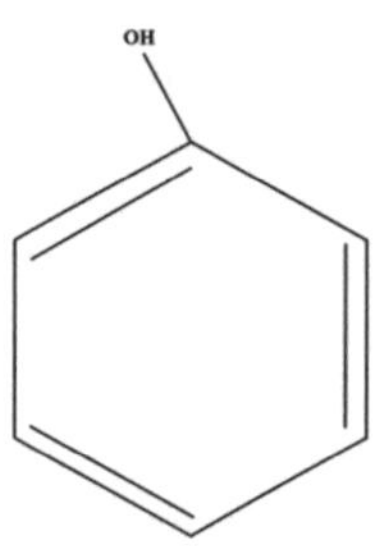

Figure 1.
Phenol.

to delocalization of the π-electrons is very much favorable and also have tendency to refurbishing aromaticity. Aromatic compounds do not easily undergo any addition reactions and a double bond is replaced by two single bonds, as we see in regular alkenes i.e., linear chains of carbon atoms containing at least one double bond. Aromatic compounds show a partially replaced the reactions, that means the replacement of atoms [9, 10].

2.2.2 Phenolic hydroxyl group

It depends upon the chemical properties of the benzene with a hydroxyl group ring and the foremost property is acidity which are generally weak acids. As compared to hydroxyl group of unsubstituted aliphatic alcohols, phenolic OH-group is more acidic [8–10].

2.3 Classification

The word phenolics includes a very wide group of chemical compound. They can be classified in many ways. Harborne and Simmonds (1964) classified these compound into groups depends upon the numeral of carbons in the molecule. (**Table 1**) [9].

An another classification represented by Swain and Bate-Smith (1962). They categorized the phenols as "common[typical]" and "less common [less typical]".

Structure	Class
C6-C1	simple phenolics
C6-C2	phenolic acids and related compounds
C6-C3	acetophenones and phenylacetic acids
C6-C3	cinnamic acids, cinnamyl aldehydes, cinnamyl alcohols
C15	coumarins, isocoumarins, and chromones
C15	chalcones, aurones, dihydrochalcones
C15	Flavans
C15	Flavones
C15	Flavanones
C15	Flavanonols
C15	Anthocyanidine
C15	Anthocyanine
C30	Biflavonyls
C6- C1- C6, C6- C2- C6	benzophenones, xanthones, &$$$;tilbenes
C6, C10, C14	Quinones
C18	Betacyanine
Lignans, neolignans	Dimers or oligomers
Lignin	Polymers
Tannins	oligomers or polymers
Phlobanhenea	Polymers

Table 1.
Classification of phenolic compounds.

Ribéreau-Gayon (1972) classified the phenols into three origins which is as follows: [10].

1. It is widely distributed in all plants or in a specific plant.

2. It is less widely distributed to known in confined number of compounds.

3. Phenolic component exists as polymers.

2.4 Biosynthesis of phenolic compounds

The biosynthesis of PC to exhibit the origin of the various families which precursors. The review of accepted pathways, newly illuminated steps in the biosynthesis for isolation of protein, gene cloning, and protein characterization.

2.4.1 Protein isolation and purification

The conventional methodology for the isolation of proteins includes the techniques for the separation of biochemical, where the protein is isolated from the other proteins based on its special chemical and physical properties. This involves molecular mass i.e. size, shape, net charge and hydrophobicity. The isolation procedure starts with the mixture of a cell extract is recognize in the course of the enzyme activity. This mostly involves crushing of the tissue in an extraction buffer and the contents of the cell [proteins] become accessible. To avoid proteolytic degradation of the enzyme, add protease inhibitors i.e. phenylmethyl sulphonylfluoride (PMSF), in the mixture of the extraction buffer. The first step is a centrifugation where the enzyme is precipitated or otherwise ends up in the supernatant which is bound to the cell wall or the cell membrane, Soluble proteins are usually separated from one another by their solubility in high-salt solutions with supported variation. Add a salt, followed by centrifuge to remove the precipitated proteins that will be fruitful strategy for proteins to separate from each other. During saturation, ammonium sulfate is usually used to precipitate proteins, as most of the proteins in the solution of this salt. The quantity of ammonium sulfate must be mixed in order to avoid subset of the proteins in the extract which can be determined by enzyme activity assays on the various fractions. Likewise, other salts, trifluoroacetic acid, protamine sulfate, polyethylene glycol, apolar solvents etc. are often used for the proteins precipitation. The next step used for further purifying the enzyme is chromatography such as High-performance liquid chromatography (HPLC), Hydrophobic interaction chromatography (HIC), Ion exchange chromatography(IEC) Bio-affinity chromatography (BAC), fast protein liquid chromatography (FPLC) and Gel filtration or size exclusion chromatography (GFC). When a fraction is obtained is the only protein during this procedure then purification is completed to homogeneity, or any contaminants are remained below the level of observation [8, 10, 11].

2.4.2 Gene cloning

It depends upon the phenolic content or its composition that are present in mutants that can altered incapable cloning of the mutated gene and cloning of the wild type of the gene [normal]. If a plant lacks in a particular phenolic compound as which is the results of a mutation, the wild type of the gene will indicate as the wild-type allele that plays an important role in the compound for biosynthesis process. The sequence of the protein encoded by the gene and this sequence of the gene can be used to deduce the amino acid present in the compound. The polymerase chain reaction was revolutionized the clone genes and it involves three steps:

a. Denaturation: To separate the two DNA strands of the template performed at 94 °C.

b. Primer annealing: It was performed at 45-70 °C depending on the GC-content and length of the primer.

c. Extension: It was performed at specific temperature recommended by the manufacturer of the enzyme Particularly at 72 °C, during which the polymerase synthesizes the DNA delineated by the primers. This procedure is repeated 20-40 times, and in each cycle every template strand is being replicated. PCR is the preferably used technique in plant, animal and microbial biology also as medicine. PCR is also considered for cloning of genes and cDNA's, and taken for genotyping using molecular markers [10, 11].

2.4.3 Insertional mutagenesis

On the basis of known sequence, which is also referred as tag in the gene of interest. There are two methods done for cloning purposes.

1. Tagging: A spontaneous or chemically induced mutant, which was been identified with no information accessible about the mutant gene.

2. Random Tagging: It is based on Principle of insertion of any gene. All the genes managing the trait of interest were rarely covered, until the mutation is not lethal [10, 11].

2.4.4 Map-based cloning

It involves identification of the molecular marker(s) which are associated with the mutation. The mutant plants will show mutation when closely linked with predominant marker allele from the mutant parent, whereas wild-type parent marker allele will show wild-type plant Once the accurate mapping is achieved, the gene sequence will be obtained on the availability of genome sequence [10, 11].

2.4.5 Candidate-gene approach

This approach is possible once there is establishment of protein databases and large DNA then the candidate gene is defined as defective gene that can originate the mutant phenotype. Once the candidate gene is identified the sequence databases were searched to identify DNA or protein sequences from the candidate gene [10–12].

2.4.6 Quantitative trait locus mapping

This can show the position of the identified candidate gene. QTL; the abbreviation for the quantitative trait loci), that can be defined as a genetic locus described by two molecular markers on a genetic map affecting a quantitative trait which is identified in 2 parental lines which differs from each other and are identified in F2 population. The F2 population are evaluated to separate the genetic and environmental effects on the trait in several locations and for years. Later the QTL was mapped to the sector of chromosome [10, 11].

2.4.7 Isolation and characteristics of recombinant proteins

Data of the purified recombinant proteins obtained from in-vitro assays can be interpreted much easily than data of crude or partially purified protein extracts that are stored from experiments. It can be due to these reasons:

1. No competing proteins with similar action.

2. No enzymes present which convert the enzyme of interest, and it reduce the concentration [10, 11].

2.5 Other biosynthesis of phenolic compounds

2.5.1 Carbohydrate catabolism

The carbohydrates for plants are acquired from photosynthesis process from the atmosphere, a fixed CO_2 is converted to carbohydrates from sunlight. The photosynthesis process within the cell for carbon-based metabolites including phenolic compounds helps the carbohydrates to form the building blocks. The two catabolic processes are the precursors of plant phenolic compounds within the plant cell. This includes many pathways such as glycolysis, pentose phosphate, phenylpropanoid pathways etc.

2.5.1.1 Glycolysis

It is also known as Embden-Meyerhof-Parnas pathway, carbohydrates generated during photosynthesis from the catabolic process are broken into pyruvate, and ultimately CO_2. This method plays two fundamental roles:

1. Building blocks for anabolism.

2. Oxidization for hexoses to urge ATP reductant, and pyruvate [10–14].

2.5.1.2 The pentose phosphate pathway

It is used to generate NADPH (nicotinamide adenine dinucleotide phosphate) It is to interrupt glucose, break down that can be used by plant. It provides sugar links that delivered as building blocks for nucleic acids and aromatic amino acids.

This pathway is divided into two phases:

1. Oxidative phase: In this phase, the glucose-6-phosphate is converted to ribulose5-phosphate,

2. Non-oxidative phase: in this phase, by reversible reactions two pentose-phosphate residues are transform to sugar- phosphate molecules [8, 10–13].

2.5.1.3 The shikimate pathway

It includes the biosynthesis of chorismate, which may later work as a precursor for the biosynthesis of the aromatic amino acids like tyrosine phenylalanine and tryptophan. This pathway was reviewed by Weaver and Herrmann [17] and Hermann and Weaver [18] in biochemistry. It was seen in both plants and microorganisms. Shikimate was synthesized from the substrates erythrose 4-phosphate and phosphoenolpyruvate. By glycolysis and the pentose phosphate pathway, these two precursors are extracted respectively, and by the enzyme DAHP synthase, they are condensed to 3-deoxy-D-arabino-heptulosonate 7- phosphate. This steps end in the formation of 3-dehydroquinate by the enzyme 3- dehydroquinate synthase, 3-dehydroshikimate and 3- dehydroquinate dehydratase, and finally shikimate by the enzyme shikimate dehydrogenase. Shikimate was converted to shikimate 3-phosphate by shikimate kinase, then to 5-enolpyruvylshikimate 3-phosphate by 5-enolpyruvylshikimate 3-phosphate synthase. EPSP is then obtained to chorismate by chorismate synthase. Chorismate is bifurcate tryptophan on the one hand, and phenylalanine and tyrosine on the opposite hand for the biosynthesis of aromatic amino acids [8, 10–16].

2.5.1.4 The general phenylpropanoid pathway

It was generating a substratum to phenylpropanoid compounds which includes coumarins, monolignols, hydroxycinnamic acids, flavonoids, stilbenes and sinapoyl esters. This pathway starts with phenylalanine via the shikimate pathway. It catalyzed phenylalanine by the enzyme phenylalanine ammonia lyase (PAL) and ends in cinnamic acid. Later, it was hydroxylated by cinnamic acid 4- hydroxylase (C4H) transfer to p-coumaric acid [10, 11].

2.5.2 Biosynthesis of phenolic acids

Phenolic acids are not abundant in most plants. There are in the form of gallic acid and salicylic acid. Gallic acid may be a precursor for the ellagitannins and gallotannins. Salicylic acid is an important defense property that mediates systemic acquired resistance (SAR), and it is also used as a signaling molecule to relay information on pathogen attack to other parts of the plant. After receiving the SA signal, a defense response is trigger the biosynthesis of pathogenesis-related (PR) proteins [8, 10].

2.5.3 Biosynthesis of flavonoids and condensed tannins

In the process of flavonoid biosynthesis, the identification and isolation of genes by the flavonoids which are a colored compound. Mutant phenotypes are identifiable from variation in color easily. The flavanonols can converted to anthocyanins. Condensed tannins transformed from polymerization of flavonoids [8, 11].

2.5.4 Monolignol biosynthesis

They are synthesized from pcoumaroyl-CoA via the shikimate and phenylpropanoid pathways and are the component of lignans and lignin, and some of them are serve as precursors for sinapoyl esters and hydroxycinnamic acids [8, 10, 11].

2.5.5 Lignan biosynthesis

In this process, the oxidative coupling of monolignol radicals are synthesized. *The monolignol radicals are generated through the action of laccases or peroxidases.* Lignans are active, and a typical pair of chemical compound which is present in some species and it binds in between the monolignols [regio-chemical control], thereby both the coupling sites and their position of the 2 monomers are controlled [10, 11].

2.5.6 Lignin biosynthesis

It involves many enzymes where the genes encoding of these enzymes need to be uniformly communicate. Lignin is a complex polymer obtained from the oxidizing the coupling of monolignol radicals. The plant's cell wall contains the polymerization of lignin, thus the monolignols has to be transferred from the cytosol and get synthesized in the cell wall [10, 11].

2.5.7 Hydroxycinnamic acid biosynthesis

By the action of 4CL as well as the corresponding CoA-esters, the biosynthesis of the hydroxycinnamic acids caffeic acid, ferulic acid, 5-hydroxyferulic acid, and sinapic acid from p-coumaric acid, the hydroxycinnamic acids are synthesized through the oxidation of aldehydes, instead of ring substitutions of the free acids. It suggests that the glucosides are messenger of th Coenzme A ester and the ring appears at the level of free acid in the lignin biosynthetic pathway [8, 11].

2.5.8 Biosynthesis of sinapoyl esters

In the phenylpropanoid pathway, Sinapaldehyde is acquired from the amino acid phenylalanine which is followed by a number of reactions of the hydroxylation and methylation [10, 11].

2.5.9 Coumarin biosynthesis

The coumarins and hydroxycoumarins are synthesized from trans-p-coumaric acid and trans-cinnamic acid in plants respectively, but the complete mechanism of its synthesis is still unknown. The possible way to biosynthetic route that coumarin is through hydroxylation to give coumaric acid, followed by glycosylation to result in trans-coumaric acid-2-O-glucoside [8, 11].

2.5.10 Stilbene biosynthesis

It has been a target for genetic engineering of disease resistance in plants, which appears similar to chalcone synthase and it is derived from the condensation of p-coumaroyl-CoA with three malonyl-CoA residues [10, 11].

2.5.11 Biosynthesis of gallotannins and ellagitannins

They are inventing from the hydrolysable tannins and 1,2,3,4,6-penta-O-galloyl-β-D-glucopyranose. Gallotannins consist of 10-12 gallic acid moieties per molecule. Ellagitannins are construct from the oxidative of gallic acid residues in pentagalloyl-glucose molecules which gives in the formation of C-C coupled 3,4,5,3′,4′,5′-hexahydroxydiphenoyl (HHDP) residues [10, 11].

2.6 Extraction technologies for phenolic compounds

2.6.1 Solid–liquid extraction

By using aqueous organic solvents from solids, the soluble constituents are removed. The selection of solvents should be accurate so that chemical or physical intervention should be within the matrix. In this method, variable such as temperature, pH, particle size, time, solvent polarity, solid–liquid ratio and riveting should be improved in order to obtained high yields of recovery of the compound which was selected. Few drawbacks of this method includes cost, toxicity, solvent combustible and also prolonged extraction times. It can be used to get PCs from herbal substance. Some methods in this extraction includes the use of poisonous solvents, low cost and can be used in combination with any other extraction techniques [4, 10, 17–19].

2.6.2 Soxhlet extraction

It usually contains the matrix with pure and hot solvent. In this way the extracted will be greater in the substance. This method is inexpensive related to energy, time and reactant. Soxhlet extraction were done in small scales in batches and can be adapted to continuously in industrial procedure. The main privilege than novel method, such as ultrasound assisted, microwave-assisted, fastidious fluid, and accelerated solvent extractions in terms of industrial implementation, consistency, effectiveness, and extract manipulation. The main disadvantage was the sensitivity of some compounds to the temperature conditions of extraction. The variants of this technique are: high-pressure, automated, ultrasound-assisted, and microwave-assisted Soxhlet extraction [4, 20].

2.6.3 Pressurized fluid extraction and supercritical fluid extraction

The extraction method is like Soxhlet extraction, then again, actually the solvents are utilized in tightening influences close to their supercritical area, so the raised temperature permits a more prominent dispersion and dissolvability of the solute to be extricated. At the point when the high pressing factor applied to the framework, the dissolvable beneath its limit is permitting the better focus in the network. These working conditions permit the utilization of low dissolvable volumes and lessen extraction times. The second extraction strategy comprises of the detachment of a compound (strong or fluid) from a grid, utilizing liquid as a dissolvable under supercritical conditions. Under supercritical conditions a liquid coincides in both fume and fluid states. The most ordinarily utilized liquids is carbon dioxide (CO2), which is joined with ethanol to change its extremity. The upsides of CO2 as extraction liquid are: moderate supercritical conditions (31.1 °C and 73.8 MPa), nonattendance of harmfulness, substance security, simple to reuse, and ease. The upsides of supercritical extraction will be: extraction limit like fluid natural solvents and the concentrates are cleaner. Mechanical utilization of supercritical liquid extraction was restricted since this strategy were created in detachment of other handling steps that are important to acquire an item [4, 18].

2.6.4 Ultrasound-assisted extraction

They used to remove bioactive mixtures, similar to cancer prevention agents, fundamental oils, steroids, and lipids from plants. The utilization of ultrasound improves the entrance of the dissolvable into cell materials, encouraging mass exchange and the arrival of the mixtures to be removed. The recurrence of ultrasound impacts the yield and extraction energy. At frequencies >20 kHz sound waves produce extension pressure cycles, in a fluid this outcome in the arrangement of air pockets that develop and breakdown close to the strong network, encouraging extraction [4, 21].

2.6.5 Microwave-assisted extraction

Microwaves are electromagnetic waves comprising of an electric field and an attractive field that waver oppositely to one another at frequencies somewhere in the range of 0.3 and 300GHz. The microwave energy acts straightforwardly on the particles by ionic conduction and dipole revolution, motivation behind why just polar materials can be warmed as such. The microwave-helped extraction relies upon the dielectric defenselessness of both dissolvable and network. Since the water inside the lattice assimilates microwaves, the interruption of the material is controlled by an inward overheating, which likewise improves the recuperation of the extricated compound. Microwave-helped extraction is characterized into shut and open frameworks. In a shut framework, the extractions are done in a fixed vessel under uniform warming; in this framework the high pressing factor and temperature permit fast and proficient extraction. Then again, open frameworks are more reasonable for extricating thermolabile mixtures, since they work under less extraordinary conditions [4, 22].

2.6.6 Pulsed electric field extraction

The cellular wall and cell membranes act as protective layer that prevent the bioactive compounds extraction in animal and plant tissues. The transmembrane segment of the cell lead to pores or electroporation by the application of an electric field. The power of the electrical pulses provides is changeable or unchangeable may form the electroporation. The pores are small associated to the whole area of the current or electric and its membrane breakdown may vary. On the contrary, increasing the intensity and time of the treatment, it is irreversible to the permeability of cell membrane [4, 23].

2.6.7 Enzyme-assisted extraction

An alternative method to solvent-based extraction. It depends on the enzymes to selectivity and catalyze reactions in aqueous humor. On the constituent of cell membranes, the enzymes with hydrolytic activity such as cellulases, hemicellulases, pectinases, etc. increases cell wall permeability and bioactive compounds extraction was yield such as antioxidants, pigments and compounds with pharmaceutical applications [4, 10, 11, 24].

2.7 Medicinal importance of phenolic compounds

Current studies have associated that consuming the foods are abundant in PC are beneficial in prevention of non-communicable diseases or lifestyle disorder which includes cardiovascular diseases, certain group of cancer, and diseases associated with aging [25]. The biological effects acquired from PC were trait to antioxidant properties [26].

They are as follows.

2.7.1 Antiseptic

PC have effects on human health which was revealed by Bravo in 1998 [29]. PC was used phenol as an antiseptic from ancient times. Now a day, it is no longer used due to, its side effects on living tissues that create blister formation specially on high concentrations. As an antiseptic agent, it is effective against the bacterium *Staphylococcus aureus* i.e. 5% (w/v) solution of phenol. It is used as an oral esthetic with the concentration of 1.4% in throat pastille. It is also in sunscreens lotions. It helps to prevent sunburns due to the presence of the aromatic ring which is an effective absorbance of the UV-B radiation (ranging from 280 and 315 nm) from the sun. It was widely used since the 1970's and nowadays due to the formation of skin rashes and acne the usage is reduced [10, 27].

2.7.2 Antioxidant

The oxidative damage and an imbalance to large biomolecules, like lipids, DNA, and proteins may be due to overproduction of oxidants in physical body. This damage includes the pathogenesis of many human diseases i.e. cardiovascular diseases (CVD), certain sorts of cancers and aging. Thus, it could be a crucial role for the prevention and treatment of chronic diseases by antioxidant phytochemicals which are demonstrated to have antioxidant abilities in human studies. Compounds are scavenging radicals that are referred as antioxidants. The important anti-oxidants are vitamin C and vitamin E. A lack of vitamin C in the diet leads to scurvy. The symptoms include rotten gums, purple lesions on the skin, loss of teeth etc. Vitamin E is a mixture of α-, β-, γ-, and δtocopherol in that α-tocopherol is the most effective. Vitamin E is lipid-soluble and has the ability to disrupt the chain reaction at the time of lipid peroxidation. They provided many health benefits by antioxidant activity of polyphenols [10, 28].

2.7.3 Protective against cardiovascular diseases

Polyphenols are helpful for preventing and treating CVD by antioxidant activity and also by other bioactivities such as preventing platelet aggregation. Anti-inflammation and adhesion which includes oxidative stress and other damage because they owe other physiological effects, like blood pressure reduction etc. [4, 29, 30].

2.7.4 Anti-obesity activity

This activity includes quercetin which may be mediated by mitogen-activated protein kinases signaling pathways (MAPK) and the adenosine monophosphate-activated protein kinase (AMPK), respectively in mature adipocytes and pre adipocytes [4, 31].

2.7.5 Anti-diabetic activity

Due to hyperglycaemia and hyperlipidaemia, diabetes is usually associated by expand the yielding of free radicals or oxidative stress. There is a remarkable decrease in plasma antioxidants in diabetes and its complication. The metabolic homeostasis was better, and the development of T2D and its complications was observed in Cohort studies showed that was retard or prevented by taking of whole grain foods [32, 33]. PCs such as flavonoids and phenolic acids are helpful in promoting health by decreasing the high risk of metabolic syndrome and the associated complications of type 2 diabetes [33–37].

2.7.6 Antiaging activity

An important factor in aging or age-associated degenerative diseases, the free radicals and oxidative stress have been believed as an antioxidant systems are declined during aging. Antiaging activities is explained by different mechanisms and revealed by antioxidant phytochemicals [34–37].

2.7.7 Protective action on Alzheimer's disease

It is particularly susceptible due to high concentration of free radicals without appropriate levels of anti-oxidation. In elderly people, the pathogenesis of dementia or AD shows oxidative stress. The study on walnuts shows that polyphenolic compounds helps to release the oxidant and decrease the inflammatory signs on the brain cells. It also repairs, the increased neurogenesis, inter-neuronal signaling, upgrade isolation of insoluble toxic protein accumulates, that play a role in preventing AD. Thus, by decreasing the oxidant stress and acetylcholinesterase that may protect or prevent against AD [4, 38].

2.7.8 Anti- cancer activity

A huge amount of fruits and vegetables in our diet had shown, a decrease risk of human cancers such as breast cancer, lung cancer, colon cancer and prostate cancer. It is revealed that flavonoids are of special attraction and bioactive compounds in plant providing defensing effects. A study shows that in mice, it provides protection against cancer of skin which are caused through ultraviolet radiation or chemical carcinogens by consumption of tea and its polyphenolic constituents [39–43].

2.7.9 Miscellaneous

Plant PC provide a means for preventing the side effects that fungal toxins (mycotoxins) and also serving in detoxification [41]. Many of the volatile PCs, such as the main PC of cloves i.e. eugenol (a hydroxyphenyl propene), or a typical component of oregano i.e. carvacrol (phenolic terpene), curcuminoids or Curcumin (diferuloylmethane) are which found only in the rhizomes of *Curcuma longa* [turmeric] are achieved. Curcumin as we all known plays an important role of various illnesses from cancer to autoimmune, neurological, cardiovascular, and diabetic etc. in the form of preventing and treating diseases [33, 42–45].

3. Conclusion

Phenolic compounds are widely found in plants with many functions and some act as defense elements against herbivores and pathogens. There are number of vegetables and fruits that contain PCs. They are classified in a range of groups according to their structure. The biosynthesis pathway explained its metabolism in plants which are beneficial to us in many ways. The recent studies show different new technics to extract phenolic compounds and studies are still under observation in prevention of many diseases. Its variations give them diverse characteristics, which helps to prevents many chronic and lifestyle disorders, like antioxidant activity, antiseptic properties, anti-diabetic activity, anti-aging, *Alzheimer's Disease*, anti-obesity, improves cardiac activity etc. Many studies were conducted to show an essential and effective antioxidant power of phenolic compounds and extracts, considering their bioavailability and bio-efficacy of phenolic compounds, which could influence the antioxidant power response, that are necessary to improve the health and well-being of people directly or indirectly. The large number of publications available on phenolic compound research and their extraction from plants over the past decade gives signifies the importance of this chapter.

Acronyms and abbreviations

AD	Alzheimer's Disease
AMPK	adenosine monophosphate-activated protein kinase
ATP	Adenosine triphosphate
BAC	Bio-affinity chromatography
CAT	catalase
C-C	Carbon Carbon
CVD	cardiovascular diseases
C4H	cinnamic acid 4- hydroxylase
cDNA	Complementary DNA
DNA	Deoxyribonucleic acid
EPSP	5-enolpyruvylshikimate 3-phosphate synthase
FPLC	fast protein liquid chromatography
GC	Guanine-cytosine
GFC	Gel filtration or size exclusion chromatography
GHz	Gigahertz
GPx	glutathione peroxidase
HHDP	3,4,5,3′,4′,5′-hexahydroxydiphenoyl
HIC	Hydrophobic interaction chromatography
HPLC	High-performance liquid chromatography
IEC	Ion exchange chromatography
kHz	kilohertz
MAPK	Mitogen-activated protein kinases
MPa	megapascal
NADPH	nicotinamide adenine dinucleotide phosphate
NF-κB	nuclear factor kappa light chain enhancer of activated B cells
OH	Hydroxy
PABA	para-aminobenzoic acid
PAL	phenylalanine ammonia lyase
PC	phenolic compound
PCR	polymerase chain reaction
PH	potential of hydrogen
PMSF	phenylmethyl sulphonylfluoride
PR	pathogenesis-related
QTL	Quantitative trait locus
RNS	reactive nitrogen species
ROS	Reactive oxygen
SAR	systemic acquired resistance
SOD	superoxide dismutase
T2D	Type 2 diabetes
UV	Ultra violet
4CL	4-coumarate-CoA ligase
%	percentage
(w/v)	concentration of solution

Author details

Mohd Abdul Gani[1] and Shama M[2*]

1 Kerala University of health sciences, Thrissur, Kerala

2 Rajiv Gandhi University of health sciences, Bangalore, Karnataka

*Address all correspondence to: shamagumc@gmail.com

IntechOpen

References

[1] Gan RY, Chan CL, Yang QQ, Li HB, Zhang D, Ge YY, Gunaratne A, Ge J, Corke H. Bioactive compounds and beneficial functions of sprouted grains. In Sprouted grains 2019 Jan 1 (pp. 191-246). AACC International Press. DOI: https://doi.org/10.1016/B978-0-12-811525-1.00009-9

[2] Hayat I, Ahmad A, Masud T, Ahmed A, Bashir S. Nutritional and health perspectives of beans (Phaseolus vulgaris L.): an overview. Critical reviews in food science and nutrition. 2014 Jan 1;54(5):580-592.

[3] Verspreet J, Dornez E, Van den Ende W, Delcour JA, Courtin CM. Cereal grain fructans: structure, variability and potential health effects. Trends in Food Science & Technology. 2015 May 1;43(1):32-42.

[4] Ruiz-Ruiz JC, Aldana GD, Cruz AI, Segura-Campos MR. Antioxidant Activity of Polyphenols Extracted from Hop Used in Craft Beer. In Bio Technological Progress and Beverage Consumption 2020 Jan 1 (pp. 283-310). Academic Press.

[5] Laura A, Moreno-Escamilla JO, Rodrigo-García J, Alvarez-Parrilla E. Phenolic compounds. In Postharvest physiology and biochemistry of fruits and vegetables 2019 Jan 1 (pp. 253-271). Woodhead Publishing. DOI: https://doi.org/10.1016/B978-0-12-813278-4.00012-9 © 2019 Elsevier Inc

[6] Chikezie, P.C., Ibegbulem, C.O., M bagwu, F.N. Bioactive principles from medicinal plants. Research Journal of Phyto chemistry. 2015 9, 88-115. https://doi.org/10.3923/rjphyto.2015.88.115.

[7] Vora, D.J.D., 2017. "Insight into the biochemical link between biodiversity and nutraceuticals. IOSR Journal of Environmental Science, Toxicology and Food Technology. 11, 22-25. https://doi.org/10.9790/2402-1103032225.

[8] Wang, T., Li, Q., Bi, K., 2018. Bioactive flavonoids in medicinal plants: structure, activity and biological fate. Asian Journal of Pharmaceutical Sciences. 13, 12-23. https://doi.org/10.1016/j.ajps.2017.08.004

[9] Vuolo MM, Lima VS, Junior MR. Phenolic compounds: structure, classification, and antioxidant power. In Bioactive compounds 2019 Jan 1 (pp. 33-50). Woodhead Publishing.

[10] Harborne, J.B. Secondary plant products: encyclopedia of plant physiology. New Series 1980, Volume 8 Phytochemistry. 19, 2803-2804. https://doi.org/10.1016/S0031-9422(00)83983-7.

[11] Wilfred V, Nicholson R. Phenolic compound biochemistry. Springer; 2006. 1-284.

[12] Vermerris W, Nicholson R. Phenolic compound biochemistry. Springer Science & Business Media; 2007 Feb 20.

[13] Pflieger, S., Lefebvre, V., and Causse, M. 2001, The candidate gene approach in plant genetics: a review, Mol. Breeding 7: 275-291

[14] Hopkinson SM. The chemistry and biochemistry of phenolic glycosides. Quarterly Reviews, Chemical Society. 1969;23(1):98-124.

[15] Nelson DL, Lehninger AL, Cox MM. Lehninger principles of biochemistry. Macmillan; 2011. 1-1158].

[16] Weaver, L. M., and Herrmann, K., 1997, Dynamics of the shikimate pathway in plants, Trends Plant Sci. 2: 346-351.

[17] Herrmann, K., and Weaver, L. M., 1999, The shikimate pathway, Annu. Rev. Plant Physiol. Plant Mol. Biol. 50: 473-503.

[18] Rispail N, Morris P, Webb KJ. Phenolic compounds: extraction and analysis. In Lotus japonicus handbook 2005 (pp. 349-354). Springer, Dordrecht.

[19] Baiano A, Bevilacqua L, Terracone C, Contò F, Del Nobile MA. Single and interactive effects of process variables on microwave-assisted and conventional extractions of antioxidants from vegetable solid wastes. Journal of Food Engineering. 2014 Jan 1; 120:135-145.

[20] Luthria DL. Influence of experimental conditions on the extraction of phenolic compounds from parsley (Petroselinum crispum) flakes using a pressurized liquid extractor. Food Chemistry. 2008 Mar 15;107(2):745-752.

[21] De Castro ML, Priego-Capote F. Soxhlet extraction: Past and present panacea. Journal of Chromatography A. 2010 Apr 16;1217(16):2383-2389.

[22] Khoddami A, Wilkes MA, Roberts TH. Techniques for analysis of plant phenolic compounds. Molecules. 2013 Feb;18(2):2328-2375.

[23] Chan CH, Yusoff R, Ngoh GC, Kung FW. Microwave-assisted extractions of active ingredients from plants. Journal of Chromatography A. 2011 Sep 16;1218(37):6213-6225.

[24] Xue D, Farid MM. Pulsed electric field extraction of valuable compounds from white button mushroom (Agaricus bisporus). Innovative Food Science & Emerging Technologies. 2015 May 1; 29:178-186.

[25] Puri M, Sharma D, Barrow CJ. Enzyme-assisted extraction of bioactives from plants. Trends in biotechnology. 2012 Jan 1;30(1):37-44.

[26] Scalbert A, Manach C, Morand C, Rémésy C, Jiménez L. Dietary polyphenols and the prevention of diseases. Critical reviews in food science and nutrition. 2005 Jun 1;45(4):287-306.

[27] Balasundram N, Sundram K, Samman S. Phenolic compounds in plants and agri-industrial by-products: Antioxidant activity, occurrence, and potential uses. Food chemistry. 2006 Jan 1;99(1):191-203.

[28] Bravo, L., 1998. Polyphenols: chemistry, dietary sources, metabolism, and nutritional significance, Nutr. Rev. 56: 317-333.

[29] Halliwell, B., Rafter, J., and Jenner A. Health promotion by flavonoids, tocopherols, tocotrienols, and other phenols: direct or indirect effects? Antioxidant or not? Am. J. Clin. Nutr. 2005. 81: 268S-276S (Suppl.).

[30] Prahalathan P, Kumar S, Raja B. Morin attenuates blood pressure and oxidative stress in deoxycorticosterone acetate-salt hypertensive rats: a biochemical and histopathological evaluation. Metabolism. 2012 Aug 1;61(8):1087-1099.

[31] Peterson JJ, Dwyer JT, Jacques PF, McCullough ML. Associations between flavonoids and cardiovascular disease incidence or mortality in European and US populations. Nutrition reviews. 2012 Sep 1;70(9):491-508.

[32] Dong G, Potenza MN. A cognitive-behavioral model of Internet gaming disorder: theoretical underpinnings and clinical implications. Journal of psychiatric research. 2014 Nov 1; 58:7-11.

[33] Belobrajdic DP, Bird AR. The potential role of phytochemicals in wholegrain cereals for the prevention of type-2 diabetes. Nutrition journal. 2013 Dec 1;12(1):62.

[34] Lin D, Xiao M, Zhao J, Li Z, Xing B, Li X, Kong M, Li L, Zhang Q, Liu Y,

Chen H. An overview of plant phenolic compounds and their importance in human nutrition and management of type 2 diabetes. Molecules. 2016 Oct;21(10):1374

[35] Pan MH, Lai CS, Tsai ML, Wu JC, Ho CT. Molecular mechanisms for anti-aging by natural dietary compounds. Molecular nutrition & food research. 2012 Jan;56(1):88-115.

[36] del Valle LG. Oxidative stress in aging: theoretical outcomes and clinical evidences in humans. Biomedicine & Aging Pathology. 2011 Jan 1;1(1):1-7.

[37] Lagouge M, Larsson NG. The role of mitochondrial DNA mutations and free radicals in disease and ageing. Journal of internal medicine. 2013 Jun;273(6):529-543.

[38] Vaiserman A, Koliada A, Zayachkivska A, Lushchak O. Nanodelivery of natural antioxidants: An anti-aging perspective. Frontiers in bioengineering and biotechnology. 2020 Jan 10; 7:447.

[39] Mendelson SD. Herbal Treatment of Major Depression: Scientific Basis and Practical Use. CRC Press; 2019 Oct 16.

[40] Mukhtar, H., and Ahmad, N., 2000, Tea polyphenols: prevention of cancer and optimizing health, Am. J. Clin. Nutrition 71 (suppl.): 1698S-1702S.

[41] Kris-Etherton PM, Hecker KD, Bonanome A, Coval SM, Binkoski AE, Hilpert KF, Griel AE, Etherton TD. Bioactive compounds in foods: their role in the prevention of cardiovascular disease and cancer. The American journal of medicine. 2002 Dec 30;113(9):71-88.

[42] Beekrum S., Govinden R., Padayachee T., and Odhav B., 2003, Naturally occurring phenols: a detoxification strategy for fumonisin B-1, Food Add. Contam. 20: 490-493.

[43] Amiot MJ, Riva C, Vinet A. Effects of dietary polyphenols on metabolic syndrome features in humans: a systematic review. Obesity Reviews. 2016 Jul;17(7):573-586.

[44] Cianciosi D, Forbes-Hernández TY, Afrin S, Gasparrini M, Reboredo-Rodriguez P, Manna PP, Zhang J, Bravo Lamas L, Martínez Flórez S, Agudo Toyos P, Quiles JL. Phenolic compounds in honey and their associated health benefits: A review. Molecules. 2018 Sep;23(9):2322.

[45] Vieira R, Souto SB, Sánchez-López E, López Machado A, Severino P, Jose S, Santini A, Fortuna A, García ML, Silva AM, Souto EB. Sugar-lowering drugs for type 2 diabetes mellitus and metabolic syndrome—review of classical and new compounds: Part-I. Pharmaceuticals. 2019 Dec;12(4):152.

Chapter 14

Polyphenols, Spices and Vegetarian Diet for Immunity and Anti-Inflammatory Drug Design

Utkarsh Ghate and Hema Kulkarni

Abstract

Much lower COVID-19 incidence and mortality in India compared to the Europe and northern America may relate to higher immunity possibly due to the low consumption of fast/packed food, liquour, tobacco, meat, HFSS- high fat, salt, sugar, besides higher exposure and a key blood protein. Indian spice intake is also double the world average and healthy cooking oil use such as Mustard, and may also explain it. Inflammation is the foundation for many ailments and challenges the immunity and vital in non-communicable ailments are at the centre stage in an aeing world. Polypehnols are crucial anti-inflammatory chemicals from spices that can for wellbeing and reduce adverse drug rections. We show this using Arthritis- a chronic auto-immune disorder, with the hlep of pharmacokinetic studies. Molecular Docking study was performed on the key bioactive compounds of important spices regarding COX2 active site (PDB ID 5IKR). Piperine in Black Pepper had most stability (Black Pepper, −9.99 Kcal/mol) followed by 'Apigenin' (Coriander, −9.63), and 'Curcumin' (Turmeric, −8.66) like quercetin in literature, and higher than the methotrexate (−8.6), the standard drug. Hence, their synergistic combination in fat medium such as clarified butter can lead the future drug design.

Keywords: Polyphenols, immunity, health, spice, Corona, pharmacokinetic, inflammatory, arthritis

1. Introduction

Spices and herbs have been the key to the health security of the oriental world, besides in western world also until the last century. Polyphenols such as flavanoids are aromatic organic compounds having many health benefits being highly antioxidant in nature [1]. These are vital in preventing chronic or non communicable diseases (NCD) also called as "lifestyle ailments" that are common in the western world [2] and the elderly (over 60 years of age), whose share in the world population is growing rapidly, from 10 to 20% of the total [3].

Inflammation and immunity are most important concepts in medicine today, as cause and remedy respectively [4]. Inflammation is the chief mediator behind chronic or lifestyle ailments prevalent today such as heart disease, cancer, diabetes and blood pressure triggering vigorous research on anti-inflammatory and/or immunity booster phytochemicals, for safety and efficacy (*ibid.*).

Antioxidants are able to reduce oxidative damage to the tissues and protect or restore immunity & health. Vegetarian diet especially fruits, vegetables, pulses are rich in antioxidants, including polyphenols. Spices are among the richest in these, and are anti-inflammatory. Thus, they can be useful to manage immunity disorders such as COVID-19 and Arthritis, an auto-immune disorder of the elderly.

Adverse drug reaction (ADR) is another major concern that spice/herbal medicines can reduce in principle and as experienced, leading to their growing global demand. ADR affects about 5–10% of the patients globally and can cause sever damage/expenses [5]. As spices are permitted food ingredients globally, ADR risk is low.

2. Covid-19 and vegetarian, spice diet relation

The COVID-19 burden cross top 10 infected countries (dt. 28 Feb. 2021) is shown in the **Table 1** and India is the 2nd most infected after USA but its no. of cases (incidence) and per million (7,990) and deaths (113) is the lowest. It is only 16% of the average (47,000/ 1 million) of the other 9 leading countries incidence and 10% of the death rate (1,110/1 million). This makes it worth studying.

2.1 Immunity Buster foods

It is known that the Immunity is compromised by the higher consumption fast/ packed foods, refined cards, intoxicants, higher salt, sugar, fat etc. We find the immunity stress foods consumed in 2–30 times in EU/USA (average 9 times intake) than India, as seen in **Tables 2** and **3**, **Figure 1.**

2.2 Immunity booster foods

Indian spice consumption, rich in polyphenols, is 2 times higher (2.07 kg/head/ year, **Table 4** and **Figure 2**) than the global average (1.01 kg/head/year) [7] or USA [8]. Cancer incidence (89 per 0.1 million) in India is 50% of the global average (197) 25% of the EU (363) or USA (387), indicating better immunity [9], possibly due to the higher spice consumption. Asthama incidence, a major respiratory ailment and immunity indicator is similarly low in India with below 10% population affected but higher levels in the European nations- 20 to 25% [10].

Country	Total Cases	Total Deaths	Tot Cases/1 M pop	Deaths/1 M pop
World	114,468,838	2,539,109	14,685	325.7
1. USA	29,202,966	524,670	87,886	1,579
2. India	11,097,134	157,092	7,990	113
3. Brazil	10,517,232	254,263	49,248	1,191
4. Russia	4,246,079	86,122	29,088	590
5. UK	4,170,519	122,705	61,222	1,801
6. France	3,736,016	86,332	57,153	1,321
7. Spain	3,188,553	69,142	68,180	1,478
8. Italy	2,907,825	97,507	48,140	1,614
9. Turkey	2,693,164	28,503	31,708	336
10. Germany	2,444,303	70,608	29,112	841

Source- https://www.worldometers.info/coronavirus/ dt. 28 Feb. 2021.

Table 1.
COVID-19 Incidence and death rate in top 10 countries.

Component	Unit	Ind	West	Ratio
1. Alcohol	Lit/ yr	5	10	2
2. Cigarettes	no.s	117	1400	9
3. Sugar	Kg/yr	19	35	2
5. Meat	Kg/yr	5	88	17
6. Fats	Gram/day	45	145	3
7. Salt	Gram/day	10	34	3
8. Refined carbohydrates kg	Kg/yr.	1.5	50	30
9. Packed foods	%	<5	45	10

Table 2.
Immunity Buster Foods Intake Globe per capita/yr. [2].

Item	Developing	% Share	Industrial	% share
Meat	369	17	958	35
Sugar	170	8	328	12
Pulses	99	5	37	1
Tubers	154	7	112	4
veg. Oil	239	11	494	18
Wheat	457	21	627	23
Rice	655	31	153	6
Total	2143		2709	

Table 3.
Various food items share in Calorie intake (Kcal/day/head) [6].

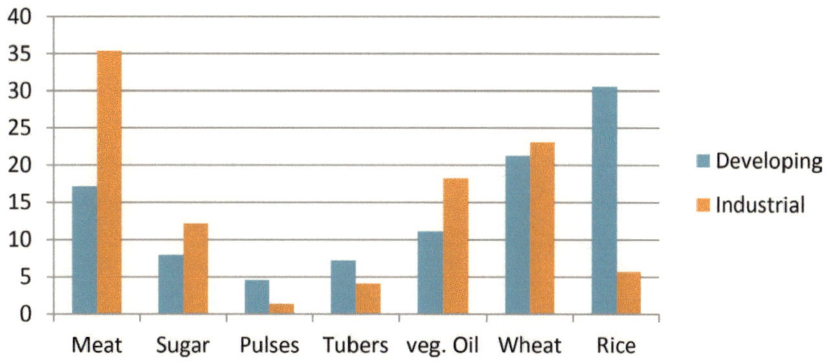

Figure 1.
Food items % in Calorie intake- nations.

However, the spice consumption in India varies greatly across states and economic strata or ethnicity and has changed over times much. Chilli for instance, has replaced black paper, common earlier for adding pungent flavor and changed the world history as European discovered India for the later. The former has no antiviral report [11] while the later is an effective antiviral [12]. Similarly, Chilli is consumed more in northern India while Black Pepper in southern India- its main producer region- and this region also shows lowest fatality rate in India- 1%. Chilli comprises nearly 20% of the 5 gram/day/head spices consumed, and Turmeric, Ginger,

Parameter- region	India [7]	USA [8]	Ratio Rest/India
Spice intake kg/head/year	1	0.6	0.6
Cancer prevalence [9]	89	363	4

Table 4.
Spice Intake & Cancer incidence- India & the world.

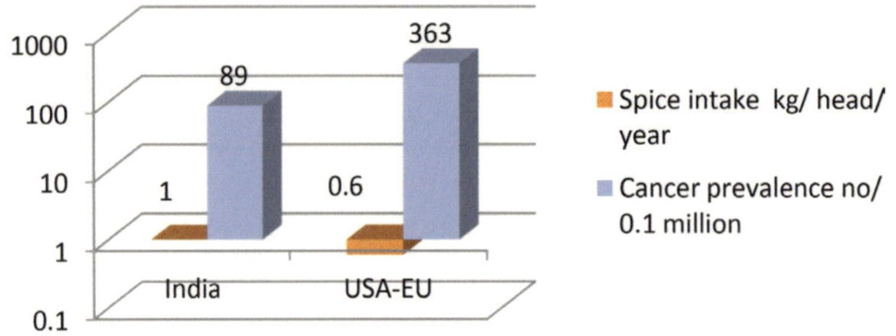

Figure 2.
Spice intake & Cancer Incidence (Note- the Y axis is in the log scale so the world and India incidence is less distinct).

Mustard, Coriander, Cumin are nearly equal, the 5 adding to 80% of the spices consumed on average [7]. More potent antiviral spices (except Ginger & Turmeric) are consumed to lesser extent viz. Cinnamon, Clove and Garlic. The potential of spice bioactive agents as possible COVID-19 remedy or relief is enlisted in **Table 5**.

Scientists have postulated other hypotheses to explain the much lower COVID-19 prevalence and mortality in India/Asia than in the Europe & America, such as the hygiene/exposure [25] and genetics [26], but the role of diet is also mentioned in both prevention and treatment with immunity focus [27]. Spices are found to be important preventive agents and immunity guards in case of the corona, based on data from 163 countries by German scientists [28].

Spices can be important immunity booster due to their bioactive compounds known to be healthy [29, 30]. They may be suppressing the inflammatory pathways

Priority	Ingredient	Effect
1. Coriander	Quercetin	Attachment, endocytosis, cell fusion [13, 14]
2. Ginger	Gingerol	Secrete IFN-β t, inhibits initiation of virus - reduces HRSV-induced plaque [15, 16]
3. Turmeric	Curcumin	Anti-inflammatory, used in rhinitis [17]
4. Pepper, Black	Piperine	Anti-proliferative activity- in vascular smooth muscle cells [18]
Ancillary		
1. Camphor	Imine	1,7,7-trimethylbicyclo [2.2.1] heptanes2-ylidene [19]
2. Cumin	EHP [1-(2-Ethyl,6-Heptyl) Phenol]	Vero cell membrane and/ or HSV-1 envelope [20]
3. Clove	Essential oil, Eugenol	Enveloped virus- HSV-1 and Newcastle [21]
4. Garlic, Onion	Allicin	Block multiplication [22]
5. Tulsi (Holy Basil)	Terpenoid, polyphenols	Non- neuraminidase inhibition [23, 24]

Table 5.
Spices with scope of COVID-19 immunity/cure.

NF-κB and STAT3 [31]. Flavonoids inhibit biosynthesis of prostaglandins (the end products of the COX and lipoxygenase pathways), which acts as a secondary messengers and are involved in various immunologic responses [32]. We illustrate below the scope of use of bioactives from spices to treat Arthritis, a common auto-immune disease with no sure modern cure yet, as drug design example to integrate the traditional wisdom with modern science and technology. But we present some other correlates of Corona intensity and mortality before that in **Table 6**.

Table 6 shows that the corona intensity (no. of cases/ million) is strongly (r = 0.7) correlated with the income per capita and even strongly (r = 0.75) with literacy. The lower corona burden in India and its poorer states may also be because the poorer, illiterate people of states such as U. P., Bihar have stronger immunity, despite higher population density. For, they mainly eat natural foods, with very low amount of packed food, bottled water so low HFSS intake. Hence, the lesser Covid-19 intensity in North & eastern India. Corona is mainly rich countries & people's ailment [33]. Lac of hygiene & exposure to microbes in the slums etc. makes people resistant to microbes, it is said [34]. Migrations improve immunity, is another hypothesis [35]. Mustard oil, common in northern India is antiviral & SARS inhibitor [36], unlike Groundnut in western India, which is an allergen. Lastly, Asians got a protein D614 mutation, making them stronger than the Europeans [37]. These 5 reasons may explain the trend besides Govt. advisory on spice decoction ("Kadha" in Hindi- https://pib.gov.in/PressReleseDetailm.aspx?PRID=1609524).

Its rationale and working mechanism is also explained by scientists [38].

2.3 Spices and herbs- global resurgence

Scientists from Russia & USA describe the health benefits of spices due to their bioactive ingredients & antioxidant nature [39]. They say rosamarinic acid content from the Mint family as beneficial (Oregano- 2,562 mg/100 gm dry weight, Sage-2,186, Mint- 1908, Sweet Basil- 1,086, & Thyme- 681). Aromatic and medicinal herbs or spices such as these and Parsley, Coriander, Onion, Cumin, Cinnamon, Bay etc. (except Chilli) protect human health due to flavanoid content e.g. Quercetin, Luteolin, Rutin, Apigenin, Myicetin. They indicate that the the herbs are found useful in treating the cancers below-

a. Turmeric- Rectal, oral, head, neck, Leukemia;

b. Saffron- Skin, rectal, hepatic,

c. Garlic- prostate, colon,

d. Onion- Gastric,

e. Mustard- Rectal, Bladder,

f. Bayleaf- melanoma.

They also mention the spices as having therapeutic effects below

1. Cardiovascular- garlic, Turmeric, Ginger
2. Neuro-degenrative- Mint, Onion
3. Antidiabetic- Cinnamon, Bayleaf, Fenugreek, Mustard

State	Popu-lation Million	Popu-lation Density[@]	Corona cases nos.	Corona – Deaths Nos.	Corona Cases/ million	Corona Fatality %[**]	Income[*]	Literacy[#]
Kerala	33	859	1111897	4539	33694	0.41	54	94
Delhi	25	7400	652742	10978	26110	1.68	111	86
Maharashtra	112	365	2600833	53795	23222	2.07	62	82
Andhra	49	303	895879	7201	18283	0.80	44	67
Karnataka	61	319	978478	12471	16041	1.27	48	75
Tamil Nadu	72	555	873219	12641	12128	1.45	54	87
West Bengal	91	1029	583839	10322	6416	1.77	35	76
Gujarat	60	308	298,596	4484	4977	1.50	55	78
Rajasthan	68	200	329595	2811	4847	0.85	31	66
M.P.	76	236	286407	3947	3769	1.38	25	69
Ttar Pradesh	199	828	610273	8783	3067	1.44	20	67
Bihar	103	1102	264604	1571	2569	0.59	13	62
Total	[***]949		[****]9486362					
Correlation coefficient				0.38	−0.07	0.71	0.76	
Correlation between				Cases-Populn. Density	Fatality-Case density	Case density- Income	Case density- literacy	

[@]no. k/ sq. km.
[***]70% of India.
[**]80% of India.
[#]2011 census data.
[*]Rs. K/head/yr- ref. RBI, 2020.
[****](deaths/ cases), M.P.- Madhya Pradesh Ref.- https://www.mohfw.gov.in/, Dt. 28-03-2021.

Table 6.
Corona intensity & socio-economic correlates across Indian states.

4. Gastrointestinal- Black Pepper, Bayleaf

5. Hypertension- Cardamom, Cinnamom,

6. Hepatic- Caraway, cardamom,

7. Endocrine- Ginger, Turmeric,

8. Obesity- Saffron, Turmeric,

9. Renal- garlic, Fennel, Ginger,

10. Alcohol abuse- Thyme, Ginger

Due to the efficacy, safety and sentiments, herbal medicines top the complimentary & alternative medicine (CAM) treatements becoming popular globally recently. The amount of money spent on CAM treatments in the U.S. has skyrocketed in recent years. As per an 2007 survey, Americans spent $33.9 billion out of pocket on CAM therapies that year [40]. About $22 billion was spent on natural products, instructional classes, and materials. Dietary supplements accounted for $14.8 billion of this amount, an expenditure equal to about 1/3rd of out-of-pocket spending or prescription drugs. The remaining $11.9 billion was spent on an estimated 354 million visits to CAM practitioners (acupuncturists, massage therapists, or chiropractors), an amount equal to about 25% of out-of-pocket costs for visits to conventional doctors. The following 10 species (3 spices, 4 herbs) prevail [40]

a. *Echinacea- 41%*,

b. Ginseng- 24%,

c. *Gingko biloba- 21%*,

d. Garlic - 10%,

e. St. John's wort- 12%,

f. Glucosamine- 12%,

g. Peppermint- 12%,

h. Fish oil/omega-3-10%,

i. Ginger - 10%,

j. Soy - 9%.

The main ailments by frequency referred to the CAM practitioners are

1. Back pain- 17%,

2. Neck pain- 6%,

3. Joint pain- 5%,

4. Arthritis- 3.5%,

5. Anxiety- 2.8%,

6. Cholesterol- 2.1%,

7. Head or chest cold- 2%,

8. Other musculoskeletal- 1.8%,

9. Severe headache or migraine- 1.6%,

10. Insomnia- 1.4%.

Reasons for the Increased Use of CAM and Dietary Supplements are

a. the increased availability of information on the Internet

b. increased contact with other cultures that traditionally use CAM.

c. Renewing d interest in formerly countercultural ideologies, such as environmentalism.

d. the perception that CAM is easier to understand, safer, and less expensive than conventional medications.

e. distrust of and frustration with the health care system.

f. a growing recognition that many factors contribute to health and well-being.

With the growing use of the herbal medicines, safety concern is emerging due to the issues such as adulteration, quality and adverse reaction [41]. Hence, World Health Organization (WHO), devised a Traditional Medicine New Strategy (2006–2013) with 3 key health priorities [42], as most countries have traditional/herbal medicine policy to mainstream it

a. mental health,

b. non-communicable diseases and

c. universal health coverage.

Nevertheless, spices and herbs have great potential in future as depicted below with in the example of an anti-inflammatory drug design excercise.

3. Arthritis drug design

There are globally 1.3 billion cases of musculoskeletal disorders and over 121,000 deaths from such disorders, as well as nearly 139 million disability-adjusted life years, or the number of years lost due to ill-health, disability or early death [43]. Globally, the proportion cases in are led by the low back pain (37%); followed by "other"

(21%); osteoarthritis (19%); neck pain (18.4%); gout (2.6%); and rheumatoid arthritis (1.3%). These proportions changed little from 1990. Surgery and the NSAID- Non steroidal anti-inflammatory drugs- are costly and cause adverse effects such as gastric and cardiovascular issues. The herbal medicines are perhaps effective and in growing demand and even exported much, as dietary supplements that are common globally and many having positive outcomes at least in the short term [44, 45]. But herbs they use are and are often adulterated such as Shalaki (Indian Frankincense/Olibanum, *Boswellia serrata*) or Guggulu (Indian Myrrh, *Commiphora wightii,* [46]. Hence, sustainable ingredients from farm are needed such as Drumstick (*Moringa olifera* tree leaves), rich in calcium and highly exported [47]. Glusosamine, a popular health supplement in arthritis is a sea shell product [48]. So it needs a vegetarian option due to the growing trend of veganism globally. Spices can address this as they contain the bio- actives to relive Arthritic pain and inflammation- Vitamin A, C & K, polypohenols, Omega 3 fatty acids, and minerals Calcium, Magnesium, Iron and Zinc [29].

Even in India, about 40% the elderly may be Arthritis affected, it is said in the study by All Indian Institute of Medical Science in metros such as Delhi AIMS [49].

Bioactive ingredient levels are low in the general market samples of spices such as Black pepper (Piperine- 2-4%), Coriander (Quercetin- 0.12%), Turmeric (Curcumin- 2-3%), as the critical agri-inputs to increase the polyphenol content e.g. Potassium for Curcumin by 50% in Turmeric [50, 51]. Higher Potassium inputs also enhances oil content in Coriander [52], for instance. So improving agro-technology can enhance the bioactive potential of the spices.

We demonstrate here the comparative advantage of spices over the standards drug for the treatment of arthritis, to illustrate the alternative approach to drug discovery. Hence, we performed docking study at Rasa Life. Co., (www.rasalsi.com) Pune during 2020 on the key bioactive compounds important spices w.r.t. COX2 (cyclooxygenase) active site (PDB ID 5IKR obtained from PDBsum). **Table 7** shows that the 4 spice ingredients have closeby values and high theraupetic potential are

Spice	Ingredient (conformation)	Site & activity (interactions)- (PDB id 5f19)	Binding energy (kcal/mol)	Remarks
High potential				
Pepper	Piperine (53)	SER530 & ARG 120 (2)	−9.99	Good activity, high probability
Coriander	Apigenin (20) Quercetin	TRP 387 & ASN 382 (2) ALA199, TRP 387 and ASN 382, TRP387, TYR385(A)	−9.94	Stable, high probability
Turmeric	Curcumin (35)	TYR 385- good activity (1)	−8.66	-“-
Less scope				
Ginger	8-shogaol (40)	Amino acid residues- SER530 & MET522 (2)	−7.51	Stable, good activity, high probability
	10- Gingerol (40)	TYR385- Good activity (1)	−7.34	-“-
Fenugreek	Diosgenin (20)	HIS214 (outside site)	−6.80	outside, not feasible
Clove	Eugenol	SER 530 & TYR 385(Chain A)	−6.33	Stable, high probability

Table 7.
Docking study results of Spices active ingredients in COX-2.

viz. Piperine (Black Pepper, −9.99 Kcal/ mol) 'Apigenin' (Coriander, −9.63) and 'Curcumin' (Turmeric, −8.66 with the stability than methotrexate (−8.6), the standard NSAID [53]. The values are also higher than the synthetically designed 'best' molecule- i.e. 4-(4-methyl-1-piperazinyl)-2-phenyl[1]benzofuro[3,2-d] pyrimidine discovered in the Saudi Arabia [53] or isatin (benzohydrazide) [54]. Ginger (−7.51 kcal/mol- 8-Shagaol), and Diosgenin from Fenugreek had the lower than the threshold values (−6.8) so are not considred here here. Other traditional medicine such as Ayurvedic top used herbs such as Behera (*Terminalia bellerica*) & Herra (*T. chebula*) have also shon high anti-inflammatory potential against COX-2 [55].

The earlier tests of herbal ayurvedic medicines in Arthritis treatment yielded encouraging results in Pune city [56], India, and in USA [57]. So this approach needs further exploration. Quercetin from Coriander & Onion, peppers has higher docking score (−12) than even the active ingredients of the commonest herbal drugs Guggul & Shalaki (<10) [58]. Quercetin is also found to be more effective than even aspirin or celecoxib in the inflammation markers cyclooxygenase (COX) that are vital in cancer biology vide studies in Russia [59] and also in India [60]. Hence, the use of these novel molecules in the arthritis context may be found safe, effective and sustainable.

4. Conclusion

Immunity decline is a major cause of the pandemics in the last century and thus inflammation control is a major challenge for healthcare system. Spices and herbs, rich in polyphenols can be vital tool in this regards as they have high antioxidant value and reduce the oxidative damage to the body. Much greater share of vegetarian diet and spices, besides less intensity of packed foods, meat, liquor, Tobacco, refined carbs, soft drinks etc. in India may be driving its higher immunity and lower burden of COVID-19. Pharmacokinetic methods such as molecular docking can be used to design drugs for immunity building and inflammation control. This is shown with the example of Arthritis where Black Pepper, Coriander & Turmeric can provide potent drugs vide docking studies with focus on piperin, quercetin/ apigenin & curcumin.

Author details

Utkarsh Ghate[1*] and Hema Kulkarni[2]

1 Gram Mooligai (Village Herbs) Co. Ltd., Durg, India

2 Government College, Jamgao-R, India

*Address all correspondence to: ughaate@gmail.com

References

[1] Del Bo' C. et al, 2019. Systematic Review on Polyphenol Intake and Health Outcomes: Is there Sufficient Evidence to Define a Health-Promoting Polyphenol-Rich Dietary Pattern? *Nutrients*, 11, 1-55.

[2] WHO, 2002. Diet, Nutrition and the Prevention of Chronic Diseases, Report of a Joint WHO/FAO Expert Consultation of the World Health Organization (WHO) and Food And Agriculture Organization Of The United Nations (FAO), HO Technical Report Series, 916, Geneva.

[3] WHO, 2018. Ageing and health, https://www.who.int/news-room/fact-sheets/detail/ageing-and-health.

[4] Aggarwal B. at al, 2011. Identification of Novel Anti-inflammatory Agents from Ayurvedic Medicine for Prevention of Chronic Diseases- "Reverse Pharmacology" and "Bedside to Bench" Approach. *Curr Drug Targets.* 2011 Oct 1; 12(11): 1595-1653.

[5] Coleman, J. J. and Sarah K Pontefract, 2016. Adverse drug reactions. *Clin Med (Lond).* ; 16(5): 481-485.

[6] Kearney, John, 2010. Food consumption trends and drivers, *Phil. Trans. R. Soc. B* 365, 2793-2807.

[7] Ferrucci, L. M. Et al., 2010. Measurement of spices and seasonings in India: Opportunities for cancer epidemiology and prevention. *Asian Pac J Cancer Prev.*; 11(6): 1621-1629.

[8] Nguyen, Ly, Lam T. Duong, Rao S. Mentreddy, 2019. The U.S. import demand for spices and herbs by differentiated sources. *Jr. Appl.Res. Medi. Arom. Pl.* DOI: 10.1016/j.jarmap 2018.12.001

[9] Bray, F.; Jacques Ferlay, Isabelle Soerjomataram, Rebecca L. Siegel, Lindsey A. Torre, Ahmedin Jemal, 2018. Global Cancer Statistics 2018: Estimates of Incidence and Mortality Worldwide for 36 Cancers in 185 Countries. *CA: Cancer J Clin.; 68: 394-424.*

[10] Anon, 2018. The Global Asthma Report 2018. Global Asthma Network, Auckland

[11] Marini E., Gloria Magi, Marina Mingoia, Armanda Pugnaloni, Bruna Facinelli, 2015' Antimicrobial and Anti-Virulence Activity of Capsaicin Against Erythromycin-Resistant, Cell-Invasive Group A Streptococci, *Front Microbiol.*; 6: 1281.

[12] Mair E., R. Liu, A G Atanasov, M Shmidtke, V M Dirsch, J M Rollinger, 2016. Antiviral and anti-proliferative in vitro activities of piperamides from black pepper. *Planta Medica* 81(S 01):S1-S381.

[13] Wu Wenjiao, Richan Li, Xianglian Li, Jian He, Shibo Jiang, Shuwen Liu and Jie Yang, 2016. Quercetin as an Antiviral Agent Inhibits Influenza A Virus (IAV) Entry, *Viruses*, 8, 6, pp. 18

[14] Dua A, Garg G, Kumar D and Mahajan R: 2014.Polyphenolic composition and antimicrobial potential of methanolic coriander (*Coriandrum sativum*) seed extract. *Int J Pharm Sci Res*; 5(6): 2302-08.

[15] Aboubakr, H A et al, 2016. In Vitro Antiviral Activity of Clove and Ginger Aqueous Extracts against Feline Calicivirus, a Surrogate for Human Norovirus. *J Food Prot* .79 (6): 1001-1012.

[16] Chang Jung San et al, 2012. Fresh ginger (*Zingiber officinale*) has anti-viral activity against human respiratory syncytial virus in human respiratory tract, *Jr. Ethnopharma.* 145(1).

[17] Utomo R. Y., Muthi Ikawati, Edy Meiyanto, 2020. Revealing the Potency

of Citrus and Galangal Constituentto Halt SARS-CoV-2 Infection. doi: 10.20944/preprints202003.0214.v1

[18] Joshi, D. R., A. C. Shrestha and N. Adhikari, 2018. A Review On Diversified Use Of The King Of Spices: *Piper Nigrum* (Black Pepper). *IJPSR, 2018; Vol. 9(10): 4089-4101.*

[19] Sokolova Anastasiya, 2015. Discovery of a New Class of Antiviral Compounds: Camphor Imine Derivatives. *Eur J Med Chem,* 105, 263-73.

[20] Mohamadein, M. M. R. M. Farrag and. A. A. I. Mekawey, 2015. Antiviral and antidermatophytic activity of a Compound from *Cuminum Cyminum* seeds, *Biomed. & Pharma. Jr.* 8(2), 573-580.

[21] Bright K. and D. H. Gilling, 2016. Natural Virucidal Compounds in Foods. *https://www.ncbi.nlm.nih.gov/pmc/articles/PMC7123660/ Viruses in Foods*. 26 : 449-469.

[22] Sharma, 2019. Efficacy of Garlic and Onion against virus. *Int.J.Res.Pharm. Sci.,*10(4),3578-3586.

[23] Ghoke, et al, 2018. Evaluation of antiviral activity of *Ocimum sanctum* and *Acacia arabica* leaves extracts against H9N2 virus using embryonated chicken egg model. BMC Complementary and Alternative Medicine volume 18, Article number: 174 pp. 1-10.

[24] Jadhav *P., L.* Hingorani *and* N. Kshirsagar' 2014. Assessment of potency of PC-complexed *Ocimum sanctum* methanol extract in embryonated eggs against Influenza virus (H1N1). *Pharmacogn Mag.* Jan-Feb; 10 (Suppl 1): S86–S91.

[25] Parveen Kumar, Bal Chander, 2020. COVID 19 mortality: Probable role of microbiome to explain disparity. *Medical Hypotheses* 144, 110209 pp 6.

[26] Chinnaswamy, Sreedhar 2020. SARS-CoV-2 infection in India bucks the trend: Trained innate immunity? *Am J Hum Biol.*; e23504.

[27] Florindo H. et al, 2020. Immune-mediated approaches against COVID-19. *Nature Nanotechnology* | VOL 15 | 630 August 2020 | 630-645.

[28] Elsayed Y. N, A. Khan* 2020. Immunity-Boosting Spices and the Novel Coronavirus. *ACS Chem. Neurosci.* 2020, 11, 12, 1696-1698.

[29] Shahidi F. and Abul Hossain, 2018, Bioactives in spices, and spice oleoresins: Phytochemicals and their beneficial effects in food preservation and health promotion, Journal of Food Bioactives, Volume 3, September 2018, pages 8-75.

[30] Jiang T. Alan, 2019. Health Benefits of Culinary Herbs and Spices, Journal of AOAC International Vol. 102, No. 2, pp. 395-411.

[31] Kunnumakkara A. B., B. L. Sailo, K. Banik, H. Choudhary , S. Prasad, S. C. Gupta, A. C. Bharti, and B. B. Aggarwal' 2018. Chronic diseases, inflammation, and spices: how are they linked? *J Transl Med*. 2018; 16: 14.

[32] Min HP, Ching SL, Chi TH. 2010. Anti-inflammatory activity of natural dietary flavonoids. *Food Funct.*; 1:15e31.

[33] Chatterjee B, R L Karandikar and S C Mande, 2020. The mortality due to COVID-19 in different nations is associated 1 with the demographic character of nations and the prevalence of autoimmunity. medRxiv preprint doi: https://doi.org/10.1101/2020.07.31.20165696;

[34] Parveen Kumar, Bal Chander, 2020. COVID 19 mortality: Probable role of microbiome to explain disparity. *Medical Hypotheses* 144, 110209 pp 6.

[35] Chinnaswamy, Sreedhar 2020. SARS-CoV-2 infection in India bucks the trend: Trained innate immunity? *Am J Hum Biol.*; e23504.

[36] Goetz Katja et al, 2020. Use of Mustard Seed Footbaths for Respiratory Tract Infections: A Pilot Study. Evidence-Based Complementary and Alternative Medicine, pp. 6.

[37] Zhang, L., Jackson, C.B., Mou, H. *et al.* SARS-CoV-2 spike-protein D614G mutation increases virion spike density and infectivity. *Nat Commun* 11, 6013 (2020).

[38] Sharma, L. 2020. Immunomodulatory Effect and Supportive Role of Traditional Herbs, Spices and Nutrients in Management of COVID- 19. doi:10.20944/preprints202009.0026.v1

[39] Yashin A, Yakov Yashin , Xiaoyan Xia and Boris Nemzer, 2017. Antioxidant Activity of Spices and Their Impact on Human Health: A Review . Antioxidants, 6, 70;

[40] Barnes PM, et al. *CDC Natl Health Statistics Rep* 2008[12]:1-24.2).

[41] Ventola, C. Lee, 2010, Current Issues Regarding Complementary and Alternative Medicine (CAM) in the United States.

[42] WHO (2013) Traditional medicine strategy: 2014-2023.

[43] Safiri *et al*, 2021. Prevalence, Deaths, and Disability-Adjusted Life Years Due to Musculoskeletal Disorders for 195 Countries and Territories 1990-2017. *Arthritis & Rheumatology*, 73(4): 702-714.

[44] Liu X, Machado GC, Eyles JP, et al. 2018. Dietary supplements for treating osteoarthritis: a systematic review and meta-analysis. *Br J Sports Med,* 52:167-175.

[45] Crawford, Cindy et al, 2019. Dietary Ingredients for Mitigating Chronic Musculoskeletal Pain. in the *Military Pain Medicine*, 20 6: 123-1247.

[46] Cunningham A B, J A Brinckmann, R N Kulloli , U Schippmann, 2018. Rising trade, declining stocks: The global gugul *Commiphora wightii* trade, *J Ethnopharmacol.* 223:22-32.

[47] Nishat Fatima*, Syeda Jabeen Fatima, 2016. Pharmacological Screening For Anti-Arthritic Activity Of Moringa Oleifera. Asian J Pharm Clin Res, 9 (3): 106-111.

[48] Benavente, M., S. Arias, L. Moreno, J. Martinez, 2015. Production of Glucosamine Hydrochloride from Crustacean Shell, Journal of Pharmacy and Pharmacology 3 1 :20-26.

[49] Singh Arvind Kumar ,M. Kalaivani, A. Krishnan, P. K. Aggarwal, and S. K. Gupta, 2014. Prevalence of Osteoarthritis of Knee Among Elderly Persons in Urban Slums Using American College of Rheumatology ACR Criteria. *J Clin Diagn Res*. 8 9 : JC09-JC11.

[50] Karthikeyan, P.K., M. Ravichandran, P. Imas, and M. Assaraf. 2009. The Effect of Potassium on the Yield and Quality of Turmeric (*Curcuma longa*). e-ifc No. 21 - Research Findings. https://www.ipipotash.org/publications/eifc-114

[51] Ghate U., H. Kulkarni, A Arunachalam, 2019. Spices in the eastern Indian laterite soil have more polyphenols? *Ind. Jr. Hill Farming*. 32(2), 236-238.

[52] Freitas, M. S. M., 2020. Potassium sources and doses in coriander fruit production and essential oil content. *Hortic. Bras.* vol.38 no.3 268-273.

[53] Shazi Shakil, Adel M Abuzenadah, Suzan M Attar, Omar Fathaldin, Rajaa Al-Raddadi, Mansour I Sulaiman, 2020. Identification of a putative anti-rheumatoid arthritis molecule by virtual

screening. *Tropical Journal of Pharmaceutical Research*; 19 (6): 1255-1261.

[54] Ravi J., K. Gangarapu, S. Manda, S. Rekulapally, 2016. Synthesis, *In Vivo* Anti-Inflammatory Activity, and Molecular Docking Studies of New Isatin Derivatives. *Intl. Jr. Medi. Chem.*, pp. 9.

[55] Shaikh R. U., M. M. Pund and R. N. Gacche. et al. 2016. Evaluation of anti-inflammatory activity of selected medicinal plants used in Indian traditional medication system. *Jr. Tradi. Compl. Medi.* 6: 355-361.

[56] Chopra Arvind, et al 2013. Ayurvedic medicine offers a good alternative to glucosamine and celecoxib in the treatment of symptomatic knee osteoarthritis: a randomized, double-blind, controlled equivalence drug trial. *Rheumatology* 52:1408_1417.

[57] Furst Daniel E. et al, 2011. Double-Blind, Randomized, Controlled, Pilot Study Comparing Classic Ayurvedic Medicine, Methotrexate, and Their Combination in Rheumatoid Arthritis. *Jr. Clin. Rheum.* 17 4 185-192.

[58] Khan MK, Ansari IA, Khan MS, Arif JM, 2013. Dietary phytochemicals as potent chemotherapeutic agents against breast cancer: Inhibition of NF-κB pathway via molecular interactions in rel homology domain of its precursor protein p105. *Phcog Mag*, 9:51-7.

[59] Manukyan A. E.and A.A. Hovhannisyan, 2020. The Quercetin And Quercetin Derivatives Interaction With Cyclooxygenase-1 And Cyclooxygenase-2. https://doi.org/10.1101/2020.12.05.413088, bioRxiv preprint.

[60] Parameswari. P, Devika. R. 2019. In silico Molecular Docking Studies of Quercetin Compound against Anti-inflammatory and Anticancer Proteins. Research *J. Pharm. and Tech.*; 12(11):5305-5309.

Chapter 15

Colon Available Bioactive Compounds Exhibits Anticancer Effect on *In-Vitro* Model of Colorectal Cancer

Poounima Patil and Suresh Killedar

Abstract

The current work was addressed to characterize gallic acid from amla fruit and quercetin from peels of pomegranate fruit and formulated into Chitosan (CS) nanoparticles and to evaluate their cytotoxicity towards human colorectal cancer (HCT 116) cell lines. Identification of the biomolecules was performed by chromatographic and spectroscopic techniques and characterization of gallic acid and quercetin loaded chitosan nanoparticles carried out by using FT-IR, X- ray diffraction, entrapment efficiency and loading content confirmed successful encapsulation of biomolecules into nanoparticles. *In vitro* drug release studies done by using simulated fluids at various pH (1.2, 4.5, 7.5, and 7.0) to mimic the GIT tract and achieved drug releases 77.56% for gallic acid 79.06% for quercetin at 24 hr. in a sustained manner. The human HCT116 cell line by MTT assay results inferred that the synthesized CS nanoparticles demonstrated shows more effective antiproliferative potential with IC_{50} value of 36.17 ug/ml than polyherbal extract 60.32ug/ml.

Keywords: gallic acid, quercetin, chitosan nanoparticles

1. Introduction

The cancer is one of the most dreaded and threatening diseases in the world, causing more than 6 million deaths a year [1]. Colon cancer is recognized as the third most common cancer worldwide with high morbidity and mortality, and the fourth common cause of death [2]. Various cytotoxic drugs are used for the treatment of colorectal cancer like 5-Fluorouracil, Oxaliplatin and Cisplatin drugs are their hydrophobic nature and their susceptibility to develop drug resistance [3, 4].

In these current work great efforts for the discovery and development of nanoformulation based on natural products on *in vitro* HCT 116 cell lines study. Here gallic acid isolated from amla fruit (*Emblica officinalis*) and quercetin isolated from peels of pomegranate fruit (*Punica granatum*). Gallic acid is a naturally available phenolic compound present in amla fruit which is water insoluble and one of the major constituent of amla which might contribute to the health effects [5, 6]. Gallic acid exerts an anticarcinogenic activity, antiproliferative and anti-apoptotic activities against pancreatic cancer cells [7–9]. Flavonoids are a group of naturally occurring class of plant secondary metabolites having polyphenolic structure in addition quercetin has anti-tumor properties, anti-inflammatory, anti-proliferative, anti-angiogenic [10–15].

There have been particular efforts to evaluate the therapeutic role of these active constituents present in plants rather than using whole extracts. The fundamental method of reasoning behind these systems contributes greatly to enhance the targeting delivery and bioavailability of phenolic and flavonoid remarkable formulation development can be made by preparation of nanotechnology products. The eco-friendly synthesis of gallic acid and quercetin loaded chitosan nanoparticles (CS nanoparticles) through green route from plant extracts have renowned a wide range of application in the field of modern science, due to increased drug efficacy and less toxicity in the nanosized mediated drug delivery model. At the same time, the use of gallic acid and quercetin in pharmaceutical formulations is limited due to its poor water solubility, poor bioavailability and instability in physiological medium [16, 17].

Lack of site specificity is one of the major reasons for the drug in reaching the target site in therapeutic concentrations in colorectal cancer [18]. Chitosan (CS), as the only naturally occurring positive charge polysaccharide, has remarkable properties including high bioavailability, super biodegradability, high biocompatibility, non-toxicity etc., On the other hand it causes sustained release of the drug from the particle in the tumor environment [19]. The main rationale behind using these types of polymers is their ability to prevent drug degradation in the gastric environment in the stomach and their ability to release the drug after entering the distal ileum [20]. Poloxamer 407 is a hydrophilic nontoxic copolymer used for its stabilizing properties and incorporation of hydrophobic drugs capability to increase the solubility of biomolecules [21]. Here combined biomolecules synergistic activity of nanotechnology approach has been developed to improve the bioavailability as to entrap these natural biomolecules into biodegradable polymeric CS nanoparticles [22].

The system of glyceryl monooleate (GMO)/chitosan is a surface-modified nanoparticulate system consisting of GMO as a lipid portion and chitosan as a coating polymer to target colonic area with poloxamer 407 as a stabilizer. Therefore, the purpose of this study was to formulate CS nanoparticles of where quercetin isolated from peels of pomegranate fruit and gallic acid isolated from amla fruit as a model hydrophobic biomolecules followed by lyophilization using probe sonicator and High Pressure Homogenization (HPH) method. CS nanoparticles prepared and optimized using Quality by design approach by using central composite factorial design. Optimized formulation further characterized for different parameters as particle size, zeta potential, FT-IR and. Release kinetic studies performed using method for conventional nanoparticle release behavior assessment.

In this study, we systematically analyzed the *In vitro* anti-cancer potential of the gallic acid and quercetin loaded chitosan nanoparticles synergy approach for combined active biomolecules and compared to their activity to combined extracts on HCT 116 human colon cancer cell lines and the mechanism of action of CS nanoparticles in regulating the growth of CRC cells. The human HCT116 cell line by MTT (3-(4,5-dimethyl-2-tiazolyl)-2,5-diphenyl-2-tetrazolium bromide) assay was exposed to cytotoxicity of polyherbal extracts, chitosan nanoparticles and cisplatin (Standard) and activity is dependent up to the concentration of 6.25-100ug/mL for 24 h followed by MTT cellular assays [23]. To determine the potential anti-cancer effect of, we synthesized CS nanoparticles using gallic acid and quercetin biomolecules, which is a phenolic and flavonoid predominantly found in amla fruit and peels of pomegranate fruit. HCT116 cells exposed to gallic acid and quercetin for 24 h exhibited significant loss of cell viability and proliferation in a dose-dependent manner.

The IC50 of polyherbal extract, chitosan nanoparticles and standard cisplatin after 48 h treatment it was found to be 60.32 and 36.17 and 8.915 ug/ml respectively. The obtained result inferred that the synthesized CS nanoparticles demonstrated shows more effective antiproliferative potential on HCT-116cell lines with IC_{50} value of 36.17 ug/ml than polyherbal extract 60.32 ug/ml were discussed briefly in this manuscript.

2. Materials and methods

2.1 Materials

Poloxamer 407 from BASF, Chitosan 90% dda obtained from CIFT Cochin, GMO from Mohini organics, standard gallic acid and quercetin purchased from Loba Chemie., 10% fetal bovine serum (Invitrogen Life Technologies USA). RPMI 1640 and McCoy's 5A medium (Fisher Scientific, Waltham, USA). All the solvents and chemicals used were procured from Himedia Laboratories, Research Lab. Mumbai.

2.2 Plant material

The sample of different parts of plant of amla and pomegranate was collected from Kolhapur district and was authenticated by Dr. Madhukar Bachulkar, Principal, Arts and Science College Peth Vadagaon Kolhapur. The voucher herbarium (PSP-1 and 2) has been deposited in the department of Pharmacognosy Bharati Vidyapeeth College of Pharmacy, Kolhapur. Amla fruit and peels of pomegranate fruit were collected in season (Feb-March), were dried under shade for 10–15 days in air.

2.3 Soxhlet extraction method

In order to extract Flavonoid and phenolics from plants with a high degree of accuracy, various solvents of differing polarities were tried as chloroform, ethanol and ethyl acetate. The dried powder of amla fruit and peels of pomegranate powder extracted with 800 ml in various solvents for 6 hours separately. All extracts were filtered and evaporated to dryness under reduced pressure at 60 °C by a rotary evaporator and to determine percentage yield for all three different solvents [24].

2.4 Phytochemical screening

2.4.1 Qualitative test

Phytochemical analysis was carried out to detect the presence of primary and secondary metabolites were used to identify the biomolecules present in the plant extract [25]. The phytochemical tests carried out for amla and peels of pomegranate extract include alkaloids, glycosides, saponins, tannins, triterpenoids, steroid, flavonoids and carbohydrate [26].

2.4.2 Quantitative tests

2.4.2.1 Determination of total phenolic content for amla extract

Phenolic compounds are important plant constituents with redox properties responsible for antioxidant activity. Folin–Ciocalteu's method (FC) is a colorimetric method based on transfer of electrons between reagents and polyphenols. Different solvent extract chloroform, ethanol and ethyl acetate of amla fruit used for determination of phenolic content. The reaction mixture was prepared by mixing 1 ml of methanolic solution of all extracts, 2.5 ml of 10% Folin–Ciocalteu's reagent dissolved in water and 2.5 ml 7.5% NaHCO3. Blank was concomitantly prepared, containing 0.5 ml methanol, 2.5 ml 10% Folin–Ciocalteu's reagent dissolved in water and 2.5 ml of 7.5% of NaHCO3. The samples were thereafter incubated in a thermostat at 45° C for 45 min. The same procedure was repeated for the standard solution of gallic acid (Standard) in methanol (10to100μg/ml) and for blank then calibration line was construed and absorbance measured at λmax 765 nm [27]. The samples were prepared in triplicate for each analysis and the mean value of absorbance was obtained.

2.4.2.2 Determination of total flavonoid content for pomegranate peel extract

Flavonoids are group of polyphenolic compound used for different activities and their potency depends on the number and position of free hydroxy groups. As a basis quantitative determination, flavonoid contents in pomegranate peel extract were determined using aluminum chloride colorimetric method with sufficient modification. In this process, flavonoid content was determined using quercetin standard (5 to 320 μg/mL) to make the calibration curve. Different solvent extracts chloroform, ethanol and ethyl acetate of pomegranate peel used for determination of flavonoid content. All procedure followed for preparation of different extracts of sample solution, blank and for standard and their corresponding absorbances were measured at 415 nm with a UV-1800 spectrophotometer [28].

3. Techniques of isolation and purification of bioactive molecules from extracts

3.1 Fractionation of bioactive compound by flash chromatographic technique

Flash chromatography instrument consisting of (Analytical technologies limited, Shanghai china) consisting of TBP2H02pump along with TBD2000 UV detector and automatic fraction collector was used for analysis. System equipped with Chromo station software was used for data monitoring during the analysis. The separation was carried out on OROCHEM OROFLO-4SiHPS column made up of silica particles [29, 30].

3.2 Gas chromatography

The gas chromatography used for estimation of residual class 3 ethyl acetate solvent in both crude extracts was performed using a Gas Chromatography system 7890 B with Agilent DB 624 column with helium gas at 1 ml/min flow mode reference solution tetrahydrofuran. GC temperature was set at 50 °C (hold for couple of min) to 250 °C at 20 °C/min. (hold up to 5 min) [31, 32].

4. Structural clarification of the bioactive molecules

The flash chromatographic fractions of amla fraction no FA004 and pomegranate fraction no FP004 were filtered, dried and kept at 4 ° C for characterization of FT-IR and ^{1}H-NMR techniques and quantitavely estimated by HPLC technique [33, 34].

4.1 FTIR spectroscopy

FTIR has proven to be a valuable tool for the characterization and identification of functional groups present in compound from plants extract. Infrared spectra was collected using IR (α-ATR Bruker Germany spectrometer) operated form 4000–600 cm^{-1} at resolution of 4 cm^{-1}. Data analyzed using Opus software.

4.2 NMR spectroscopy of the isolated compound

Only fraction A16 and Fraction P4 was additionally elucidated by ^{1}HNMR by using solvent D_6 + $CDCL_3$ MIX. The analysis was done at the BRUKER instrument of 400 MHz [35].

4.3 HPLC of isolated compounds

HPLC PU-2080 Plus (Systronics) with UV-2075 plus intelligent detector and HPLC C18 column (250 × 4.6 mm, 5 μm) was set at 270 nm for estimation of gallic acid and 259 nm for estimation of quercetin [36, 37]. HPLC PU-2080 Plus (Systronics) with UV-2075 plus intelligent detector and HPLC C18 column (250 × 4.6 mm, 5 μm) was set at 270 nm for estimation of gallic acid and 259 nm for estimation of quercetin. The mobile phase Acetonitrile and 2% Acetic Acid with ratio 40:60 used for elution of both compounds. Flow of mobile phase and injection loop was set at 1.0 ml/min and 20 μL respectively. Quantitative determination of gallic acid and quercetin content in fraction concentrations (FA004 by flash chromatography of amla extract) in the range 0.01 to 0.5 mg/ml and (Fraction FP004 by flash chromatography of pomegranate extract) in the concentration range of 0.01 to 0.5 μg/ml used.

5. Determination of solubility of isolated compound

The two isolated compounds were analyzed for their solubility in different solvents as in DMSO, ethanol, methanol and acetone.

6. Melting point determination

Melting points of two isolated biomolecule compounds was done in thermonic apparatus to determine its identity and purity. The observed melting point of isolated compound was compared with the standard melting point of respective gallic acid and quercetin.

7. Antioxidant activity by DPPH method

2, 2-diphenyl-1-picryl-hydrazyl (DPPH) is a stable organic free radical used to estimate the antioxidant activity of various compounds. The scavenging action on DPPH radical from amla fruit and isolated fraction (FA004 by flash chromatography) and peels of pomegranate extract and isolated fraction (FP004 by flash chromatography of peels of pomegranate extract) determined by following method [38, 39]. As of different concentrations was mixed with an aliquot of DPPH (1 ml, 0.004% w/v) and analyzed at 517 nm. Then the scavenging capacity was calculated using equation number (1).

$$\text{Scavenging activity}(\%) = \frac{(\Delta A517 \text{ of control} - \Delta A517 \text{ of sample})}{\Delta A517 \text{ of control}} \times 100 \quad (1)$$

8. Formulation of nanoparticles

An o/w nanoemulsion of gallic acid and quercetin was prepared by using a GMO/chitosan framework as reported by with slight modifications. Briefly, isolated gallic acid (100 mg) and quercetin (100 mg) were dissolved in molten GMO (2 g), then add 12.5 ml of 0.1% poloxamer 407 sonicated at 18 W for 3 min in probe sonicator. To this emulsion, dropwise 12.5 ml of 2.4% chitosan solution was added again using probe sonicator at 16 W for 4 min [40, 41]. Finally this phase was subjected to twelve cycles of HPH at 15,000 psi to give the nanoemulsion. Then, lyophilized with 2% mannitol as a cryoprotectant for 48 hr. Central composite design [42] was applied to examine the combined effect two variables, each at 2 levels and the possible 9 combinations of CS nanoparticles.

9. Characterization of polyherbal nanoparticles

9.1 Particle size and zeta potential

Average particle size and zeta potential of the CS nanoparticles were determined by Particle Size Analyzer (Zetasizer Ver System; Malvern Instruments Ltd., Malvern, UK). To analyze particle size, nanosuspension was diluted with filtered (0.22 lm) ultra pure water [43–45].

9.2 By FTIR spectroscopy

FTIR has proven to be a valuable tool for identification of functional groups present in compound from plants. Attenuated total reflection/Fourier transform infrared spectroscopic (ATR/FTIR) spectra was collected at room temperature by coupling ATR accessory to an FTIR spectrometer (Perkin Elmer, Spectrum 100).

9.3 *In vitro* release studies

CS nanoparticles were tested in various simulated fluids at different pH to evaluate the release of nanoparticles at particular pH and also to determine the drug release [46]. Four milligrams of CS nanoparticles were dispersed in a freshly prepared phosphate-buffered saline (PBS; pH = 2.0, 4.5, 6.8, 7.4) as a release medium in a dialysis membrane sac (mw cut-off 12 kDa; Sigma Aldrich) to simulate ileo-colon conditions for 24 hr [47]. The enclosed dialysis sac was immersed in a beaker containing 50 mL of the release medium. The beaker was placed in a shaking incubator at 37 °C under mild agitation (90–100 rpm) PBS; pH = 2.0 for first four hour, pH = 4.5 for next five to nine hour, pH = 6.8 for next ten to thirteen hour and finally pH = 7.4 for fourteen to twenty-four hour. The supernatant 5 ml withdrawn at specified time intervals and assayed for drug release in UV spectrophotometrically gallic acid at 270 nm and quercetin at 259 nm.

10. *In vitro* anticancer activity (cytotoxicity) by MTT assay

10.1 Cell culture

A human colorectal adenocarcinoma cell line (HCT116) were cultured with RPMI 1640 and McCoy's 5A medium (Fisher Scientific, Waltham, MA, USA), respectively [48]. All cell culture mediums contained 10% fetal bovine serum. Cells were incubated in a CO_2 incubator at 37 °C with 5% CO_2. After reaching confluency, cells were isolated from the dish with Trypsin–EDTA. The cell suspension was centrifuged at 1000 r/min for 5 min and then re-suspended in growth medium for further experiments.

10.2 Cell viability assay

Cell viability was studied using an MTT assay. Cells were grown in a medium containing 10% FBS, seeded in 96-well plates at a density of $2x10^5$ cells/well, and incubated at 37 °C in CO_2 incubator with 5% CO_2 for 24 h [49]. Then, polyherbal extract, CS nanoparticles and standard cisplatin were added (final concentrations of 6.25, 12.5, 25, 50, and 100 ug/ml) to the mono-layers of cells, which were subsequently incubated for at 24 and 48 h, media were aspirated and MTT solution at a concentration of 5 mg/ml in phosphate-buffered saline (PBS) buffer was added 20 ml/well. After further incubation (3 h), the media was removed and replaced with 100 ml of DMSO. Plates were washed with 1% acetic acid, air-dried, and then 10 mM Tris base pH 7.4 (150 μl) was added to the wells to solubilize the dye. The plates were shaken vigorously for 5 min and color absorbance was measured at 540 nm using an ELISA microplate reader. (ELISA reader Denver Jasco Model 7800 UV/VIS Spectrophotometer Jasco Tokyo, Japan) Untreated cells were used as positive controls with 100% viability and cells without assay reagents were used as a blank.

11. Result and discussion

11.1 Soxhlet extraction method

Soxhlet extraction method carried out for extraction of amla fruit and peels of pomegranate fruit by using three solvents as chloroform, ethanol and ethyl acetate separately. Ethyl acetate solvent gives highest yield 42.51% (Amla fruit) and 42.89% (Pomegranate peels) hence used for extraction of phenolics and flavonoids [50].

11.2 Phytochemical screening

11.2.1 Qualitative tests

Qualitative tests of Phytochemical screening of amla fruit and pomegranate peel extract gives positive test for presence of flavonoids, alkaloids, tannins and carbohydrate.

11.2.2 Quantitative test

11.2.2.1 Total phenolic content

Calibration curve of standard gallic acid showed linear equation at $y = 0.014x + 0.395$, $R2 = 0.996$ The content of phenolics in different solvents was as 25.73 ± 0.21, 42.09 ± 0.19 and 63.76 ± 0.29 mg GAE/g for chloroform, ethanol and ethyl acetate respectively. As compare to other solvent ethyl acetate gave more yields hence this is suitable solvent for extraction of phenolics.

11.2.2.2 Total flavonoid content

The concentration of flavonoid standard quercetin on the calibration line was based on the calculated absorbance at $y = 0.017 \times +0.412$, $R2 = 0.990$ (**Figure 1**)

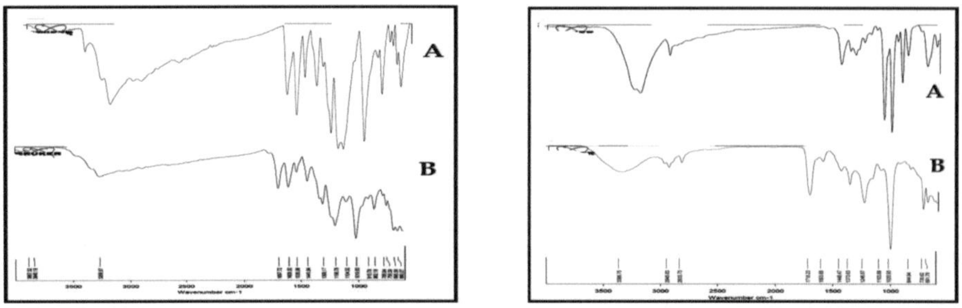

[A] [B]

Figure 1.
IR spectra of isolate of [A] Amla fruit [B] pomegranate peel extract.

then, the content of flavonoids in pomegranate peels of different solvent extract was expressed in terms as mg QE/g. The content of flavonoid in different solvents was as 32.88 ± 0.26, 42.11 ± 0.29 and 70.8 ± 0.1732 mg QE/g for chloroform, ethanol and ethyl acetate respectively. As compare to other solvents ethyl acetate gave more yields.

11.3 Techniques of isolation and purification of bioactive molecule from amla fruit and pomegranate peel extract

11.3.1 Fractionation of bioactive compound by flash chromatographic technique

The mobile phase used as ethyl acetate: methanol 100:0 to 0:100 with flow rates were kept at 4 ml/min with wavelength for amla fruit at 270 nm and pomegranate peel extract at 263.5 nm. Column was loaded with 8.0 gm slurry (3 g extract +5 g silica gel) in 25 gm of silica gel (200–400 mesh size).

1 gm of amla extract and peels of pomegranate powder extract separately mixed with 3 gm of silica gel and triturated properly in mortar and pestle. Then, properly mixed extract samples were loaded in sample holder. The separation was completed in 15 minutes only. The Five fractions were isolated by linear gradient with peak tube volume was 14 ml and run time was 15 min. Different fractions no. FA001 to FA005 from amla extract and FP001 to FP005 from peels of pomegranate extract were isolated and dried on buchi roto evaporator (R-210 water bath B-491) for dryness.

Among all five fractions of amla extract fraction number the UV spectra of fraction no FA004 phytoconstituent which gives absorbance at 270.5 nm and this absorbance confirmed with standard gallic acid solution spectra at 272 nm. (**Figure 2A**) The percentage yield of fraction FA004 was found to be 33.4 mg/gm. The five fractions of peels of pomegranate extract fraction no FB004 gives maximum absorbance at 263.5 nm and also this absorbance confirmed by standard quercetin sample absorbance at 345 nm (**Figure 2B**) scanning with UV Spectrophotometry summarized in **Tables 1** and **2** . The percentage yield of fraction FP004 was found to be 42.6 mg/gm. Further these two isolated fractions no FA004 and FP004 characterized for IR, H1NMR, HPLC and HPTLC techniques for better results.

11.3.2 Gas chromatography

The study represented a simple gas chromatographic method for estimation of ethyl acetate contents in both amla and pomegranate extract. The GC analysis of the crude

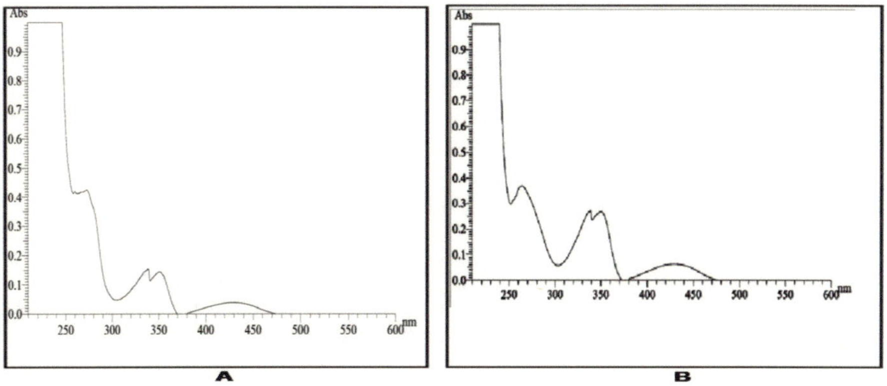

Figure 2.
Flash chromatogram of [A] Amla fruit [B] pomegranate peel extract.

Peak	Start (nm)	Apex (nm)	End (nm)	Height (Abs)	Area
FA001	600.0	430.5	379.0	0.039	1.257
FA002	379.0	350.5	340.5	0.144	2.708
FA003	340.5	339.5	304.0	0.154	3.471
FA004	304.0	270.5	262.0	0.426	11.114
FA005	262.0	239.0	210.0	1.971	70.223

Table 1.
Flash chromatography of amla extract.

Peak	Start (nm)	Apex (nm)	End (nm)	Height (Abs)	Area
FP001	600.0	429.5	379.0	0.064	2.745
FP002	379.0	350.0	340.5	0.266	5.440
FP003	340.5	335.0	302.5	0.272	5.920
FP004	302.0	263.5	252.0	0.369	12.189
FP005	252.0	230.5	210.0	1.893	53.258

Table 2.
Flash chromatography of peels of pomegranate extract.

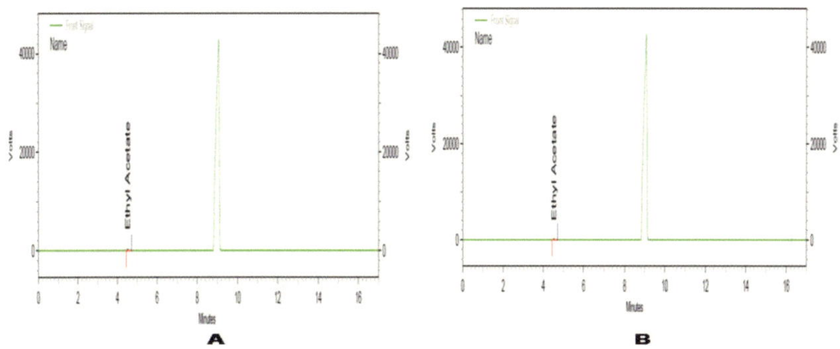

Figure 3.
Gas chromatogram of [A] Amla fruit [B] pomegranate peel extract.

ethyl acetate extracts of amla gives retention time at 4.526 min (**Figure 3A**) and for pomegranate fruit at 4.528 min. (**Figure 3B**) Ethyl acetate concentration in amla fruit was found to be 1305.376 ppm and in pomegranate fruit was found to be 1538.440 ppm. Excellent results were obtained within the worldwide accepted validation reference values and particularly taking into account the low concentration levels investigated [51].

11.4 Structural clarification of the bioactive molecules

The isolated compounds (Fraction No. FA004 from amla extract and FP004 from peels of pomegranate extract by flash chromatography) was characterized by using FT-IR, ^{1}H-NMR and quantitatively estimated by using HPLC technique [52].

11.4.1 FTIR spectroscopy of the isolated compound

FT-IR spectra of isolate of amla fruit extract resulted in presence of functional groups hydroxyl (-OH) stretch, C-H stretch of alkenes, C=O stretch for acid and aromatic benzonoid ring (**Figure 1A**) and FT-IR spectra of isolate of Pomegranate peel extract resulted in presence of functional groups hydroxyl (-OH) stretch at 3366 cm^{-1}, C-H stretch of alkenes at 2945 cm^{-1}, C=O stretch for lactone and aromatic benzonoid ring 1020 cm^{-1} (**Figure 3**).

11.4.2 NMR spectroscopy of the isolated compound

The analysis was done at the BRUKER instrument of 400 MHz d 9.136 (1H, H-7, s), 7.08 (1H, H-2, H-6, s) and 5.011 (1H, H-3, H-4, H-5, s). ^{1}H NMR of isolate of amla fruit showed the aromatic proton, acidic proton and hydroxyl proton and presence of 7 carbons in structure (**Figure 4A**) given molecular formula as $C_7H_6O_5$ [53] ^{1}H-NMR signals of isolate of Pomegranate peel extract shows signals at 12 (S 1H OH Pyran), 6.2 (S 2H Aromatic OH), 6.9 (S1H Aromatic OH), 7.1 (S1H Aromatic OH), 7–8 (S Aromatic proton) ^{1}H NMR showed the aromatic proton and hydroxyl proton and presence of 15 carbons (**Figure 4B**) in structure given molecular formula as $C_{15}H_{10}O_7$.

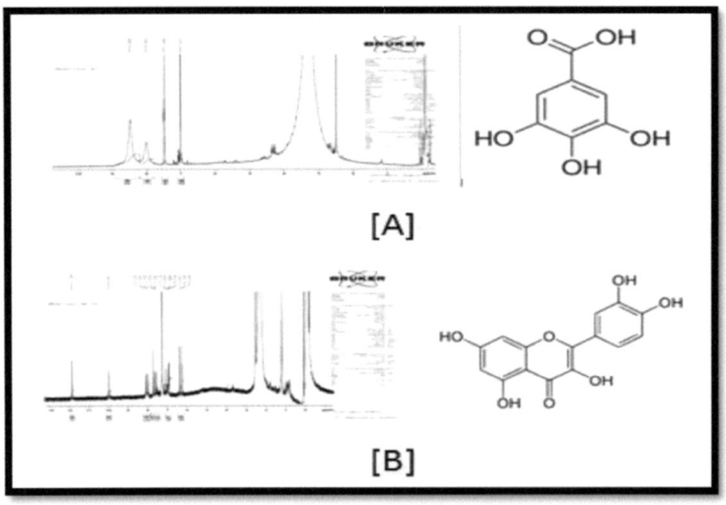

Figure 4.
NMR spectra of isolated compound and structure of compounds [A] Amla fruit (Gallic acid) [B] pomegranate peel extract (quercetin).

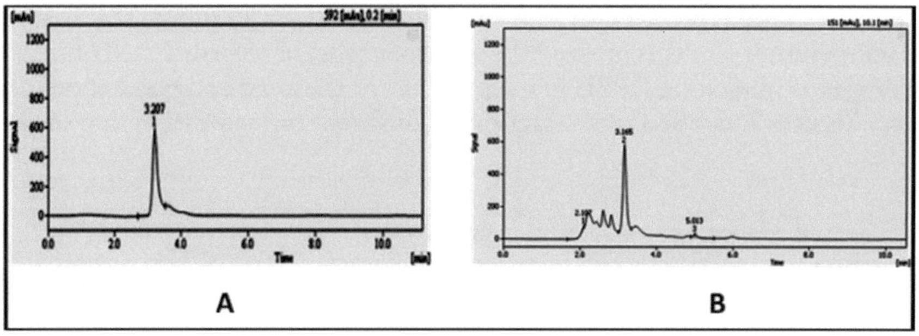

Figure 5.
Chromatogram of [A] standard Gallic acid [B] isolated fraction of amla extract.

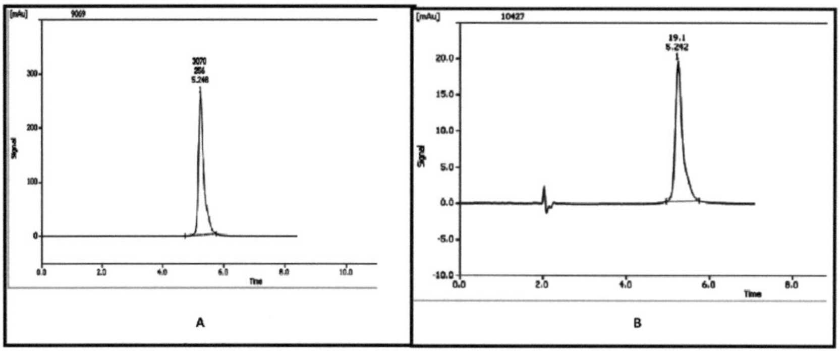

Figure 6.
Chromatogram of [A] standard quercetin [B] isolated fraction of pomegranate peel extract.

11.4.3 HPLC analysis of isolated compounds

A comparison between the spectra of fruits of amla extract (Fraction no A004 by flash chromatography) peak at 3.165 min confirmed with that of standard gallic acid peak at 3.207 min respectively (**Figure 5A** and **B**). A good linearity was found from 5–15 μg/mL gallic acid, and the linear regression equation was y = 8008x-397.0 (rc = 0.999) where y is the peak height. The gallic acid from amla fruit extract was fractionated by HPLC of which 27.15 ± 0.001 μg/mg GAE equivalent by HPLC method were characterized.

Same comparison between the spectra of peels of pomegranate (Fraction no B004 by flash chromatography) peak at 5.242 min with that of standard quercetin confirmed that the retention time of the analyte was 5.248 min respectively (**Figure 6A** and **B**). Linearity for the developed method was found over the concentration range 3–18 μg/ml with a linear regression equation was y = 16.01x + 25628 where y is the peak height correlation coefficient of 0.999.

11.5. Antioxidant activity

11.5.1 Antioxidant activity by DPPH method

The DPPH is a stable free radical, which has been widely accepted as a tool for estimating free radical-scavenging activities of antioxidants. The scavenging activity on 2, 2-diphenyl-1-picryl-hydrazyl (DPPH) radical of both the fruits extract and isolated fractions was determined by following method. The extracts of different concentrations were mixed with an aliquot of DPPH (1 ml, 0.004% w/v) [54].

The mixtures were vigorously shaken and left to stand for 30 min in the dark at room temperature. For this method the absorbance were recorded at 517 nm. The percentages of remaining DPPH in the presence of the amla and pomegranate peel extract (**Figures** 7 and **8**) and its fractions at different concentrations are shown in **Table** 3.

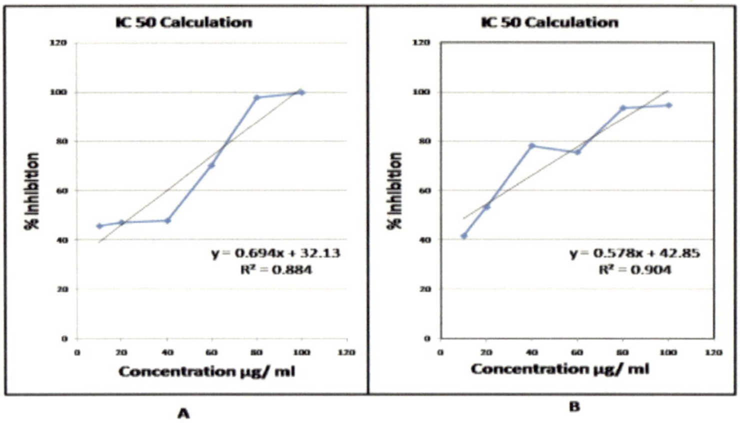

Figure 7.
DPPH radial scavenging activity (A) Amla extract (B) isolated fraction of FA004 by flash chromatography.

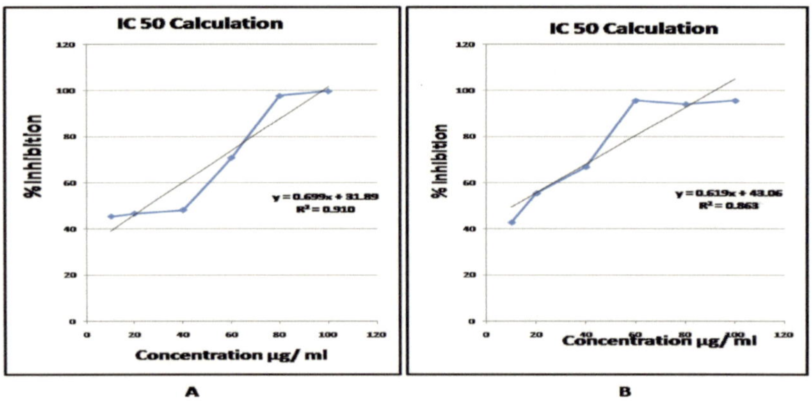

Figure 8.
DPPH radial scavenging activity (A) peels of pomegranate extract (B) isolated fraction no FP004 by flash chromatography.

Sample	R^2	IC_{50}
Ascorbic acid (Standard)	0.996	8.98 µg/ml
Amla extract	0.884	25.74 µg/ml
Isolated Fraction [FA004]	0.904	14.44 µg/ml
Pomegranate peel extract	0.863	29.89 µg/ml
Isolated Fraction [FP004]	0.910	11.21 µg/ml

Table 3.
Antioxidant activity of amla and pomegranate extract by DPPH.

11.6 Determination of solubility of isolated compound

The isolated compound was analyzed for their solubility in different solvents. White colored powder of amla extract (FA004 Flash chromatography) which is soluble in ether, ethanol, methanol, glycerol and acetone. Yellow colored crystalline powder of pomegranate extract (FP004 Flash chromatography) practically insoluble in water and soluble in DMSO, ethanol, methanol and acetone.

11.7 Melting point determination

Melting point of compound was done in thermonic apparatus to determine its identity and purity. The observed melting point of isolated compound of amla extract (FA004 Flash chromatography) was 255–257 ° C compared with the standard melting point (260 °C) of respective isolated gallic acid. The observed melting point of isolated compound of pomegranate extract (FP004 Flash chromatography) was 313–316 °C compared with the standard melting point (316 °C) of respective isolated quercetin.

11.8 Formulation of CS nanoparticles

In this study the goals for optimization were to minimizing particle size and maximum Zeta potential. Desirability ramp showing optimum conditions to formulate CS nanoparticles as chitosan 2.4%, and Poloxamer (407) 0.1% to achieve particle size 218.33 nm and zeta potential11.50 mV with desirability 1.000.

11.9 Characterization of CS nanoparticles

11.9.1 Analysis of particle size and zeta potential

A mean diameter of particle size of CS nanoparticles was found to be 214.2 ± 1.28 nm with +14.7 mV zeta potential [14, 55, 56]. Chitosan on the other hand has a positive charge in acidic solutions due to the presence of protonated amino groups which was appropriate adhere negatively charged intestinal mucus layer. This explains that outer coating of nanoparticles was CS only.

11.9.2 FTIR of CS nanoparticles

The characteristic groups of chitosan at (**Figure 9A**) 3285.15 cm^{-1} for O-H stretching 2875.66 cm^{-1} for C-H stretching and 1415.23 cm^{-1} for amide C-N stretching. The bands at 1150.54 cm^{-1} for asymmetric stretching of the bond C-O-C and 1062.04 and 1023.35 cm^{-1} for vibrations involving the C-O bonds of primary alcohols [57]. The carbon chain of poloxamer 407 (**Figure 9B**) at 2881.11 cm^{-1} aliphatic C-H stretching, plane O-H bend at 1365.12 cm^{-1} and 1242.02 cm^{-1}, C-O stretch at 1096.99 cm^{-1}, $CH{=}CR_2$ at 840.46 cm^{-1}. The C=O functionality of GMO (**Figure 9C**) was seen with a strong peak at 1738 cm^{-1}. In the spectrum of gallic acid (**Figure 9D**) there is a broad band at 3194.61 cm^{-1} related to OH stretching and hydrogen bonds between phenolic hydroxyl groups. The COOH stretch/bend is observed at 1255.93 cm^{-1} Aromatic ring stretching is observed at 1454.44 cm^{-1} [58]. C- O stretching is at 1021.45 cm^{-1} In the spectrum of quercetin (**Figure 9E**) there is a broad band at 3194.61 cm^{-1} related to OH stretching and hydrogen bonds between phenolic hydroxyl groups. O-H stretch at 3190.38 cm^{-1}, =C-H stretch at 2935.23 cm^{-1}, aromatic C=C stretch at 1454.09 cm^{-1} and aromatic C-O stretch at

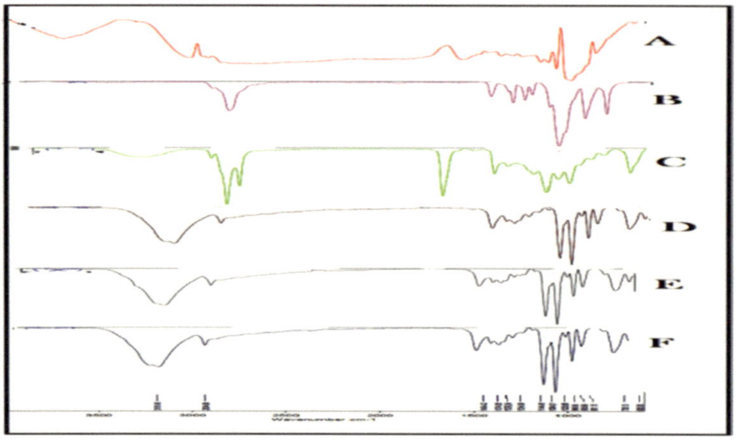

Figure 9.
IR spectra of CS nanoparticles (a) chitosan (B) Poloxamer 407 (C) GMO (D) quercetin (E) Gallic acid (E) CS nanoparticles.

1145.06 cm^{-1}. The COOH stretch/bend is observed at 1255.93 cm^{-1} The spectra of the gallic acid and quercetin loaded CS nanoparticles showed that O-H stretch of gallic acid and quercetin was disappeared (**Figure 9F**). Here all data of FTIR results conclude that encapsulation of gallic acid and quercetin into CS nanoparticles with intermolecular hydrogen bonding occurred in the nanoformulation which correlated with the less crystalline compared to both pure biomolecules.

11.9.3 In vitro release studies

As a result CS nanoparticles have indicated improved drug releases 77.56% for gallic acid 79.06% for quercetin at 24 hr. respectively. So the CS nanoparticles can be considered as a potential barrier, which can release the biomolecules at colonic pH [59]. By engineering chitosan approach gallic acid and quercetin biomolecules achieved sustained and controlled release and also benefitted by its targeting property to colonic region. To describe the mechanism of gallic acid and quercetin release from the CS nanoparticles, [60] the data was plotted into a few kinetic models and best fitted information into the Korsmeyer–Peppas power law model.

11.10 Methods of anticancer activity determination

11.10.1 In vitro cytotoxicity by MTT assay

After 24 hours of incubation, cell viability was determined by the MTT assay. The nanoparticles induced cell cytotoxicity in a concentration dependent manner, as illustrated. Cytotoxicity of polyherbal extracts, CS nanoparticles and cisplatin (Standard) was dose on HCT 116 cell lines and activity is dependent up to the concentration of 6.25–100 ug/mL. The IC50 of polyherbal extract, chitosan nanoparticles and standard after 48 h treatment it was found to be 60.32 and 36.17 and 8.915 ug/ml respectively summarized in **Table 4**.

The antiproliferative potential of all samples shown as cytotoxicity of standard cisplatin (**Figure 10A**) CS nanoparticles (**Figure 10B**) polyherbal extract (**Figure 10C**) was done on HCT 116 cell lines and activity is dependent up to the concentration of 6.25–100 ug/mL. MTT assay determined the cytotoxic effect of all samples by decreasing the cell viability of HCT116 colon cancer cells with different serial dilutions. The half maximal inhibitory concentration (IC50) was evaluated to determine the effectiveness of CS nanoparticles in inhibiting

Name of Samples	Concentrations ug/ml					
	Untreated	**6.25**	**12.5**	**25**	**50**	**100**
Standard (A)	1.157	0.680	0.582	0.19	0.0885	0.06
	IC_{50} = 10.55 ug/ml					
Polyherbal extract (B)	1.157	1.677	1.301	0.794	0.478	0.182
	IC_{50} = 60.32 ug/ml					
CS nanoparticles (C)	1.157	0.906	0.892	0.79	0.141	0.028
	IC_{50} = 36.173 ug/ml					

Table 4.
MTT data analysis of HCT-116 cell lines.

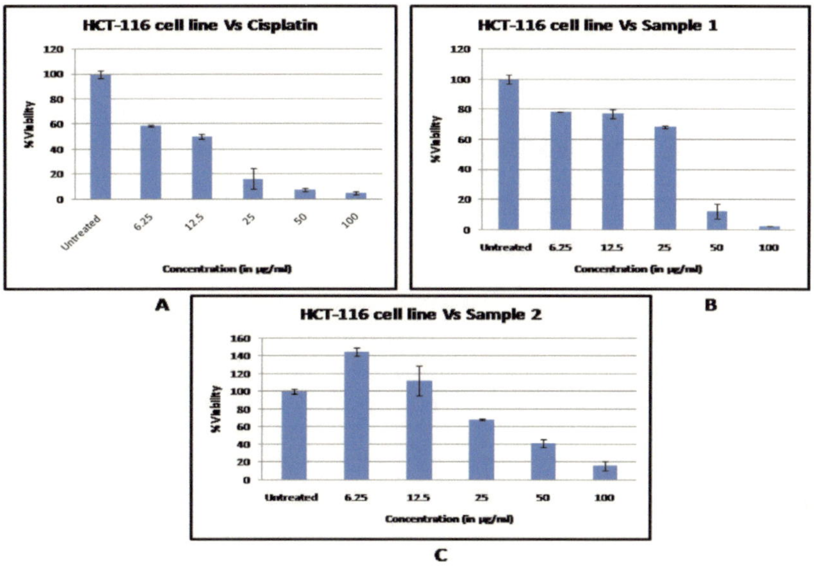

Figure 10.
Concentrations used for MTT assay (A) standard (cisplatin) (B) Polyherbal extract (C) CS nanoparticles.

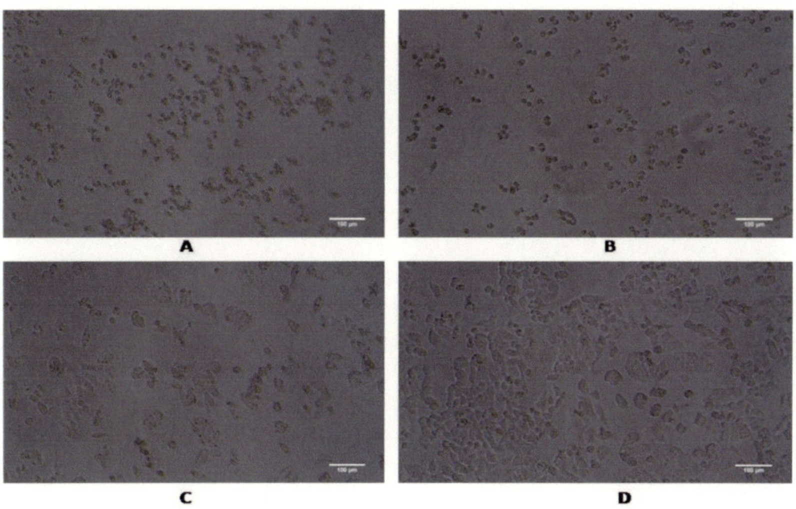

Figure 11.
Microscopy imaging of cellular uptakes (A) standard (B) CS nanoparticles (C) Polyherbal extract (D) untreated HCT116 cell lines.

biological or biochemical functions. CS nanoparticles shows a higher cytotoxic effect on HCT116 cells with low concentrations (IC50 = 36.173 μg/ml) than polyherbal extract (IC50 = 60.32 μg/ml) that might be due to the active biomolecules capped to the nanoparticles.

HCT 116 cell lines considered to have more prominent take-up for CS nanoparticles and more stable even at low concentrations and longer interval than polyherbal extract. Microscopy imaging of cellular uptakes shows as standard cisplatin (**Figure 11A**) CS nanoparticles (**Figure 11B**) polyherbal extract (**Figure 11C**) and untreated HCT116 cell lines (**Figure 11D**) HCT 116 cell lines subjectively were deemed to have had greater uptake for CS nanoparticles and more stable even at low concentrations than polyherbal extract expected to be longer interval than polyherbal extract.

12. Conclusion

In conclusion, the presence of phenolic compound (gallic acid) and flavonoid (quercetin) could be one of the contributing factors for mechanism of *in vitro* studies on HCT 116 cell lines. Model hydrophobic biomolecules with nanoparticle size range, positive charge on particle with good value and sustained in-vitro releases of gallic acid and quercetin especially in wide pH range of entire gastrointestinal tract from nanoparticles were special findings associated with colonic site. Therefore, discovery and development of new nanoformulation based on natural products have been the reported to have a controlled effect on cancer cell lines; therefore, they have the potential to be used as an important therapeutic anticancer biomolecules. Further studies are warranted to decipher the probable mechanism by which gallic acid and quercetin nanoparticles exert anticancer effect.

Author details

Poounima Patil[1*] and Suresh Killedar[2]

1 Department of Pharmaceutical Chemistry, Bharati Vidyapeeth College of Pharmacy, Kolhapur, Maharashtra, India

2 Shree Sant Gajanan Maharaj College of Pharmacy, Gadhiglaj, Mahagaon, Maharashtra, India

*Address all correspondence to: poournima6@gmail.com

References

[1] Ramanathan, R.K., et al., *Safety and toxicity analysis of oxaliplatin combined with fluorouracil or as a single agent in patients with previously treated advanced colorectal cancer.* Journal of clinical oncology, 2003. **21**(15): p. 2904-2911.

[2] Haggar, F.A. and R.P. Boushey, *Colorectal cancer epidemiology: incidence, mortality, survival, and risk factors.* Clinics in colon and rectal surgery, 2009. **22**(4): p. 191.

[3] Lage, H., *An overview of cancer multidrug resistance: a still unsolved problem.* Cellular and molecular life sciences, 2008. **65**(20): p. 3145.

[4] Jain, A., et al., *Design and development of ligand-appended polysaccharidic nanoparticles for the delivery of oxaliplatin in colorectal cancer.* Nanomedicine: Nanotechnology, Biology and Medicine, 2010. **6**(1): p. 179-190.

[5] Park, W.H., *Gallic acid induces HeLa cell death via increasing GSH depletion rather than ROS levels.* Oncology reports, 2017. **37**(2): p. 1277-1283.

[6] Dorniani, D., et al., *Graphene oxide-gallic acid nanodelivery system for cancer therapy.* Nanoscale research letters, 2016. **11**(1): p. 491.

[7] Sánchez-Carranza, J.N., et al., *Gallic acid sensitizes paclitaxel-resistant human ovarian carcinoma cells through an increase in reactive oxygen species and subsequent downregulation of ERK activation.* Oncology reports, 2018. **39**(6): p. 3007-3014.

[8] Bahorun, T., et al., *The effect of black tea on risk factors of cardiovascular disease in a normal population.* Preventive medicine, 2012. **54**: p. S98-S102.

[9] Cedó, L., et al., *Gallic acid is an active component for the anticarcinogenic action of grape seed procyanidins in pancreatic cancer cells.* Nutrition and cancer, 2014. **66**(1): p. 88-96.

[10] Gibellini, L., et al., *Quercetin and cancer chemoprevention, Evid.* Based Complement. Altern. Med, 2011. **591356**.

[11] Heim, K.E., A.R. Tagliaferro, and D.J. Bobilya, *Flavonoid antioxidants: chemistry, metabolism and structure-activity relationships.* The Journal of nutritional biochemistry, 2002. **13**(10): p. 572-584.

[12] Kim, H.-S., et al., *Quercetin enhances hypoxia-mediated apoptosis via direct inhibition of AMPK activity in HCT116 colon cancer.* Apoptosis, 2012. **17**(9): p. 938-949.

[13] Liu, K.C., et al., *The roles of endoplasmic reticulum stress and mitochondrial apoptotic signaling pathway in quercetin-mediated cell death of human prostate cancer PC-3 cells.* Environmental toxicology, 2014. **29**(4): p. 428-439.

[14] Ramadon, D., E. Anwar, and Y. Harahap, *In vitro penetration and bioavailability of novel transdermal quercetin-loaded ethosomal gel.* Indian Journal of Pharmaceutical Sciences, 2018. **79**(6): p. 948-956.

[15] Vijayakumar, A., et al., *Quercetin-loaded solid lipid nanoparticle dispersion with improved physicochemical properties and cellular uptake.* Aaps Pharmscitech, 2017. **18**(3): p. 875-883.

[16] Lesser, S. and S. Wolffram, *Oral bioavailability of the flavonol quercetin a review.* Current Topics in Nutraceutical Research, 2006. **4**(3/4): p. 239.

[17] Ader, P., A. Wessmann, and S. Wolffram, *Bioavailability and metabolism of the flavonol quercetin in the pig.* Free Radical Biology and Medicine, 2000. **28**(7): p. 1056-1067.

[18] Kumari, A., S.K. Yadav, and S.C. Yadav, *Biodegradable polymeric*

nanoparticles based drug delivery systems. Colloids and surfaces B: biointerfaces, 2010. **75**(1): p. 1-18.

[19] Gulbake, A. and S.K. Jain, *Chitosan: a potential polymer for colon-specific drug delivery system.* Expert opinion on drug delivery, 2012. **9**(6): p. 713-729.

[20] Mladenovska, K., et al., *5-ASA loaded chitosan–Ca–alginate microparticles: Preparation and physicochemical characterization.* International journal of pharmaceutics, 2007. **345**(1-2): p. 59-69.

[21] Talasaz, A.H., et al., *In situ gel forming systems of poloxamer 407 and hydroxypropyl cellulose or hydroxypropyl methyl cellulose mixtures for controlled delivery of vancomycin.* Journal of applied polymer science, 2008. **109**(4): p. 2369-2374.

[22] Gurunathan, S., et al., *Nanoparticle-mediated combination therapy: two-in-one approach for cancer.* International journal of molecular sciences, 2018. **19**(10): p. 3264.

[23] Kuppusamy, P., et al., *In vitro anticancer activity of Au, Ag nanoparticles synthesized using Commelina nudiflora L. aqueous extract against HCT-116 colon cancer cells.* Biological trace element research, 2016. **173**(2): p. 297-305.

[24] Zhang, Q.-W., L.-G. Lin, and W.-C. Ye, *Techniques for extraction and isolation of natural products: a comprehensive review.* Chinese medicine, 2018. **13**(1): p. 20.

[25] Tepal, P., *Phytochemical screening, total flavonoid and phenolic content assays of various solvent extracts of tepal of Musa paradisiaca.* Malaysian Journal of Analytical Sciences, 2016. **20**(5): p. 1181-90.

[26] Hussain, J., et al., *Qualitative and quantitative comparison of rutin, quercetin and gallic acid concentrations in Syrian Capparis spinosa. L Leaves.* Journal of Pharmacognosy and Phytochemistry, 2017. **6**(4): p. 407-415.

[27] Sánchez-Rangel, J.C., et al., *The Folin–Ciocalteu assay revisited: improvement of its specificity for total phenolic content determination.* Analytical Methods, 2013. **5**(21): p. 5990-5999.

[28] Chang, C.-C., et al., *Estimation of total flavonoid content in propolis by two complementary colorimetric methods.* Journal of food and drug analysis, 2002. **10**(3).

[29] Silver, J., M. Drooby, and R. Lewis, *Purification of phenolic flavanoids with flash chromatography.* Planta Medica, 2012. **78**(11): p. PJ102.

[30] Arora, S. and P. Itankar, *Extraction, isolation and identification of flavonoid from Chenopodium album aerial parts.* Journal of traditional and complementary medicine, 2018. **8**(4): p. 476-482.

[31] Lee, K.J., et al., *Isolation and bioactivity analysis of ethyl acetate extract from Acer tegmentosum using in vitro assay and on-line screening HPLC-ABTS+ system.* Journal of analytical methods in chemistry, 2014. **2014**.

[32] Gad, M., et al., *Static headspace gas chromatographic method for the determination of residual solvents in cephalosporins.* RSC Advances, 2015. **5**(22): p. 17150-17159.

[33] Thakker, V.Y., et al., *Simultaneous estimation of gallic acid, curcumin and quercetin by HPTLC method.* J Adv Pharm Educ Res, 2011. **1**: p. 70-80.

[34] Vijayalakshmi, R., and and R. Ravindhran, *Comparative fingerprint and extraction yield of Diospyrus ferrea (willd.) Bakh. root with phenol compounds (gallic acid), as determined by uv–vis and ft–ir spectroscopy.* Asian Pacific Journal of Tropical Biomedicine, 2012. **2**(3): p. S1367-S1371.

[35] Blunder, M., et al., *Efficient identification of flavones, flavanones and their glycosides in routine analysis via off-line combination of sensitive NMR and HPLC experiments.* Food chemistry, 2017. **218**: p. 600-609.

[36] Sawant, L., B. Prabhakar, and N. Pandita, *Quantitative HPLC analysis of ascorbic acid and gallic acid in Phyllanthus emblica.* J. Anal. Bioanal. Tech, 2010. **1**(2).

[37] Bansal, V., et al., *Rapid HPLC method for determination of vitamin c, phenolic acids, hydroxycinnamic acid, and flavonoids in seasonal samples of emblica officinalis juice.* Journal of liquid chromatography & related technologies, 2015. **38**(5): p. 619-624.

[38] Li, Y., et al., *Evaluation of antioxidant properties of pomegranate peel extract in comparison with pomegranate pulp extract.* Food chemistry, 2006. **96**(2): p. 254-260.

[39] Liu, X., et al., *Identification of phenolics in the fruit of emblica (Phyllanthus emblica L.) and their antioxidant activities.* Food chemistry, 2008. **109**(4): p. 909-915.

[40] Pandit, A.A., *Solid Lipid Nano-particulate Formulation for ifosfamide: Development and Characterization*, 2011, Creighton University.

[41] Riegger, B.R., et al., *Chitosan nanoparticles via high-pressure homogenization-assisted miniemulsion crosslinking for mixed-matrix membrane adsorbers.* Carbohydrate polymers, 2018. **201**: p. 172-181.

[42] Vozza, G., et al., *Application of Box-Behnken experimental design for the formulation and optimisation of selenomethionine-loaded chitosan nanoparticles coated with zein for oral delivery.* International journal of pharmaceutics, 2018. **551**(1-2): p. 257-269.

[43] Basha, M., et al., *Benzocaine loaded solid lipid nanoparticles: formulation design, in vitro and in vivo evaluation of local anesthetic effect.* Current drug delivery, 2015. **12**(6): p. 680-692.

[44] Naik, J., et al., *Formulation and evaluation of poly (L-lactide-co-[epsilon]-caprolactone) loaded gliclazide biodegradable nanoparticles as a control release carrier.* International Journal of Drug Delivery, 2013. **5**(3): p. 300.

[45] El-Badry, M., G. Fetih, and M. Fathy, *Improvement of solubility and dissolution rate of indomethacin by solid dispersions in Gelucire 50/13 and PEG4000.* Saudi Pharmaceutical Journal, 2009. **17**(3): p. 217-225.

[46] Tığlı Aydın, R.S. and M. Pulat, *5-Fluorouracil encapsulated chitosan nanoparticles for pH-stimulated drug delivery: evaluation of controlled release kinetics.* Journal of Nanomaterials, 2012. **2012**.

[47] Pasparakis, G. and N. Bouropoulos, *Swelling studies and in vitro release of verapamil from calcium alginate and calcium alginate–chitosan beads.* International journal of pharmaceutics, 2006. **323**(1-2): p. 34-42.

[48] Gnanavel, V., V. Palanichamy, and S.M. Roopan, *Biosynthesis and characterization of copper oxide nano-particles and its anticancer activity on human colon cancer cell lines (HCT-116).* Journal of Photochemistry and Photobiology B: Biology, 2017. **171**: p. 133-138.

[49] Sufyani, A., et al., *Characterization and Anticancer Potential of Silver Nanoparticles Biosynthesized from Olea chrysophylla and Lavandula dentata Leaf Extracts on HCT116 Colon Cancer Cells.* Journal of Nanomaterials, 2019. **2019**.

[50] Lezoul, N.E.H., et al., *Extraction Processes with Several Solvents on Total*

Bioactive Compounds in Different Organs of Three Medicinal Plants. Molecules, 2020. **25**(20): p. 4672.

[51] Guideline, I.H.T. *Impurities: Guideline for residual solvents*. in *International Conference on Harmonisation Q3C, Geneva*. 1997.

[52] Upadhyaya, K., et al., *Identification of gallic acid based glycoconjugates as a novel tubulin polymerization inhibitors.* Organic & biomolecular chemistry, 2016. **14**(4): p. 1338-1358.

[53] Kamatham, S., N. Kumar, and P. Gudipalli, *Isolation and characterization of gallic acid and methyl gallate from the seed coats of Givotia rottleriformis Griff. and their anti-proliferative effect on human epidermoid carcinoma A431 cells.* Toxicology Reports, 2015. **2**: p. 520-529.

[54] López-Martínez, L.M., et al., *A 1 H NMR Investigation of the interaction between phenolic acids found in mango (Manguifera Indica Cv Ataulfo) and papaya (Carica papaya Cv Maradol) and 1, 1-diphenyl-2-picrylhydrazyl (DPPH) free radicals.* PloS one, 2015. **10**(11): p. e0140242.

[55] Smitha, B., S. Sridhar, and A. Khan, *Chitosan–sodium alginate polyion complexes as fuel cell membranes.* European Polymer Journal, 2005. **41**(8): p. 1859-1866.

[56] Mittal, A.K., S. Kumar, and U.C. Banerjee, *Quercetin and gallic acid mediated synthesis of bimetallic (silver and selenium) nanoparticles and their antitumor and antimicrobial potential.* Journal of colloid and interface science, 2014. **431**: p. 194-199.

[57] Rampino, A., et al., *Chitosan-pectin hybrid nanoparticles prepared by coating and blending techniques.* European Journal of Pharmaceutical Sciences, 2016. **84**: p. 37-45.

[58] Hassani, A., et al., *Preparation, characterization and therapeutic properties of gum arabic-stabilized gallic acid nanoparticles.* Scientific Reports, 2020. **10**(1): p. 1-18.

[59] Yu, S.-H., et al., *Preparation and characterization of radical and pH-responsive chitosan–gallic acid conjugate drug carriers.* Carbohydrate polymers, 2011. **84**(2): p. 794-802.

[60] Siddiqui, I.A., et al., *Nanochemo-prevention: sustained release of bioactive food components for cancer prevention.* Nutrition and cancer, 2010. **62**(7): p. 883-890.

Chapter 16

Cysteine in Broiler Poultry Nutrition

Iyakutye Jacob Nte and Hollinshead Holly Gunn

Abstract

The SAAs are limiting in the major poultry feed ingredients, ranking first and fifth in soya bean meal and maize, respectively. Feed ingredients rich in protein, in particular and other nutrients, enhance Energy supply and protein accretion. Modern commercial broilers have reduced maintenance needs and high amino acid requirements, and are more responsive to protein (amino acids) than energy. Cysteine is a semi-essential amino acid belonging to the SAAs. It plays essential roles in protein synthesis, structure and function, causing growth depressing effects in broiler chicks when there is methionine:cysteine imbalance. Genetically predetermined amino acid sequences in proteins are essential for production of adequate quantities of meat, milk and eggs. Therefore, ideal amino acid ratios which conform to the requirements of broilers should be utilized. In nutrition, amino acids are equivalent to proteins, hence the shift in focus from proteins to individual amino acids, expressed as ideal ratios to lysine. The SAAs are practically relevant and have critical nutritional roles in animal nutrition with over 90% production being used to fortify animal (particularly poultry) diets. A balance in the methionine:cysteine ratio is necessary to ensure efficient utilization of the SAAs for proper growth and development in broiler poultry.

Keywords: Cysteine, Sulfur-containing amino acids, Methionine-cysteine ratio, Broiler poultry

1. Introduction

The diets of broilers, and indeed most poultry, consists of maize and soya bean meal, primarily. These ingredients are limiting in certain amino acids, with the sulfur-containing amino acids (SAAs) ranking first and fifth in soya bean meal and maize, respectively [1]. Sulfur-containing amino acids are amino acids containing at least one sulfur atom, and therefore are considered as a group of sulfur bioactive molecules [2]. Generally, they affect protein metabolism, like other amino acids, leading to reduced protein synthesis when they are deficient in animal diets [3]. Among the four common sulfur-containing amino acids, namely methionine, cysteine, homocysteine, and taurine, only methionine and cysteine are incorporated into proteins [4]. However, all amino acids as constituents of proteins are a-amino acids, in which the molecular structure has the amino group attached to the same carbon atom as the carboxyl group [5], and only such amino acids are relevant for animal nutrition [1]. Apart from the ideal protein concept proposed for different categories of poultry, ideal amino acid ratios have also been proposed for broiler chickens [6]. Since methionine and lysine are the first and second limiting amino acids in poultry diets, their supplementation enhances the efficiency of protein utilization and hence, excretion of nitrogen.

The remarkable increase in the growth potentials of broilers in recent times, following their genetic improvement, has been attributed to artificial genetic selection [7], resulting in increased appetite and early attainment of market weight. However, some authors [8, 9] have suggested that other factors in combination with genetic make-up, such as nutrition, environment, age, sex, management, and health care, account for the successes achieved in managing dietary energy intake of broilers. Modern broilers perhaps eat to their physical capacity or adjust their feed intake in response to several factors including dietary energy [10], and increased nutrient density results in a linear improvement in weight gain and feed efficiency, without reduction in intake [11]. According to [12] constant intake of feeds high in protein and other nutrients increases supply of energy and results in a linear increase in protein accretion in tissues, until a maximum rate - a genetically defined term, is reached. Although the commercial objective in meat production is fostered by protein accretion, increased supply of energy beyond the "maximum rate", would merely translate into an excess of body fat [13], which is undesirable in terms of energetic efficiency [12].

There is the notion that today's broilers are more responsive to dietary protein (amino acids) and less to energy concentration due to reduced maintenance needs. This is occasioned by the significant reduction in market age and increased amino acid requirement, as driven by increase in the lean (muscle) growth as a percent of body weight [12]. It was reported by [14] that as little as 0.10% supplemental cysteine is growth depressing in chicks fed methionine deficient diets. This creates an imbalance in cysteine:methionine ratio, which affects the efficiency of DL-2-hydroxy-4-(methylthio) butyric acid, a precursor of methionine [6]. Apart from this imbalance, bioavailability of amino acids in proteins, which implies metabolism after digestion and absorption, is important in ensuring that they are absorbed in suitable chemical forms that can enhance protein synthesis [15–17]. Consequently, there is a dire need to ensure a balance in amino acid content of feeds using the ideal amino acid ratio, under the assumption that the ratio should remain largely unaffected by the variables that affect amino acid requirements [18]. It is also essential to supply dietary amino acids in their required profile conforming to the requirements of poultry [19].

2. Amino acids nutrition in poultry

It is well accepted that amino acids, as nutrients, are building blocks of proteins and play essential roles in the nutritional composition of all feed stuffs and vital physiological roles in the body of all livestock [20, 21]. The fact that it would have been very difficult, if not impossible, to produce the quantity of meat, milk, eggs, and fish demanded by consumers, without amino acids, accentuates their importance. The series of amino acids within the protein molecule, referred to as the amino acid sequence, is genetically predetermined, and they are essential nutrients which are a vital integral part of animal feeding regimens [1]. From direct hydrolysis of common nutritional feed proteins, about 20 different amino acids have been identified. These are known as the twenty canonical amino acids [2, 22]. In poultry nutrition, 21 amino acids are needed to form body proteins [23].

As functional and structural units of proteins, they are nutritionally classified into two groups: non-essential (synthesized in the body) and essential (cannot be synthesized rapidly and in sufficient quantities to meet their metabolic requirements) [24]. However, a number of non-essential amino acids can only be synthesized from essential amino acids (EAAs) and are called semi-essential amino acids (**Table 1**). It was further noted by [1] that the classification of amino acids into essential and non-essential should not be taken to imply that non-essential amino acids are not required for the synthesis of proteins. Consequently, [23] opined that a sufficient amount of non-essential amino acids must also be supplied alongside the essential amino acids by the diet, to prevent the conversion of essential into non-essential amino acids. To undertake such amino acid inter-conversions the animal requires sources of carbohydrates and suitable nitrogen compounds [1].

The amino acids relevant in animal nutrition are only α-amino acids, which exist as two optically active isomers i.e. the L-forms and D- forms, with one being a mirror image of the other (**Figure 1**). However, only the L-forms are found in proteins. Consequently, if both forms are supplied in animal diets in a 47:47 ratio, known as a "racemic mixture", the D-forms will be converted to the L-forms for ease of metabolism. There are differences in recommended essential amino acid levels in various guidelines, which raise concerns for the poultry sector [24]. Nevertheless, in the diet of poultry, amino acids must be balanced to avoid loss of energy that can be diverted to the synthesis of fats [25].

Amino acids chemically bound in proteins must be separated from the parent protein unit, before they can pass from the lumen of the gut across the intestinal wall (absorption) into the bloodstream. This separation occurs with the help of proteolytic digestive enzymes (proteases). The absorbed amino acids are transported via the hepatic portal vein into the liver, which is the principal organ for the metabolism of amino acids. The metabolism of proteins is made up of two opposing processes which occur simultaneously: accretion of proteins (anabolism i.e. synthesis) and breakdown of proteins (catabolism i.e. proteolysis) [1]. They further noted that whereas in mature animals a balance is reached between synthesis and proteolysis with no increase in the mass of the muscle but with continuous turnover, synthesis supersedes proteolysis in young growing animals, building up the proteins into muscle. Although, broilers are able to compensate for deficiencies of non-essential amino acids within certain limits through auto-synthesis, protein synthesis is terminated if one of the essential amino acids is lacking because some amino acids (the essential ones) cannot be synthesized by the organism [1]. Therefore, they opined that since the amino acid sequence of a protein is genetically predetermined, all the required amino acids must be present at the same time for individual amino acids to be synthesized (synchronous synthesis).

Once digested and absorbed, amino acids are used as the building blocks of structural proteins (muscle, skin, ligaments), metabolic proteins, enzymes, and precursors of several body components. Because body proteins are constantly being synthesized and degraded, an adequate amino acid supply is critical to support growth or egg production [23]. Broilers, like other poultry, are believed to develop better immune function when adequate levels of dietary amino acids are provided.

Essential	Arginine[1], Histidine, Isoleucine, Leucine, Lysine, Methionine, Phenylalanine, Threonine, Tryptophan, Valine
Semi-essential	Cystine (Cysteine)[2], Tyrosine
Non-essential	Alanine, Asparagine, Aspartic Acid, Glutamic Acid, Glutamine, Glycine, Proline, Serine

[1]*In swine, Arginine is essential only in young animals.*
[2]*Cystine = dimer of cysteine.*
Source: Dalibard et al. [1].

Table 1.
Essentiality of amino acids in pigs and poultry nutrition.

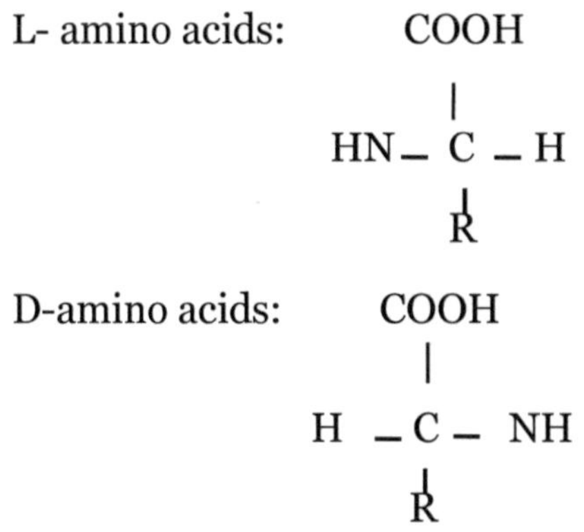

Figure 1.
General structure of L - D- isomers. Source: Dalibard et al. [1].

Since the health status and productivity of poultry are directly related to their immune status, there will be an increased demand for amino acids, particularly essential amino acids, under conditions of immune stress [23]. This is because amino acids are indispensable in the production of antibodies and cytokine, and hence in immune function [26–28].

Amino acids are analogous to proteins, from the standpoint of the fundamentals of nutrition. The main emphasis in the nutrition of animals has therefore, been shifted from a focus on protein as a whole to a focus on individual amino acids [1]. The authors also noted that great importance is therefore attached to the concept of amino acid flux - the continuous supply of free amino acids from the feed into the animal's metabolism and in the ideal ratios. These should be taken into account, when supplementing amino acids to mixtures of feed. In modern practical feeding systems, amino acid supplementation has been proven to be an effective method to continuously balance the amino acid supply at the site of protein synthesis. Therefore, the knowledge of digestible amino acid requirements and their digestibility in common feed ingredients fed to poultry are viewed as important tools in advancing knowledge in amino acid nutrition and metabolism of poultry [4]. However, according to [29–31] there are variations in the utilization efficiencies of individual essential amino acids.

2.1 Ideal amino acid ratios

A myriad of dietary, genetic and environmental factors impinge on the amino acid requirements of all livestock. The general notion nowadays is that, poultry requirement for any amino acid is proportionally linked to the requirement for the others. The indication is that the supply of one amino acid will improve performance only if no other amino acid is limiting [32]. Consequently, they also noted that poultry and swine nutritionists use lysine as a reference point in the ideal amino acid concept, and express the requirement for other amino acids as a percentage of the requirement for lysine. However, this was first established for swine for different weight categories [33]. The choice of lysine as the reference amino acid was based on a number of conditions namely, its position as the second limiting amino acid and ease of supplementation in commercial diets; its exclusive post-absorption use in protein accretion, maintenance and lack of a precursor role; relative ease of analysis in feedstuffs; and availability of a large pool of data on responses under different dietary concentrations, body compositional and varying environmental conditions [34, 35].

It has been well recognized that the requirements for amino acids by poultry cannot be valid under all dietary, sex and body compositional scenarios [36, 37]. A way out of this challenge, in order to obtain reliable amino acid requirements, is to express all amino acid requirements as ideal ratios to lysine. The ideal amino acid ratio utilizes the concept that the ideal ratios of the absolute or indispensable amino acids to lysine as published by [38] are slightly altered by drastic deviations in their requirements occasioned by genetic or environmental factors. Normally, the ideal amino acid profile only includes provisions for essential amino acids implying that the diet supplies sufficient non-essential amino acids. The non-essential amino acids should make up about half of the dietary protein with the remainder supplied by essential amino acids [39–41]. Ideal amino acid profiles should be based on digestible amino acids, particularly when diet formulation is done with other ingredients other than maize and soya bean [34]. If the amino acids supplied are in the proper, or ideal, ratio in relation to the needs of the animal, then amino acids in excess of the least limiting amino acid will be deaminated [42] and likely used as a source of energy rather than towards body protein accretion.

The overall benefits of the concept of ideal amino acid ratios are two fold namely, it enables the calculation of the requirements for the indispensable amino acids after an accurate determination of lysine requirement, and it allows the most efficient and economical use of proteins in diet formulation to allow for maximum utilization and minimum excretion of nitrogen [34].

2.2 Sulfur-containing amino acids

Sulfur is an abundant element in biological systems, which plays an important role in processes essential for life as a constituent of proteins, vitamins and other crucial biomolecules [22]. Sulfur-containing amino acids (SAAs) are amino acids which contain a sulfydryl group and are considered to be non-polar and hydrophobic [43, 44]. Generally, they play crucial roles in protein structure, metabolism, immunity, and oxidation [2, 44–46]. As noted earlier, there are four common sulfur-containing amino acids namely, methionine, cysteine, homocysteine, and taurine (**Figure 2**), but only methionine and cysteine are incorporated into proteins [5]. On this account, they are deemed as the principal or primary Sulfur-containing amino acids, although homocysteine and taurine also play important physiological roles. They are, therefore, classified as proteinogenic, canonic amino acids incorporated into the structure of proteins [22].

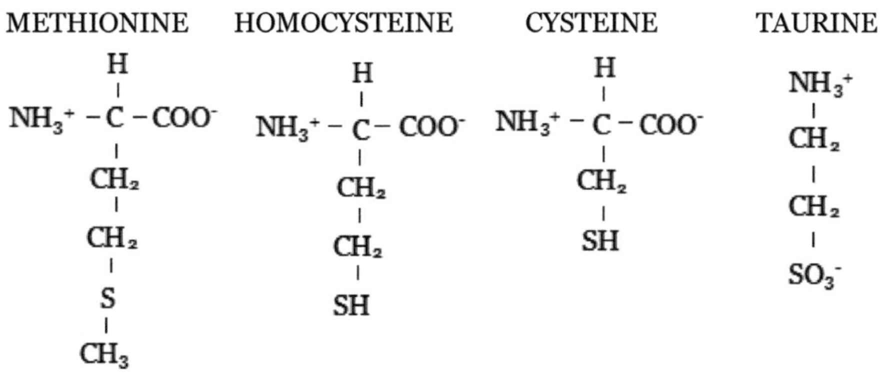

Figure 2.
Structures of the sulfur-containing amino acids. Source: Brosnan and Brosnan [4].

2.2.1: Primary SAAs: Methionine and cysteine are generally considered to be non-polar and hydrophobic, and are present in animal and plant proteins in varying proportions. Methionine is one of the most hydrophobic amino acids and is almost always found in the interior of proteins. Cysteine, which is genetically encoded by two possible codons (nucleotide triplets of mRNA) UGU and UGC [45], ionizes and readily forms disulfide linkages because of the ease with which it dissociates to yield a thiolate anion. Cysteine is also confined to the interior of proteins because the thiol group can be easily oxidized to form disulphide bonds. Methionine is an essential amino acid whereas cysteine is semi-essential because it can be synthesized from methionine and serine by trans-sulfuration [47]. Both methionine and cysteine are gluconeogenic, but methionine is a neutral amino acid while cysteine is basic [48].

Depending on the species of animal, cysteine may be responsible for up to 47% of the dietary methionine requirement, and this proportion has been shown to be lower in high performance animals [1]. The requirement for SAAs in the diet of animals is assessed on the basis of the content of methionine and cysteine [43]. When fed at supplemental levels well above the dietary requirement, methionine causes more serious growth depressing effects than other essential amino acids, but not much is known about responses of broilers to excess dietary cysteine [6, 49]. However, [50] suggested that excess dietary L-cysteine causes acute metabolic acidosis in chicks but not in pigs and rats. According to [6], no other amino acid, even at far higher doses, is known to elicit such lethality as observed with excess L-cysteine.

Sulfur-containing amino acids play critical roles in protein synthesis, structure and function. Sulfur amino acids are involved in the synthesis of intracellular anti-oxidants such as glutathione and N-acetyl cysteine, and represent a powerful part of cell antioxidant system [22]. Thus, they are essential in the maintenance of normal cellular functions and health. In addition to their worthy antioxidant action, sulfur-containing amino acids may offer a chelating site for heavy metals. Accordingly, they may be supplemented during chelating therapy, providing beneficial effects in eliminating toxic metals [22]. When animals are fed cysteine deficient diets, the SAAs and their derivatives (L-cysteine, L-cystine, N-acetyl-L-cysteine, and L-methionine) are not known to have depressing effects on their growth at iso-sulfurous levels [51]. L-cysteine and L-cystine can partially replace or reduce the metabolic requirement for methionine in different species of animals to the level of 38–77%, and are known as "spare amino acids" [52]. Research efforts into SAAs is of great practical relevance to animal nutrition in that well over 90% of their production is used to fortify diets for animals, particularly poultry [53]. He further indicated that poultry diets around the world are based on corn and soybean meal, and these diets for poultry, without fortification, are deficient in SAAs. Therefore, the nutritional roles of SAAs are of critical importance to the animal nutritionist as well as to the metabolic scientist.

This chapter focuses on cysteine which is biosynthesized from methionine, plays critical roles in protein structure, apparently irreplaceable by any other amino acid, amidst its role in broiler chicken nutrition – particularly the need for a balance between it and other amino acids to foster the growth, development and overall productivity of broilers.

2.2.2: Forms and derivatives of cysteine: Cysteine is special among coded amino acids because it contains a reactive sulph-hydryl group. Cysteine therefore, easily undergoes oxidation and, like methionine, it is confined to the interior of proteins. In the process it reacts with itself to form a disulphide bond, or with other thiols (Sulfur-containing compounds), yielding cystine [48]. Cystine is therefore a dimer of cysteine. In the plasma, and in fact the extracellular space, cysteine occurs primarily as cystine [54], and these are the two primary forms of cysteine relevant to animal nutrition. From its metabolic pathway it produces few intermediate substrates and derivatives namely, cystathionine, homocysteine, γ-glutamylcysteine, glutathione, hypotaurine and taurine.

The levels of cysteine and cystine in the cell milieu are maintained by adjustments in the ratio of L-cystine to L-cysteine by cellular control of their efflux and uptake. According to [50], the intracellular ratio of L-cystine to L-cysteine is improved by the efflux and uptake of L-cysteine and L-cystine from and by the cells, respectively. Conversely, their extracellular ratio is increased by uptake of L-cysteine and its oxidation to L-cystine, and efflux of L-cystine by the cells. This is illustrated in **Figure 3**.

3. Cysteine requirements

An animal's amino acid requirement is made up of the total requirement for protein accretion and maintenance. Due to faster growth rate and earlier market age of modern commercial broilers, the requirement for maintenance function becomes reduced [53]. Consequently, the relative need for protein accretion to maintenance varies for individual amino acids. Therefore, the requirement for cysteine and other amino acids with high maintenance needs relative to lysine will reduce [35]. In broiler poultry nutrition, optimum amino acid density must be maintained when considering the balance between energy and proteins in their feed, indicating a higher ratio of essential amino acids to energy in modern broilers [13]. This conforms to the fact that modern commercial broilers are different from those offered by the poultry industry in the 90s when the NRC nutrient requirement for poultry was published. Genetic selection, management practices, and changes in feed are believed to be partially responsible for this [55, 56].

The requirement for total SAAs recommended by [37] were 0.9, 0.69, and 0.57 for 0–3, 3–6, and 6–8 weeks, respectively, as against weekly requirements of 0.94, 0.9, 0.82, 0.78, 0.74, 0.71, and 0.67% for 1–7 weeks of age in modern broilers as presented by [57]. However, proper assessment of amino acid requirements have remain unresolved owing to the difficulty posed by underestimation and overestimation by the oxidation and nitrogen balance methodologies, respectively [51]. Biosynthesis of cysteine occurs in animals and plants via the trans-sulfuration pathway from methionine, in the presence of adequate nitrogen and sulfur [58]. However, since cysteine is synthesized from methionine via the trans-sulfuration pathway, its requirement is usually considered together with methionine [59].

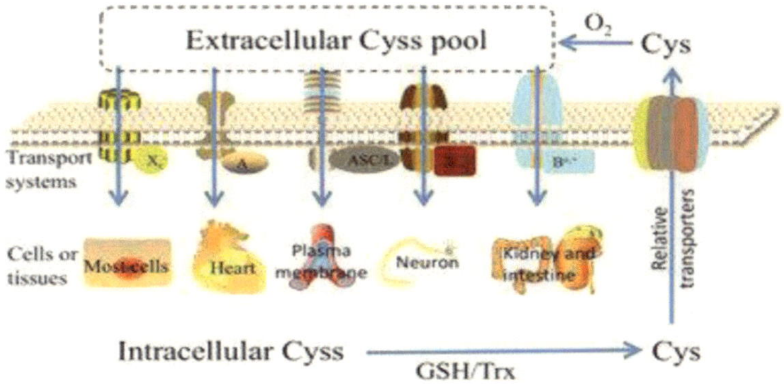

Figure 3.
Extracellular and intracellular L-cysteine/L-cystine balance and L-cysteine/L-cystine transport systems. Glu, L-glutamate; Cyss, L-cystine; Cys, L-cysteine; GSH, glutathione; Trx, thioredoxin. Source: Baker [50].

Although there is adequate physiological concentrations of cysteine, many cells still rely on the trans-sulfuration pathway for a minimum of 47% of their cysteine requirements [60]. Cysteine requirement is therefore subsumed in the total SAAs requirement as captured by [35, 55] for different age ranges in poultry. Commercial diets are traditionally formulated to meet broiler requirements for methionine + cystine (Met + Cys), based on the assumption that amounts of dietary Met are converted into Cys [61].

4. Cysteine digestibility and bioavailability

Following a combination of heat treatment and alkaline food processing some alterations occur in the chemical nature of cysteine leading to some effects in its digestibility and subsequent absorption. These two processes are vital in ensuring the assimilation of amino acids by broilers. Heat processing causes the oxidation of a significant portion of protein-bound cysteine to cystine, which has lower digestibility [62]. This may probably be due to the formation of disulphide bridges during the transformation process. Dietary cystine is also converted to lanthionine under the influence of heat and alkali treatment [63]. The reduced SAA activity of lanthionine results in reduced availability of protein-bound cysteine [64]. Since protein metabolism continues even when no protein is being consumed, some of the amino acids released are oxidized and are not available for re-synthesis of new proteins. Feeding a protein-free diet to broilers, therefore, elicit a cysteine response (reduction in body weight loss and improves nitrogen balance) [65], indicating that it could be substantially depleted in the body pool making it the first limiting amino acid for endogenous protein synthesis [53].

All the nutrients ingested by an animal via its diet cannot be utilized by the animal because some are undigested. Furthermore, some are either absorbed in forms that cannot be utilized for physiological and metabolic functions in the body or are not absorbed at all. The available nutrients refer to the portion of the nutrients that are digested, absorbed and metabolized [20]. The same is true of amino acids, which are bioavailable if they occur in forms that can be utilized by the cells for maintenance or production. Digestibility of amino acids, therefore, is the digestion of the amino acids consumed in the diets and their subsequent absorption from the lumen of the small intestine into the bloodstream [66]. The portion of the absorbed amino acids present in chemical forms amenable to protein synthesis indicates their bioavailability [17]. The concept of digestible amino acids is critical to establish ideal protein ratios [67], and broiler diets are now formulated based on digestible proteins and amino acids [68].

5. Vital roles of cysteine in broiler poultry nutrition

Cysteine can be synthesized from methionine and serine by trans-sulfuration [47]. As one of the naturally occurring biogenic amino acids, cysteine plays crucial roles at all the levels of protein structure because it is easily oxidized to cystine, a feature that is very vital for the analysis of the primary structure of proteins; for effects on changes in secondary structure and for stabilization of tertiary and quartenary structure of proteins [69]. It was further noted that it possesses a sulfhydryl group in its side chain, according it a special position that cannot be replaced by any other amino acid. Cysteine, by virtue of its ability to form inter- and intra-chain disulfide bonds, plays a crucial role in protein structure and in protein-folding pathways. Such bonds, known as disulphide linkages, are common in proteins destined for export or residence on the plasma membrane [70]. He also noted that any mismatched disulfide bonds are rearranged to ensure the correct protein folding under the influence of protein disulfide isomerase (an endoplasmic reticulum protein).

Basically, cysteine (and methionine too) is incorporated into structural proteins, and it is also required for normal growth. The two are major protein constituents of feathers and hair, with methionine occurring in greater percentage in muscle while cysteine is higher in feather keratin [61]. Cysteine has been reported to be thirteen (13) times higher in broiler feathers than methionine [71, 72]. This indicates their importance in growth and feather development of broiler poultry. Reduced feed intake and weight gain have been associated with L-cysteine supplementation in young animals. Its anoretic effects have been reported to manifest as reduced final body weight, body weight gain, feed intake, and feed efficiency in rats [73]. This was attributed to the bitter taste imparted by L-cysteine. However [74], reported that reduced (Cys) and oxidized (cystine) forms of cysteine support animal growth equally when provided in a cyst(e)ine deficient and methionine adequate diet. Since L-cysteine is a spare amino acid for Methionine, as the adverse effects of L-methionine deficiency can be ameliorated by L-cysteine supplementation in the diet of animals [75]. In the opinion of [54], whole body protein synthesis and physiological Homeostasis can be maintained by dietary supplementation of L-cysteine under conditions of impaired L-methionine catabolism.

Cysteine is involved in the biosynthesis of methionine by accelerating the pathway leading to the formation of pheomelanin, thereby blocking the formation of eumelanin that produces dark colors [76]. Cysteine itself is a powerful antioxidant and has the potential to trap reactive oxygen species (ROS) [5]. It plays a central role in the antioxidant protection system of the body such as glutathione (GSH), by functioning as a precursor of some constituents [77, 78]. GSH is a potent antioxidant which protects the body against toxic effects of elevated levels of endogenous and exogenous electrophiles [79]. Taurine, another SAA, and Hydrogen sulphide are also produced from dietary L-cysteine and they play vital roles in the reduction of oxidative stress and protection against several environmental toxins [80].

It has the capacity to improve intestinal histomorphometric indices of broiler chickens with a consequent increase in absorption of nutrients [81]. High L-cysteine concentration has been observed in proteins and mucins that contribute to the maintenance of gut integrity and plays key roles in intestinal structure and function [82, 83]. The indications are, therefore, that L-cysteine deficiency causes certain degrees of intestinal distortions and is essential in the maintenance of intestinal integrity and function.

Lipid metabolism is also mediated by L-cysteine and its derivatives such as S-methyl L-cysteine with hypoglyemic and antihyperlipidemic characteristics, through reduction in fasting blood sugar and total triglycerides [84], and N-acetlycysteine which improves lipid metabolism by affecting serum cholesterol, triglycerides, Very High Density Lipoproteins (VHDL), and High Density Lipoproteins (HDL) levels [85]. The mode of action of L-cysteine underlying these effects are not clear, but it is believed to be partially accounted for by its target on gene expression of certain biochemical substances such as element-binding protein and fatty acid synthase [85]. Its roles in lipid metabolism and positive correlation with fat mass are eloquent testimonies [75].

Cysteine is, in fact, a rather potent reducing agent in addition to its capacity of being capable of either chelating or complexing trace elements [6]. Its reducing agent bio-activity when supplemented in diets at 0.38%, is capable of converting pentavalent to trivalent organic arsenic, which is up to 100 times more toxic and of great significance in poultry and animal nutrition [86]. The authors also opined that this has great implication in the use of certain poultry drugs e.g. coccidiostats containing pentavalent organic arsenic, whose toxicity is accentuated by pharmacologic cysteine. It is well established that modest excesses of SAAs, particularly cysteine, can have marked pharmacologic effects on trace-mineral utilization, but far less is known about effects of excess cysteine ingestion [87].

6. Methionine-cysteine balance

Methionine and cysteine are closely related in that the latter is endogenously synthesized from the former via the trans-sulfuration pathway by L-methionine degradation [43]. In this pathway, methionine is converted to homocysteine, which in turn donates a sulfur group to serine (a non-essential amino acid) to ultimately form cysteine. The production of cysteine accounts for 47% of methionine dietary requirement [48]. L-cysteine can furnish up to 47% and 77% of the requirements for SAAs in young and older animals, respectively [88]. Nevertheless, the practice of formulating commercial diets to contain adequate methionine+cysteine, with the assumption that dietary methionine is converted to cysteine, is common. This may lead to reduced efficiency of amino acid utilization, since methionine will be supplied in excess. This can be addressed by adequate knowledge of methionine:cysteine ratio in relation to total sulfur amino acids (TSAAs), and the quantity of methionine converted to cysteine [61]. Another condition of imbalance is created when excess cysteine is provided in methionine deficient diets, with growth depressing effects in chicks [14]. Such imbalances need to be addressed to ensure efficient utilization of the SAAs by broilers.

As broilers age or increase in weight, maintenance needs for amino acids, including methionine and cysteine, and ideal amino acid ratios will alter. However, not much is known about the methionine:cysteine ratio, although a ratio of 52:43 [89] and a minimum of 49:45 [61] has been recommended for poultry and growing broilers, respectively. Also, little is known about the effects of excess cyst(e)ine on chicks, but among the EAAs, excess methionine is known to have the most adverse effects on growth [71, 90]. Variations in the ratio of these amino acids affect growth responses in broilers, and the utilization and efficacy of hydroxyl analogues of methionine or its precursors [6]. Therefore, determination of the optimum methionine:cysteine ratio in relation to TSAAs is necessary to foster proper growth and development of broilers.

7. Conclusions

The primary ingredients used in broiler poultry nutrition are limiting in SAAs in particular. Methionine and cysteine as the major SAAs, are α-amino acids which are important in animal nutrition. The recent increase in growth potentials of modern commercial broilers has been attributed to genetic improvement resulting in increased appetite and early market weight. They are more responsive to proteins (amino acids) than to energy concentration due to reduced maintenance need. Amino acids play vital nutritional and physiological roles in all livestock and must be supplied in appropriate ratios to foster protein accretion, the major objective in broiler poultry production. Nutritionally, amino acids are equivalent to proteins, warranting a shift in focus from proteins to individual amino acids.

Dietary amino acids must be in the ideal ratios for efficient use of proteins in diet formulation, maximum utilization and minimum excretion of nitrogen. Among the SAAs, methionine and cysteine are gluconeogenic and need to be balanced to circumvent the lethality associated with excess cysteine, an apparently irreplaceable amino acid. However, biosynthesis of cysteine occurs in animals and plants via the trans-sulfuration pathway from methionine, and its requirement is subsumed in the TSAAs requirement in poultry. Basically, cysteine is incorporated into structural proteins, and is required for normal growth. A balance in the methionine:cysteine ratio is therefore, necessary to ensure efficient utilization of the SAAs, so as to foster proper growth and development in broiler poultry.

Acknowledgements

We are deeply indebted to the omniscience God, the Fountain of all wisdom, who we worship and adore, for endowing and equipping us with the necessary wisdom and skills to write this chapter. We also thank our colleagues friends and well-wishers for their encouragement. Special appreciation to Joan Business Services, Azikiwe Road, Port Harcourt, Nigeria, for skillfully typesetting the manuscript under short notice.

Finally, we wish to thank our wives and children, very specially, for their encouragement, and for having to bear with our apparent momentary unavailability during the preparation of the manuscript: For Nte - Ruth (wife) and Awaji-inyiemi (son); and for Gunn - Regina (wife), Ewomazino (daughter), Oghenemaro (daughter), and Okgariorieze (son). We love you all unconditionally.

Conflict of interest

The authors declare no conflict of interest.

Author details

Iyakutye Jacob Nte* and Hollinshead Holly Gunn
Department of Animal Science, Faculty of Agriculture, Rivers State University, Port Harcourt, Nigeria

*Address all correspondence to: nte.iyakutye@ust.edu.ng

References

[1] Dalibard P, Hess, V, LE Tutour, L, Peisker, M, Per, S, Perojo Gutierrez, A, Redshaw, M. Amino acid in animal nutrition. Belgium: Felana; 2014. 98p.

[2] Reddy MK. Amino acid. Encyclopaedia Britanica. https://www.britanica.com/science/amino-acid.

[3] Tesseraud S, Coustard SM, Collin A, Seilez I. Role of sulphur amino acids in controlling nutrient metabolism and cell functions: implications for nutrition. British Journal of Nutrition. 2009; 101(8):1132-1139.

[4] Brosnan JT, Brosnan ME. The sulfur-containing amino acids: an overview. Journal of Nutrition. 2006: 136(6):1636S-1640S. DOI:10.1093/jn/136.6.1636S.

[5] Oshimura E. Sakamoto K. Amino Acids, Peptides, and Proteins. Cosmetic Science and Technology: Theoretical Principles and Applications. Elsevier Inc. 2017. http://dx.doi.org/10.1016/B978-0-12-802005-0.00019-7

[6] Baker DH. Advances in protein—amino acid nutrition of poultry. Amino Acids. 2009: 37:29-41. DOI:10.1007/s00726-008-0198-3

[7] Havenstein HB, Ferket PR, Qureshi MA. Growth, livability, and feed conversion of 1957 versus 2001 broilers when fed representative 1957 and 2001 diets. Poultry Science. 2003; 82: 1500-1508.

[8] Ferket PR, Gernat AG. Factors that affect feed intake of meat birds: A review. International Journal of Poultry Science. 2006;5(10):905-911.

[9] Rosa PS, Faria Filho DE, Dahlke F, Vieira BS, Macan M, Furlan RL. Effect of energy intake on performance and carcass composition of broiler chickens from two different genetic groups. Revista Brasileira de Ciência Avicola. 2007;9(2):117-122.

[10] Ahiwe, EU, Omede, AA, Abdallh MB, Iji PA. Managing Dietary Energy Intake by Broiler Chickens to Reduce Production Costs and Improve Product Quality, Animal Husbandry and Nutrition. Yücel B, Takin T. IntechOpen, 2018. DOI:10.5772/intechopen.76972. Available from: https: //www.intechopen.com/books/animal-husbandry-andnutrition/managing-dietarv-energv-intake-by-broiler-chickens-to-reduceproduction-costs-and-improve-product-qu.

[11] Lamot D. First week nutrition of broiler chickens: effects on growth, metabolic status, organ development, and carcass composition [PhD thesis]. Wageningen NL. 2017.

[12] Aftab U. Energy and amino acid requirements of broiler chickens: keeping pace with the genetic progress. World's Poultry Science Journal. 2019;75(4):507-514. DOI:10.1017/S0043933919000564

[13] De Lange CFM, Swanson KC. Genetic influences in nutrient utilization in growing farm animals. In: Mosenthin R, Zentek Z, Zebrowska T, editors. Biology of Nutrition in Growing Animals. Elsevier, 2006. p. 541-558.

[14] Dilger RN, Baker DH. DL-Methionine is as efficacious as Lmethionine, but modest L-cystine excesses are anorexigenic in sulfur amino acid-deficient purified and practical-type diets fed to chicks. Poult. Sci. 2007; 86:2367-2374.

[15] Dalibard P, Paillard E. Use of digestible amino acid concept in formulating diets for poultry. Anim. Feed Sci. Technol. 1995; 53: 189-204.

[16] Batterham ES. Availability and utilization of amino acids for growing pigs. Nutr. Res. Rev. 1992; 5: 1-18

[17] Lewis AJ, Bayley HS. Amino acid bioavailability. In: Ammerman C.B, Baker D.H, Lewis AJ, editors. Bioavailability of nutrients for animals, amino acids, minerals and vitamins. New York, USA: Academic Press, 1995. p. 35-65.

[18] Baker DH, Han Y. Ideal amino acid profile for chicks during the first three weeks posthatching. Poultry Science. 1994; 73: 1441-1447.

[19] Ullah Z, Au M, Nisa MU, Sarwar M. Digestible amino acids: significance and prospects in poultry: a review. Int. J. Agric. Biol. 2015; 17: 851-859

[20] Bortoluzzi C, Rochell SJ, Applegate TJ. Threonine, arginine, and glutamine: Influences on intestinal physiology, immunology, and microbiology in broilers. Poult. Sci. 2018; 97(3):937-945.

[21] Debnath BC, Biswas P, Roy B. The effects of supplemental threonine on performance, carcass characteristics, immune response and gut health of broilers in subtropics during pre-starter and starter period. J. Anim. Physiol. Anim. Nutr. 2019; 103(1):29-40.

[22] Brosnan JT, Brosnan ME. The sulfur-containing amino acids: an overview. Journal of Nutrition. 2006; 136(6): 1636S-164oS.

[23] Colovic MB, Vasic VM, Djuric, DM, Krstic, DZ. Sulphur-containing Amino Acids: Protective Role Against Free Radicals and Heavy Metals. Current Medicinal Chemistry. 2018, 25(3):324-335. DOI: https://doi.org/10.2174/09298 67324666170609075434

[24] Applegate TJ. Factors affecting feed intake – What do we know? In: Proceedings of the Arkansas. Nutrition Conference. The Poultry Federation; Rogers, Arkansas, USA. 2012

[25] Alagawany M, Elnesr SS, Farag MR, Tiwarid R, Yatoo MI, Karthik K, Michalak I, Dhama K. Nutritional significance of amino acids, vitamins and minerals as nutraceuticals in poultry production and health — a comprehensive review. Veterinary Quarterly. 2021; 41(1): 1-29.

[26] Le Floc'h N, Meichior D, Obled C. Modifications of protein and amino acid metabolism during inflammation and immune system activation. Livest. Prod. Sci. 2004; 87(1):37-45.

[27] Kidd MT. Nutritional modulation of immune function in broilers. Poult. Sci. 2004; 83(4):650-657.

[28] Li P, Yin YL, Li D, Kim SW, Wu G. Amino acids and immune function. Br. J. Nutr. 2007; 98(2):237-252.

[29] Han K, Lee JH. The role of synthetic amino acids in monogastric animal production-review. Asian Australasian Journal of Animal Sciences. 2000; 13(4): 543-560.

[30] Baker DH, Fernandez SR, Parsons CM, Edwardslll HM, Emmert JL, Webel DM Maintenance requirement for valine and efficiency of its use above maintenance for accretion of whole body valine and protein in young chicks. J. Nutr. 1996; 126:1844-1851

[31] Edwards III HM, Baker DH, Fernandez SR, Parsons CM. Maintenance threonine requirement and efficiency of its use for accretion of whole- body threonine and protein in young chicks. Br. J. Nutr. 1997; 78:111-119.

[32] Edwards III HM, Fernandez SR, Baker DH. Maintenance lysine requirement and efficiency of using lysine for accretion of whole-body lysine

and protein in young chicks. Poult. Sci. 1999; 78:1412-1417

[33] Schutte JB, De Jong J. Ideal amino acid profile for poultry. In: Brufau J, Tacon A, editors. Feed manufacturing in the Mediterranean region: Recent advances in research and technology. Zaragoza: CIHEAM, 1999. p. 259-263 (Cahiers Options Méditerranéennes; n. 37)

[34] Chung TK, Baker DH. Ideal amino acid pattern for 10-kilogram pigs. Journal of Animal Science. 1992; 70:3102-3111.

[35] Mack S, Bercovici D, De Groote G, Leclercq B, Lippens M, Pack, M, Schutte JB, Van Cauwenberghe S. Ideal amino acid profile and dietary lysine specification for broiler chickens of 20 to 40 days of age. British Poultry Science. 1999; 40(2): 257-265. http://dx.doi.org/10.1080/0007166q987683

[36] Emmert JL, Baker DH. Use of the protein concept for precision formulation of amino acid levels in broilers diets. J. Appi. Poult. Res. 1997; 6: 462-470.

[37] Taherkhani R, Shivazad M, Zaghari M, Shahneh ZA. Male and female response to different amino acid ratios during the third and fourth week posthatch. International Journal of Poultry Science. 2005; 4(8): 563-567.

[38] Nutrient Requirements of Poultry, 9th rev. ed. Washington, DC: National Research Council, National Academy Press, 1994. p155.

[39] Bregendahl K, Roberts S, Kerr B, Hoehler D. Ideal Amino Acid Profile for 28-to-34-Week-Old Laying Hens. Animal Industry Report: 2008; AS 654, ASL R2332. DOI: https://doi.org/1o.31274/ans_air-18o814-192 Available at: https://lib. dr.iastate.edu/ans air/vol654/iss 1/81.

[40] Heger J, Frydrych Z, Fronek P. The effect of non-essential nitrogen on the utilization of dietary protein in the growing rat. Journal of Animal Physiology and Animal Nutrition. 1987; 57:130-139.

[41] Heger J, Mengesha S, Vodehnal D. Effect of essential:nitrogen ratio on protein utilization in the growing pig. British Journal of Nutrition. 1998; 80:537-554.

[42] Lenis NP, van Diepen HT, Bikker P, Jongbloed AW, van der Meulen J. Effect of the ratio between essential and nonessential amino acids in the diet on utilization of nitrogen and amino acids by growing pigs. Journal of Animal Science. 1999; 77:1777-1787.

[43] Van Milgen J, Dourmad JY. Concept and application of ideal protein for pigs. J Animal Sci Biotechnol. 2015; 6:15. https://doi.org/10.1186/s40104-015-0016-1

[44] Bin P, Huang R, Zou X. Oxidation Resistance of sulfur amino acids: Methionine and cysteine. Biomed. Research International. 2017; 1-6. DOI:10.1155/2017/9584932.

[45] Kim, SW. Amino Acids and Immune Function. Vienna, Austria: Springer, 2003.

[46] Grimble RF. The effects of sulfur amino acid intake on immune function in humans. Journal of Nutrition. 2006; 136(6):1660S-1665S.

[47] WHY CYSTEINE IS SPECIAL? Written by Jacek Leluk. Available from: http://www.cryst.bbk.ac.uk/pps97/assignments/projects/leluk/project.htm Assessed on 3-02-2021.

[48] Stipanuk MH. Sulfur amino acid metabolism: pathways for production and removal of homocysteine and cysteine. Annu Rev Nutr. 2004;24:539-577.

[49] Muramatsu K, Odagiri H, Morishita S, Takeuchi H. Effect of excess levels of individual amino acids on growth of rats fed casein diets. J. Nutr. 1971;101:1117-1125

[50] Baker DH. Comparative species utilization and toxicity of sulfur amino acids. J. Nutr. 2006; 136:1670S-1675S.

[51] Dilger RN, Baker DH. Cyst(e)ine imbalance and its effect on methionine precursor utilization in chicks. J. Anim. Sci. 2008; 86:1832-1840.

[52] Baker DH. Utilization of precursors for L-amino acids. In: D'Mello JPF (ed) Amino acids in farm animal nutrition. Wallingford: CAB International; 1994. p. 37-64

[53] Baker DH. Comparative nutrition and metabolism: Explication of open questions with emphasis on protein and amino acids. PNAS. 2005; 102(50): 17897-17902

[54] Yin J, Ren W, Yang G, Duan J, Huang X, Fang R, Li C, Li T, Yin Y, Hou Y, Kim SW, Wu G. Mol. L-Cysteine metabolism and its nutritional implications Nutr. Food Res. 2016; 60: 134-146. DOI:10.1002/mnfr.201500031

[55] Havenstein, H.B., Ferket, P.R. and Qureshi, M.A. Growth, livability, and feed conversion of 1957 versus 2001 broilers when fed representative 1957 and 2001 diet. Poultry Science. 2003; 82: 1500-1508.

[56] Baker DH, Han Y, Ideal amino acid profile for chicks during the first three weeks posthatching. Poultry Science. 1994; 73:1441-1447.

[57] Dozier WA, Kidd MT, Corzo A. Amino acid responses of broilers. J. Appl. Poult. Res. 2008; 17:157-167

[58] Hesse H, Nikiforova V, Gakiere B, Hoefgen, R. Molecular analysis and control of cysteine biosynthesis: integration of nitrogen and sulphur metabolism. Journal of Experimental Botany. 2004; 55(401): 1283-1292. DOI: 10.1093/jxb/erhl36

[59] Goulart CC, Costa FGP, Silva JHV, Souza JG, Rodrigues VP, Oliveira CFS. Requirements of digestible methionine+cystine for broiler chickens at 1 to 42 days of age. R Bras Zootec. 2011; 40(4):797-803.

[60] Mosharov E, Crawford MR, Banerjee R. The quantitatively important relationship between homocysteine metabolism and glutathione synthesis by the trans-ulfuration pathway and its regulation by redox changes. Biochemistry. 2000; 39:13005-11.

[61] Pacheco LG, Sakomura NK, Suzuki RM, Dorigam JCP, Viana GS, Van Milgen J, Denadai JC. Methionine to cystine ratio in the total sulfur amino acid requirements and sulfur amino acid metabolism using labelled amino acid approach for broilers. BMC Vet. Res. 2018; 14: 364. DOI: 10.1186/s12917-018-16778

[62] Miller EL, Huang YX, Kasinathan S, Rayner B, Luzzana U, Moretti VM, Valfr F, Torrissen KR, Jensen HB, Opstvedt J. Heat-damaged protein has reduced ileal true digestibility of cysteine and aspartic acid in chicks. J. Anim. Sci. 2001;79(1):65.

[63] Robbins KR, Baker DH, Finley JW. Studies on the utilization of lysinoalanine and lanthionine. J. Nutr. 1980;110:907-915.

[64] Parsons CM, Hashimoto K, Wedekind KJ, Han Y, Baker DH. Effect of over processing on availability of amino acids and energy in soybean meal. Poult. Sci. 1992; 71: 133-140

[65] Yoshida A, Moritoki K. Nitrogen sparing action of methionine and

threonine in rats receiving a protein free diet. Nutr Rep Int. 1974; 9:159-168.

[66] Fuller M. Amino acid bioavailability-A brief history. In: Proceeding of 9th International Symposium University of Alberta, Alberta, Canada. 2003. pp: 183-198

[67] Moore DT, Baker K, Firman JD. Digestible sulfur amino acid requirements for male turkeys to five weeks of age. J. Appl. Poult. Res. 2001; 10: 363-370

[68] Huang KH, Ravindran V, Li X, Bryden WL. Influence of age on the apparent ileal amino acid digestibility of feed ingredient for broiler chickens. Brit. Poul. Sci. 2005; 46: 236-245.

[69] Wu G. Amino acids: metabolism, functions, and nutrition. Amino Acids. 2009; 37(1): 1-17.

[70] Jessop CE. Oxidative folding in the mammalian endoplasmic reticulum. Biochem Soc Trans. 2004; 32:655-8.

[71] Emmans GC, Fisher C. Problems in nutritional theory. Nutrient requirements of poultry and nutritional research. 1986; 1:9-39.

[72] Talpaz H, De La Torre JR, Sharpe PJH, Hurwitz S. Dynamic optimization model for feeding of broilers. Agric Syst. 1986; 20(2):121-32.

[73] Lee, S, Han, KH, Nakamura, Y, Kawakami, S. et al., Dietary Lcysteine improves the antioxidative potential and lipid metabolism in rats fed a normal diet. Biosci. Biotechnol. Biochem. 2013; 77: 1430-1434.

[74] Baker DH. Comparative species utilization and toxicity of sulfur amino acids. J. Nutr. 2006; 136:1670S-1675S

[75] Elshorbagy AK, Nurk E, Gjesdal CG, Tell GS. Homocysteine, cysteine, and body composition in the Hordaland Homocysteine Study: does cysteine link amino acid and lipid metabolism? Am. J. Clin. Nutr. 2008; 88: 738-746.

[76] Solano F. Melanins: skin pigments and much more types, structural models, biological functions, and formation routes. New J Sci. 2014; 1e28, 498276.

[77] Mari M, Morales A, Colell A, Garcıa-Ruiz C, Fernandez Checa JC. Mitochondrial glutathione, a key survival antioxidant. Antioxid Redox Signal. 2009;11(11):2685-2700.

[78] Jong CJ, Azuma J, Schaffer S. Mechanism underlying the antioxidant activity of taurine: prevention of mitochondrial oxidant production. Amino Acids. 2012; 42(6): 2223-2232.

[79] Chen Y, Dong H, Thompson DC, Shertzer HG. Glutathione defense mechanism in liver injury: insights from animal models. Food Chem. Toxicol. 2013; 60, 38-44.

[80] Das J, Roy A, Sil PC. Mechanism of the protective action of taurine in toxin and drug induced organ pathophysiology and diabetic complications: a review. Food Funct. 2012; 3:1251-1264.

[81] Elwan HA, Elnesr SS, Xu Q, Xie C, Dong X, Zou X. Effects of in ovo methionine-cysteine injection on embryonic development, antioxidant status, IGF-I and TLR4 gene expression, and jejunum histomorphometry in newly hatched broiler chicks exposed to heat stress during incubation. Animals. 2019; 9(1):25.

[82] Badaloo A, Hsu JW, Taylor-Bryan C, Green C. Dietary cysteine is used more efficiently by children with severe acute malnutrition with edema compared with those without edema. Am. J. Clin. Nutr. 2012; 95, 84-90.

[83] Bauchart-Thevret C, Stoll B, Chacko S, Burrin DG. Sulfur amino acid

deficiency upregulates intestinal methionine cycle activity and suppresses epithelial growth in neonatal pigs. Am. J. Physiol. Endocrinol. Metabol. 2009; 296:E1239-E1250.

[84] Senthilkumar GP, Thomas S, Sivaraman K, Sankar P. Study the effect of S-methyl L-cysteine on lipid metabolism in an experimental model of diet induced obesity. J. Clin. Diagn. Res. 2013;2449-2451.

[85] Sit M, Yilmaz EE, Tosun M, Aktas G. Effects of Nacetyl cysteine on lipid levels and on leukocyte and platelet count in rats after splenectomy. Niger. J. Clin. Pract. 2014; 17, 343-345.

[86] Czarnecki GL, Baker DH, Garst JE. Arsenic-sulfur amino acid interactions in the chick. J. Anim. Sci. 1984; 59:1573-1581.

[87] Baker DH, Czarnecki-Maulden GL. Pharmacologic role of cysteine in ameliorating or exacerbating mineral toxicities. J. Nutr. 1987; 117:1003- 1010.

[88] Buttery PJ, D'Mello JPF. Amino acid metabolism in farm animals: an overview. In: D'mello JPF, editor, Amino Acids in Farm Animal Nutrition. Wallingford, UK, 1994, 1-10. Avaiable from: https://www.researchgate.net/publication/299105295_Influence_of_excessive_lysine_andor_methionine_supplementatio_on_growth_performance_and_carcass_traits_in_broiler_chicks. Assessed on 11-03-2021.

[89] Baker DH. Comparative species utilization and toxicity of sulfur amino acids. J. Nutr. 2006; 136:167oS-1675S.

[90] Edmonds MS, Baker DH. Comparative effects of individual amino acid excesses when added to a corn—soybean meal diet: effects on growth and dietary choice in the chick. J. Anim. Sci. 1987; 65:699-705.